Trends in Mathematics

Trends in Mathematics is a series devoted to the publication of volumes arising from conferences and lecture series focusing on a particular topic from any area of mathematics. Its aim is to make current developments available to the community as rapidly as possible without compromise to quality and to archive these for reference.

Proposals for volumes can be submitted using the Online Book Project Submission Form at our website www.birkhauser-science.com.

Material submitted for publication must be screened and prepared as follows:

All contributions should undergo a reviewing process similar to that carried out by journals and be checked for correct use of language which, as a rule, is English. Articles without proofs, or which do not contain any significantly new results, should be rejected. High quality survey papers, however, are welcome.

We expect the organizers to deliver manuscripts in a form that is essentially ready for direct reproduction. Any version of TEX is acceptable, but the entire collection of files must be in one particular dialect of TEX and unified according to simple instructions available from Birkhäuser.

Furthermore, in order to guarantee the timely appearance of the proceedings it is essential that the final version of the entire material be submitted no later than one year after the conference.

For further volumes:
http://www.springer.com/series/4961

Julien Barral • Stéphane Seuret
Editors

Further Developments in Fractals and Related Fields

Mathematical Foundations and Connections

Birkhäuser

Editors
Julien Barral
LAGA - Institut Galilée
Université Paris 13
Villetaneuse, France

Stéphane Seuret
Laboratoire d'Analyse et de Mathématiques Appliquées
Université Paris-Est Créteil - Val de Marne
Creteil, France

ISBN 978-0-8176-8399-3 ISBN 978-0-8176-8400-6 (eBook)
DOI 10.1007/978-0-8176-8400-6
Springer New York Heidelberg Dordrecht London

Library of Congress Control Number: 2012956553

Mathematics Subject Classification (2010): 11Kxx, 26Axx, 28Axx, 28Dxx, 30xx, 37xx, 40Axx, 42Axx, 60Gxx

Printed on acid-free paper

Birkhäuser is part of Springer Science+Business Media (www.birkhauser-science.com)

In memoriam

Benoît Mandelbrot,

whose friendship

was precious to us,

and whose scientific legacy

is so important to our community.

Preface

This volume is a collection of 13 peer-reviewed chapters consisting of expository/survey chapters and research articles on fractals. Many of these chapters were presented at the second edition of the international conference "Fractals and Related Fields," held on Porquerolles Island, France, in June 2011. The success of this event proved the dynamism of the mathematical activity in the numerous branches connected to fractal geometry.

The selected chapters cover the following topics:

- Geometric measure theory
- Ergodic theory, dynamical systems
- Harmonic analysis
- Multifractal analysis
- Number theory
- Probability theory

The three surveys are written by famous experts in their respective fields. The other chapters are either original contributions or accessible expositions of very recent developments, also written by leaders in their respective domains.

This book naturally follows the previous one, "Recent Development in Fractals and Related Fields" which was published after the first conference, "Fractals and Related Fields." It is intended for researchers and graduate students wishing to discover new trends in fractal geometry.

Villetaneuse, France — Julien Barral
Créteil Cedex, France — Stéphane Seuret

Contents

The Rauzy Gasket

Pierre Arnoux and Štěpán Starosta

Abstract We define the Rauzy gasket as a subset of the standard two-dimensional simplex associated with letter frequencies of ternary episturmian words. We prove that the Rauzy gasket is homeomorphic to the usual Sierpiński gasket (by a two-dimensional generalization of the Minkowski ? function) and to the Apollonian gasket (by a map which is smooth on the boundary of the simplex). We prove that it is also homothetic to the invariant set of the fully subtractive algorithm, hence of measure 0.

1 Introduction

Strict episturmian ternary words, also called Arnoux–Rauzy words, are a natural generalization of Sturmian words (see Sect. 2 for the definitions). Each such word is uniquely ergodic, and in particular, its letters have a well-defined frequency; one can prove that these frequencies completely define the minimal symbolic system associated with such a word.

These dynamical systems are associated with a particular family of interval exchange transformations (see [1]). It is known that some of these systems (in particular those defined by a substitution) can be represented by a toral rotation,

P. Arnoux (✉)
IML - Institut de Mathématiques de Luminy, UPR-Cnrs 9016, Campus de Luminy case 907, 13288 Marseille Cedex 9, France
e-mail: arnoux@iml.univ-mrs.fr

Š. Starosta
Department of Mathematics, FNSPE, Czech Technical University in Prague, Trojanova 13, 120 00 Prague 2, Czech Republic

Department of Theoretical Computer Science, FIT, Czech Technical University in Prague, Thákurova 9, 160 00 Prague 6, Czech Republic
e-mail: stepan.starosta@fit.cvut.cz

J. Barral and S. Seuret (eds.), *Further Developments in Fractals and Related Fields*, Trends in Mathematics, DOI 10.1007/978-0-8176-8400-6_1,

Fig. 1 The Rauzy gasket

and in particular, they have pure discrete spectrum (see [13]); on the other hand, it is known that some examples of Arnoux–Rauzy words are not balanced (see [3]); hence they cannot be represented by a toral rotation. It would be interesting to understand what is the general behavior; a preliminary step would be to find a "good" measure on the parameter set of the family of episturmian systems.

Arnoux–Rauzy words admit a natural renormalization process, which acts on the frequencies; this renormalization can be considered as a kind of generalized continued fraction; indeed, the equivalent renormalization on Sturmian words is a symbolic version of the classical additive continued fraction. It is then natural to look for an invariant measure (a Gauss measure) for the renormalization.

In this chapter, we define the Rauzy gasket as the set of admissible vectors of letter frequencies for episturmian ternary words; it is a compact subset of the two-dimensional simplex (see Fig. 1). The Rauzy gasket parametrizes the set of episturmian systems, and there is a generalized continued fraction algorithm acting on it. We would like to find a Gauss measure for this algorithm; however, since this set is fractal, one would first need to compute its Hausdorff dimension and Hausdorff measure.

We are far from reaching this goal, and some preliminary investigations are given in this chapter. We prove that the Rauzy gasket is homeomorphic to the usual Sierpiński gasket by a map which is a generalization of the Minkowski ? function; this map is not differentiable, and not absolutely continuous, so the Hausdorff dimensions of both sets have no reason to be the same. We also prove that the Rauzy gasket is homeomorphic to the Apollonian gasket by a quite regular map, since it is smooth on the boundary of the complementary triangles, although it is not a diffeomorphism on the rational points. We then show that the same set occurs in a classical two-dimensional continued fraction algorithm, the fully subtractive algorithm. We deduce from the proof of [11] that the Rauzy gasket has zero Lebesgue measure.

In Sect. 2, we give the necessary definitions for episturmian words and explain the origin of the problem. In Sect. 3, we define the Rauzy gasket and the related continued fraction. In Sect. 4, we prove that the Rauzy gasket is homeomorphic to the classical Sierpiński gasket by a generalization of the Minkowski ? function. In Sect. 5, we prove that it is homeomorphic to the Apollonian gasket. In Sect. 6, we show that the continued fraction associated with the Rauzy gasket is conjugate by a linear change of coordinates to the induction of the fully subtractive algorithm on the central part of the simplex, and we deduce that its Lebesgue measure is 0. In the last section, we give a few remarks and open questions.

2 Preliminaries

2.1 Background: Complexity and Sturmian Words

Let $\mathscr{A}$ denote an *alphabet*, a finite set of *letters*. A finite (infinite) sequence of letters is called a *finite (infinite) word*. We say that a finite word $\mathbf{w} = w_0 w_1 \ldots w_n$, where $w_i \in \mathscr{A}$, is a *factor* of a word $\mathbf{v} = v_0 v_1 \ldots$ (finite or infinite) if there exists an index k such that $w_0 w_1 \ldots w_n = v_k v_{k+1} \ldots v_{k+n}$. Furthermore, such an index k is called an *occurrence* of $\mathbf{w}$ in $\mathbf{v}$. By $|\mathbf{v}|_{\mathbf{w}}$ we denote the *number of occurrences* of $\mathbf{w}$ in $\mathbf{v}$.

The *language* of an infinite word $\mathbf{u}$ is the set of factors of $\mathbf{u}$. We say that this language is *closed under reversal* if, for any factor $\mathbf{w} = w_0 w_1 \ldots w_n$ of $\mathbf{u}$, its *reverse* word $w_n w_{n-1} \ldots w_0$ is also a factor of $\mathbf{u}$.

The *shift map* on $\mathscr{A}^\infty$ associates to any infinite word $\mathbf{u}$ the word $\mathbf{v}$ defined for all i by $v_i = u_{i+1}$. The dynamical system associated with a word $\mathbf{u}$ is the closure of its orbit by the shift; it is also the set of words whose language is contained in the language of $\mathbf{u}$. Hence it is completely determined by this language.

Let $\mathbf{u} = u_0 u_1 \ldots$ be an infinite word over $\mathscr{A}$ and $\mathbf{w}$ be a factor of $\mathbf{u}$. Let $f_{\mathbf{u}}(\mathbf{w})$ denote the limit $\lim_{n \to +\infty} \frac{|u_0 u_1 \ldots u_{n-1}|_{\mathbf{w}}}{n}$, if it exists. Such a number is then called the *frequency* of the factor $\mathbf{w}$ in $\mathbf{u}$.

The *factor complexity* of $\mathbf{u}$, denoted $\mathscr{C}(n)$, is the mapping which associates to an integer n the number of factors of $\mathbf{u}$ of length n. As usual, the *empty word* is counted as a factor and we have $\mathscr{C}(0) = 1$ and $\mathscr{C}(1) = \#\mathscr{A}$.

A factor $\mathbf{w}$ of $\mathbf{u}$ is said to be left (right) special, if there exist two distinct letters a and b such that $a\mathbf{w}$ and $b\mathbf{w}$ ($\mathbf{w}a$ and $\mathbf{w}b$) are factors of $\mathbf{u}$.

It is well-known that the complexity of a non eventually periodic word is strictly increasing; hence such a word has complexity $\mathscr{C}(n) \geq n+1$. Aperiodic words of minimal complexity are of particular interest.

Definition 1. An infinite word is called a *Sturmian word* if it is of minimal complexity $\mathscr{C}(n) = n+1$.

Let $\lfloor x \rfloor$ (resp. $\lceil x \rceil$) denote the *floor* function, i.e., the largest integer $n \leq x$ (resp. the *ceiling* function, that is the smallest integer $n \geq x$).

Definition 2. An infinite word $\mathbf{u}$ is a *rotation word* if there exist $\alpha, \beta \in [0,1]$ such that for all n, $u_n = \lfloor (n+1)\alpha + \beta \rfloor - \lfloor n\alpha + \beta \rfloor$ or for all n, $u_n = \lceil (n+1)\alpha + \beta \rceil - \lceil n\alpha + \beta \rceil$ (the difference between $\lfloor x \rfloor$ and $\lceil x \rceil$ is irrelevant unless $n\alpha + \beta$ takes integer values for some n).

One can prove that a rotation word is periodic if and only if α is rational, that the Sturmian words are the aperiodic rotation words, and that the closure in $\{0,1\}^{\mathbb{N}}$, for the product topology, of the set of Sturmian words is the set of rotation words.

Since, in a Sturmian word, there are three different factors of length 2, and words 10 and 01 must occur in a not eventually constant sequence, one of the words 00 or 11 does not occur. Hence, in a Sturmian word, one of the letters is *isolated*. Sturmian words admit a renormalization process by erasing the letter following the isolated letter, which gives a new Sturmian word; it is a symbolic counterpart of the classical continued fraction algorithm acting on the angle of the corresponding rotation. We will look more closely at this renormalization process in Sect. 4.1.

2.2 Arnoux–Rauzy Words and Episturmian Words: Definition

It is natural to look for good generalizations of Sturmian words, and one such family is the set of strict episturmian words:

Definition 3. Strict episturmian words on three letters, also called ternary Arnoux–Rauzy (AR for short) words, are infinite words of complexity $\mathscr{C}(n) = 2n+1$ such that for each n there is only one left special factor and one right special factor of length n.

Since $\mathscr{C}(1) = 3$, they are words on three letters; they were first described in [14] and were further studied in [1], where a geometric representation was introduced. AR words code a specific family of six-interval exchange transformations.

One inconvenience of this definition is that the set of AR words is not closed in $\mathscr{A}^{\infty}$ for the natural topology; this is already the case for Sturmian words, whose closure contains periodic rotation words. It is then natural to consider the closure; the language of any AR word, like the language of a Sturmian word, is closed under reversal, and this leads to the definition of a wider family of infinite words known as *episturmian* words (see [4, 6, 7] for a survey) defined on an arbitrary alphabet.

Definition 4. Episturmian words are the infinite words having their language closed under reversal and at most one left special factor of each length.

It is an easy consequence of the definition that the set of episturmian words is closed in $\mathscr{A}^\infty$. Some episturmian words, those who can be extended on the left in several ways, are of particular interest.

Definition 5. An aperiodic episturmian word is *standard* (or *characteristic*) if each of its prefix is a left special factor.

Remark 1. Since every prefix of a left special factor is a left special factor, the definition implies that every aperiodic episturmian word has the same set of factors as a unique standard episturmian word. In other words, the subshift generated by an aperiodic episturmian word contains a unique standard word, which can be used as canonical representative of the system.

If an episturmian word is periodic, a factor longer than the period cannot be special, so the definition does not apply. In that case, we will say by abuse of language that a periodic episturmian word is standard if any prefix is a special left factor whenever there exists a special left factor of the same length. One can prove that the finite subshift generated by a periodic nonconstant episturmian word contains exactly two standard words with this definition.

By a result of Boshernitzan [2], episturmian words are uniquely ergodic; hence the frequencies of factors in episturmian words exist and are positive. On frequencies of AR words see more in [16].

The next section gives a classification of ternary episturmian words.

2.3 Ternary AR Words: Renormalization

In what follows we will consider the alphabet to be fixed $\mathscr{A} = \{1,2,3\}$. For all $i \in \mathscr{A}$, we define the morphism σ_i on the free monoid $\mathscr{A}^*$ by

$$\sigma_i(j) = ij \quad \text{if} \quad j \neq i \qquad \text{and} \qquad \sigma_i(i) = i.$$

We denote by $\mathscr{S}$ the set $\{\sigma_1, \sigma_2, \sigma_3\}$. The following is a restatement of claims in [14].

Proposition 1. *Let* $\mathbf{u}$ *be an AR word and* σ *a morphism from* $\mathscr{S}$*. Then* $\sigma(\mathbf{u})$ *is an AR word.*

This leads to the following claim using our notation.

Corollary 1. *Let* $\mathbf{u}$ *be an AR word and* $i \in \mathscr{A}$*. Then for all letter* $j \in \mathscr{A}$ *we have*

$$f_{\sigma_i(\mathbf{u})}(j) = \frac{f_{\mathbf{u}}(j)}{2 - f_{\mathbf{u}}(i)} \quad \textit{for} \quad i \neq j \qquad \textit{and} \qquad f_{\sigma_i(\mathbf{u})}(i) = \frac{1}{2 - f_{\mathbf{u}}(i)}.$$

What is less obvious is that, conversely, any standard AR word can be renormalized by using one of the morphisms in $\mathscr{S}$. This is the content of the next proposition, which is again a restatement of claims in [14] in our terms.

Proposition 2. *Let* $\mathbf{v}$ *be a standard AR word; then there exist an index* $i \in \mathscr{A}$ *and a standard AR word* $\mathbf{u}$ *such that* $\mathbf{v} = \sigma_i(\mathbf{u})$.

Furthermore, we have

$$f_{\mathbf{u}}(j) = \frac{f_{\mathbf{v}}(j)}{f_{\mathbf{v}}(i)} \quad \text{for} \quad i \neq j \quad \text{and} \quad f_{\mathbf{u}}(i) = \frac{2f_{\mathbf{v}}(i) - 1}{f_{\mathbf{v}}(i)}.$$

Let $(\zeta_1, \zeta_2, \zeta_3)$ denote the vector of frequencies of letters of the AR word $\mathbf{v}$. (Clearly, $\zeta_1 + \zeta_2 + \zeta_3 = 1$.) It follows from [1] that ζ_1 and ζ_2 are rationally independent irrational numbers. Furthermore, it follows from the previous proposition that one of the frequencies is always strictly greater than the sum of the two others, i.e., one letter is always *dominating* the word. Moreover, the dominating letter is *separating* (see [4], Lemma 4), i.e., all factors of $\mathbf{v}$ of length 2 contain at least one occurrence of the separating letter. The renormalization procedure is very simple: to obtain the renormalized word, it suffices to erase the letter following each non-separating letter. In other words, the index i from the last proposition is clearly the separating letter of $\mathbf{v}$ and $\mathbf{u}$ is the renormalized word. Moreover, the previous proposition ensures that, for a standard AR word, this procedure can be infinitely iterated.

In this way, one can associate to any AR subshift an infinite sequence of morphisms in $\mathscr{S}$; this sequence can be seen as a symbolic version of a generalized continued fraction expansion on the set of frequencies, as we show in the next section. This sequence is also the *$\mathscr{S}$-adic expansion* of the subshift (see [5]).

This construction has been extended to episturmian words; the following theorem summarizes results in [6, 7].

Theorem 1. *Let* $\mathbf{u}$ *be a standard episturmian word; then, one can find a standard episturmian word* $\mathbf{v}$ *and a morphism* $\sigma_i \in \mathscr{S}$ *such that* $u = \sigma_i(v)$. *Furthermore, the morphism* σ_i *is uniquely defined by the frequencies of the letters of* $\mathbf{u}$, *unless* $\mathbf{u}$ *is periodic of period* 2.

It follows immediately that, to any episturmian system, one can associate by iterating this construction a sequence σ_{i_n} *of morphisms (the word* $(i_n)_{n\in\mathbb{N}}$ *is called the directive word); three cases are possible:*

1. *Every letter in* $\mathscr{A}$ *occurs infinitely often in* (i_n)*; then the word* $\mathbf{u}$ *is an AR word and* (i_n) *is uniquely defined.*
2. *One letter occurs a finite number of times in* (i_n)*; then the word* $\mathbf{u}$ *is the image by a morphism (finite composition of elements of* $\mathscr{S}$*) of a Sturmian word and* (i_n) *is uniquely defined.*
3. *The word* (i_n) *is eventually constant; then the word* $\mathbf{u}$ *is periodic, and if the word* (i_n) *is not constant, there are two such possible words, one ending in* ij^∞ *and the other in* ji^∞.

The basic element of the proof is that the morphism σ_i is determined by the largest frequency ζ_i. Since this largest frequency satisfies $\zeta_i \geq \zeta_j + \zeta_k$, it is uniquely defined, unless we have $\zeta_k = 0$ and $\zeta_i = \zeta_j$, which corresponds, up to a permutation of indices, to the frequencies $(1/2, 1/2, 0)$ and a periodic word of period 2, and to the directive words 12^∞ and 21^∞, depending on whether we choose to erase letter 1 or 2. It is easy to check that this phenomenon occurs during the renormalization of a word if and only if the vector of frequencies is rational.

Remark 2. This phenomenon already occurs for rotation words; in that case, it is linked to the fact that an irrational number has only one continued fraction expansion, but a rational number has two finite continued fraction expansions.

3 The Rauzy Gasket

We are interested in the set of all possible vectors of letter frequencies $(\zeta_1, \zeta_2, \zeta_3)$ of episturmian words. For convenience, we will speak of the set of frequencies. We can now define the Rauzy gasket and show some of its properties.

Definition 6. The Rauzy gasket, denoted by $\mathbf{R}$, is the set of frequencies of episturmian words.

We will also be interested in the following subsets of the Rauzy gasket:

Definition 7. We denote by $\mathbf{R}_{\text{aper}}$ the set of frequencies of aperiodic episturmian words and by $\mathbf{R}_{\text{AR}}$ the set of frequencies of AR words.

Lemma 1. *We have* $\mathbf{R}_{\text{AR}} \subset \mathbf{R}_{\text{aper}} \subset \mathbf{R} = \overline{\mathbf{R}_{\text{AR}}}$.

Proof. The only nontrivial fact is that $\mathbf{R} \subset \overline{\mathbf{R}_{\text{AR}}}$. But every episturmian word can be approached arbitrarily close by an AR word: it suffices to take a long prefix of the directive word of the word and to compose it with the directive word $(123)^\infty$ to get an AR word. Hence the frequency vector of the given episturmian word can be approximated as closely as we want by the frequency vector of an AR word. □

The elements of $\mathbf{R}_{\text{aper}}$ are exactly the irrational elements of $\mathbf{R}$. Indeed, the frequencies completely characterize an episturmian system. If the word is periodic, its frequencies are rational. On the other hand, if the frequencies are rational, the height of the frequency vector (defined as the sum of coefficients of the smallest collinear integer vector) is strictly decreasing under renormalization, unless the two smallest coordinates are 0; this implies that, starting from any rational frequency, a finite number of renormalizations changes the word to a constant word, so an episturmian word with rational frequencies is periodic.

One can prove that the elements of $\mathbf{R}_{\text{aper}}$ which are not in $\mathbf{R}_{\text{AR}}$ are irrational vectors which satisfy one rational relation, since they are the images of a Sturmian word, with only two letters, by a morphism. One could conjecture that the elements of $\mathbf{R}_{\text{AR}}$ are completely irrational, but we do not know a proof of this.

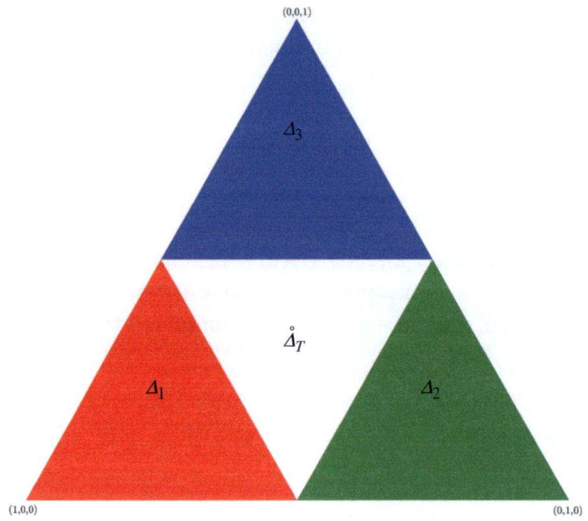

Fig. 2 The standard simplex Δ partitioned into four subsets $\Delta_1, \Delta_2, \Delta_3$, and $\mathring{\Delta}_T$

3.1 The Rauzy Gasket as an Iterated Function System

Let Δ denote the convex span of $\{e_1, e_2, e_3\}$, i.e.,

$$\Delta := \left\{(x_1,x_2,x_3) \in \mathbb{R}^3_{\geq 0} \mid \sum x_i = 1\right\}.$$

Let $\mathring{\Delta}_T$ denote the open set of triplets that satisfy the triangular inequalities $x_i < x_j + x_k$; it is the interior of the convex span of the centers of the sides of Δ:

$$\mathring{\Delta}_T := \left\{ (x_1,x_2,x_3) \in \Delta \,\middle|\, \forall j, x_j < \sum_{i \neq j} x_i \right\}.$$

Furthermore, let us denote for all j

$$\Delta_j := \left\{ (x_1,x_2,x_3) \in \Delta \,\middle|\, x_j \geq \sum_{i \neq j} x_i \right\}.$$

One has $\mathring{\Delta}_T = \Delta \setminus \bigcup \Delta_i$; see Fig. 2.

We consider the linear mapping $\tilde{F}$ defined on the set of strictly positive vectors which do not satisfy the triangle inequality by subtracting the two smaller coordinates form the larger one as follows:

$$\tilde{F} : (x_1,x_2,x_3) \mapsto \begin{cases} (x_1 - x_2 - x_3, x_2, x_3) & \text{if} \quad x_1 \geq x_2 + x_3, \\ (x_1, x_2 - x_1 - x_3, x_3) & \text{if} \quad x_2 \geq x_1 + x_3, \\ (x_1, x_2, x_3 - x_1 - x_2) & \text{if} \quad x_3 \geq x_1 + x_2, \end{cases}$$

Note that the definition is not consistent on the set of vectors of the form $(x,x,0)$, $(x,0,x)$, or $(0,x,x)$; we make an arbitrary choice in these three cases, which correspond to periodic words of period 2. As we have seen above, the renormalization operation is not well defined for these words.

This linear map, acting on the positive cone, gives rise to a projective map acting on $\Delta \setminus \mathring{\Delta}_T$; it will be denoted by $F : \Delta \setminus \mathring{\Delta}_T \mapsto \Delta$.

If $\mathbf{v}$ is any standard episturmian word and $\mathbf{u}$ is the corresponding renormalized word, one can now rewrite the formula from Proposition 2:

$$(f_{\mathbf{u}}(1), f_{\mathbf{u}}(2), f_{\mathbf{u}}(3)) = F\left(f_{\mathbf{v}}(1), f_{\mathbf{v}}(2), f_{\mathbf{v}}(3)\right).$$

The map $\tilde{F}$ is 3-to-1, and the inversion $\tilde{F}^{-1}$ has three branches, denoted $\tilde{f}_i$, for $i \in \mathscr{A}$. These branches define projective maps $f_i : \Delta \to \Delta_i$. These maps correspond to linear maps given by matrices M_i, defined as:

$$M_1 = \begin{pmatrix} 1 & 1 & 1 \\ 0 & 1 & 0 \\ 0 & 0 & 1 \end{pmatrix}, \qquad M_2 = \begin{pmatrix} 1 & 0 & 0 \\ 1 & 1 & 1 \\ 0 & 0 & 1 \end{pmatrix}, \qquad M_3 = \begin{pmatrix} 1 & 0 & 0 \\ 0 & 1 & 0 \\ 1 & 1 & 1 \end{pmatrix}.$$

We will use these inverse branches to find the set where F can be iterated infinitely many times; remark that $f_i(\Delta) = \Delta_i$ and $f_j(\Delta) = \Delta_j$, $i \neq j$ are disjoint except for one point.

It follows from Propositions 1, 2 that $\mathbf{R}$ satisfies

$$\mathbf{R} = \cup_{i \in \mathscr{A}} f_i(\mathbf{R})$$

so that $\mathbf{R}$ is the solution of an iterated function system. The function f_i being projective defined by positive matrix is contracting on the standard simplex; however, e_i, the ith vector of the standard basis of $\mathbb{R}^3$, is an indifferent fixed point for f_i, so this iterated function system is only weakly contracting, and a little care must be taken to apply the theorem of Hutchinson to prove unicity of the solution of this equation.

We define the family of maps $f_{i,j,n}$ by $f_{i,j,n} = f_i^n f_j$, for $i,j \in \mathscr{A}$, $i \neq j$, and $n \in \mathbb{N}^+$.

Lemma 2. *There exists a constant $c < 1$ such that the maps $f_{i,j,n}$ are strict contractions with contraction ratio less than c.*

Proof. Since the f_i, defined by positive matrices on the positive simplex, are weak contractions, it is enough to prove it for $n = 1$. Direct computation of the Jacobian matrix of $f_1 f_2$ shows that, on the simplex, it is everywhere contracting by a contraction factor at least $\frac{2}{3}$. The result follows by symmetry for all $f_i f_j$. □

Hence, the set of all the maps $f_{i,j,n}$ forms an infinite strictly contracting iterated function system; we can apply a modified version of Hutchinson's theorem. Let $H(\Delta)$ denote the set of all nonempty compact subsets of Δ, equipped with Hausdorff metric. Define $\Phi : H(\Delta) \mapsto H(\Delta)$ as

$$\Phi(X) = \overline{\bigcup_{i,j\in\mathscr{A},i\neq j,n\in\mathbb{N}^+} f_{i,j,n}(X)}.$$

It is clear from the above lemma that Φ is a strict contraction on $H(\Delta)$; hence it has a unique fixed point, which is $\mathbf{R}$. Indeed, the analysis of the previous section showed that any nonconstant episturmian word $\mathbf{v}$, having a nonconstant directive word, can be renormalized as $\mathbf{v} = \sigma_i^n \sigma_j(\mathbf{u})$; restated in terms of frequency, this means that the Rauzy gasket satisfies:

$$\mathbf{R} = \{e_1, e_2, e_3\} \cup \bigcup_{i,j,n} f_{i,j,n}\mathbf{R}$$

from which it follows immediately that $\Phi(\mathbf{R}) = \mathbf{R}$.

We can now prove the main result of this section:

Theorem 2. *The Rauzy gasket* $\mathbf{R}$ *is the unique nonempty compact subset of the standard simplex which satisfies the equation:*

$$\mathbf{R} = \cup_{i\in\mathscr{A}} f_i(\mathbf{R}).$$

Proof. The only thing to prove is the uniqueness. Let X be another solution; from $X = \cup_{i\in\mathscr{A}} f_i(X)$, we obtain that $f_{i,j,n}(X) \subset X$; hence $\Phi(X) \subset X$. Let x be any element of X; by definition of X, we can find a sequence x_n in X, with $x_0 = x$, and an infinite word $(i_n)_{n\in\mathbb{N}}$, such that $x_n = f_{i_n}(x_{n+1})$. If the word is not constant, x is in some $f_{i,j,n}(X)$. If it is constant, x is in $\cap_{n\in\mathbb{N}} f_i^n(X)$; hence $x = e_i$. Hence $X = \{e_1, e_2, e_3\} \cup (\cup_{i,j\in\mathscr{A},i\neq j,n\in\mathbb{N}^+} f_{i,j,n}X) \subset \Phi(X)$, so $\Phi(X) = X$ and $X = \mathbf{R}$. □

We note the following proposition, which might be useful to compute the Hausdorff dimension of $\mathbf{R}$:

Proposition 3. *The family* $\{f_{i,j,n}\}$ *satisfies the open set condition; that is, there is an open set* U *such that all the* $f_{i,j,n}(U)$ *are disjoint and contained in* U.

Proof. Let $\mathring{\Delta}$ be the interior of Δ. It is clear that $f_{i,j,n}(\mathring{\Delta}) \subset \mathring{\Delta}$; the explicit coordinates are easily computed, and the corresponding triangles are disjoint. Figure 3 shows the disposition of the triangles $f_{1,2,n}(\mathring{\Delta})$. □

3.2 *Symbolic Dynamics for the Rauzy Gasket*

The map F gives us a symbolic word associated with the elements of $\mathbf{R}$, unique except for the rational points. The easiest proof relies on the following lemma:

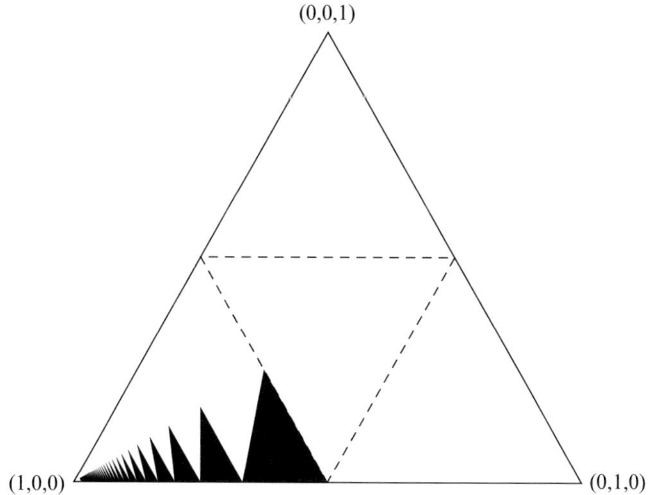

Fig. 3 The triangles $f_{1,2,n}(\mathring{\Delta})$

Lemma 3. *Let (i_n) be any infinite word in $\mathscr{A}^\infty$. The set $\cap_{n\in\mathbb{N}} f_{i_0} f_{i_1} \cdots f_{i_n}(\Delta)$ contains exactly one point.*

Proof. Remark first that $f_1^n(\Delta)$ is the triangle with vertices $f_1^n(e_1) = (1,0,0)$, $f_1^n(e_2) = (\frac{n}{n+1}, \frac{1}{n+1}, 0)$, $f_1^n(e_3) = (\frac{n}{n+1}, 0, \frac{1}{n+1})$, whose diameter tends to 0; the intersection of these triangles is the point $(1,0,0)$. Similarly $f_j^n(\Delta)$ converges to $\{e_j\}$.

Hence, if the word (i_n) is eventually constant, i.e., for all $n > N$ we have $i_n = j$ for some $j \in \mathscr{A}$, the limit of the corresponding set is reduced to the point $f_{i_0} f_{i_1} \cdots f_{i_N}(e_j)$.

If the word is not eventually constant, it can be decomposed in a unique way as a product of $f_{i,j,n}$; hence the diameter of the images goes to zero, so the intersection of this sequence of decreasing compact sets is reduced to a point. □

Definition 8. The symbolic coding for the Rauzy gasket is the map, $\pi_F : \mathscr{A}^\infty \to \mathbf{R}$, which associates to any infinite word (i_n) the unique point defined by the previous lemma.

This map is one-to-one, except on eventually constant words, corresponding to rational points, where it is 2-to-one. In the previous setting, the reciprocal map is easy to describe. Let ν be the coding map: $\cup_{i\in\mathscr{A}} \Delta_i \to \mathscr{A}$ defined by $\nu(x) = i$ if $x \in \Delta_i$ (this map is ill-defined on the three middle points, elements of $\Delta_i \cap \Delta_j$, $i \neq j$); the coding word of x is the word $(\nu(F^n(x))_{n\in\mathbb{N}}$. It is well defined on the irrational points, and one could easily and tediously give the descriptions of the two coding words for the rational points.

Remark that π_F is obviously continuous; it is true, but less obvious, that the reciprocal map is continuous where it is well defined.

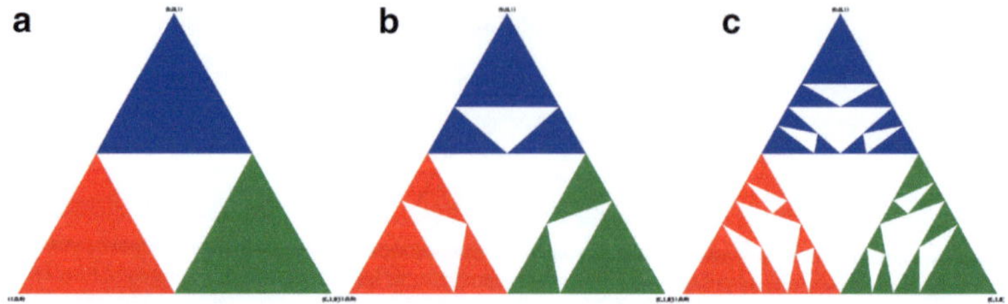

Fig. 4 (**a**) The set K_1. (**b**) The set K_2. (**c**) The set K_3

Remark 3. To get an explicit approximation of $\mathbf{R}$, a first possibility is to start with $V_0 = \{e_1, e_2, e_3\}$ and to build an increasing sequence. Define $\Psi(X) = \cup_{i\in\mathscr{A}} f_i(X)$ and consider the recurrent sequence given by $V_{n+1} = \Psi(V_n)$; the union of V_n for all n is the set of rational points of $\mathbf{R}$, and its closure is the Hausdorff limit $\mathbf{R}$ of the sequence V_n.

One can of course obtain a decreasing sequence by removing triangles. Denote $K_0 = \Delta$ and $K_n = \Psi^n(K_0)$. We get a decreasing sequence which converges to $\mathbf{R}$. Figure 4 shows K_1, K_2, and K_3.

Each set K_n can bee seen as a union of 3^n triangles. The set V_n as defined above is the set of the vertices of those triangles.

One easily shows that the sets $\mathbf{R}_{\text{aper}}$ and $\mathbf{R}_{\text{AR}}$ are (noncompact) solutions to the equation $\Psi(X) = X$; they also satisfy the equation $X = \cup_{i,j,n} f_{i,j,n}(X)$ (without taking the closure here). One can also show that, if $\mathring{\Delta}$ is the interior of Δ, we have $\mathbf{R}_{\text{AR}} = \cap_n \Psi^n(\mathring{\Delta})$.

4 Relation with the Sierpiński Gasket and a Generalization of the Question Mark Function

The above properties, in particular the approximation by the sets K_n and the three types of points in $\mathbf{R}$ corresponding to periodic, non-strict episturmian, and AR words, are reminiscent of the topology of the Sierpiński gasket. We will show that this set is in fact homeomorphic to the Sierpiński gasket; we first recall basic facts about the Minkowski question mark function.

4.1 The Minkowski Question Mark Function

Dynamical systems generated by rotation words are completely determined by the frequency x of the letter 0. As we recalled in Sect. 2.1, such a system can be

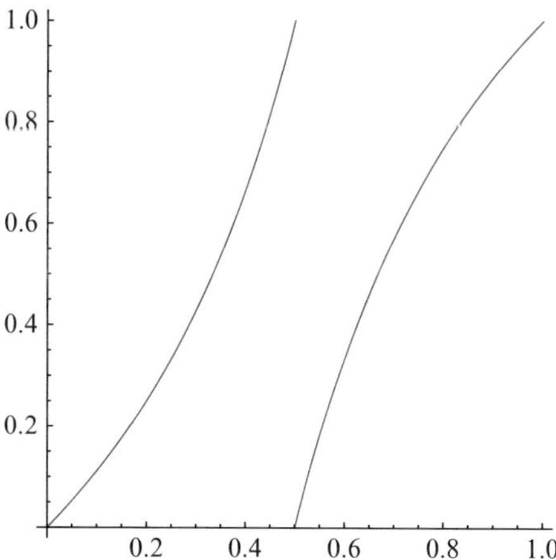

Fig. 5 The additive continued fraction map

renormalized by erasing any occurrence of the most frequent letter following the other letter; a simple computation shows that the frequency of 0 in the new system is $\phi(x)$, where ϕ is defined as

$$\phi : [0,1] \to [0,1] : \qquad x \mapsto \begin{cases} \frac{x}{1-x} & \text{if} \quad x < \frac{1}{2}, \\ 2 - \frac{1}{x} & \text{if} \quad x \geq \frac{1}{2}. \end{cases}$$

This map, represented in Fig. 5, is an exotic version of the usual additive continued fraction map; it has two indifferent fixed repelling points in 0 and 1, and it is ill-defined in $\frac{1}{2}$ (as we will see, the choice of the value 0 or 1 for $\phi(\frac{1}{2})$ is irrelevant; we have chosen here $\phi(\frac{1}{2}) = 0$).

By using the coding function ν defined by $\nu(x) = 0$ if $x < \frac{1}{2}$ and $\nu(x) = 1$ if $x \geq \frac{1}{2}$, we can associate to any $x \in [0,1]$ a coding word $\nu_\phi(x) = (\nu(\phi^n(x)))_{n\in\mathbb{N}}$. This defines a map $\nu_\phi : [0,1] \to \{0,1\}^{\mathbb{N}}$, which is one-to-one, and avoids only words which are eventually constant of value 1.

It is easy to prove that this map is increasing for the usual lexicographic order on $\{0,1\}^{\mathbb{N}}$, and that there is a reciprocal function which is increasing and one-to-one except on the set of eventually constant words.

One can do exactly the same thing with the function $\gamma : [0,1] \to [0,1] : x \mapsto 2x \mod 1$ and define a coding map $\nu_\gamma : [0,1] \to \{0,1\}^{\mathbb{N}} : x \mapsto (\nu(\psi^n(x)))_{n\in\mathbb{N}}$. This is the usual binary expansion of real numbers, which is also increasing and whose reverse map is defined, for any binary sequence $(\varepsilon_n)_{n\in\mathbb{N}}$, by $\nu_\gamma^{-1}(\varepsilon) = \sum_{n=0}^{\infty} \frac{\varepsilon_n}{2^{n+1}}$.

Definition 9. The question mark function ? of Minkowski is defined by ?: $[0,1] \to [0,1] \quad x \mapsto ?(x) = \nu_\gamma^{-1}\left(\nu_\phi(x)\right)$.

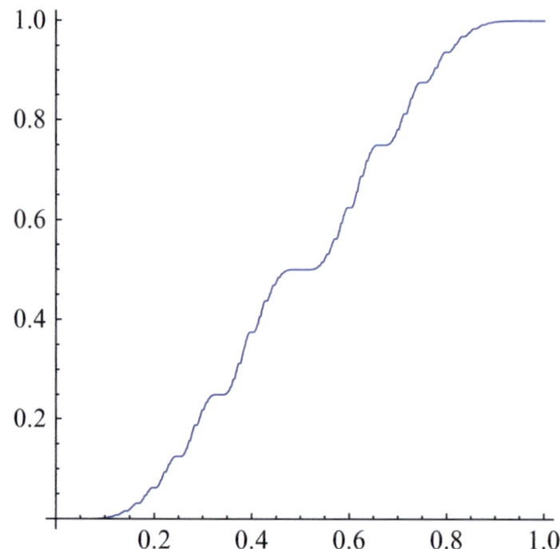

Fig. 6 The Minkowski ? function, also known as the slippery devil's staircase

The graph of this function is given in Fig. 6. The following properties of the function ? are easy to prove:

- The function ? is an increasing homeomorphism from $[0,1]$ to itself.
- It takes all rational numbers to dyadic numbers.
- It takes all quadratic numbers to rational numbers.
- It conjugates ϕ and γ: $\phi = ? \circ \gamma \circ ?^{-1}$.
- It has derivative 0 in 0 and in all rational numbers.

One can also prove that it is not absolutely continuous. Another way to define ? is to send the Farey set of order n to the set of all dyadic rationals between 0 and 1 such that their denominator in completely reduced form equals 2^k for some $0 \leq k \leq n$, preserving the order, and to show that this extends by continuity to an homeomorphism of $[0,1]$ to itself.

4.2 The Sierpiński Gasket

We consider the iterated function system $\{g_1, g_2, g_3\}$, where g_i is defined on Δ by $g_i(x_1, x_2, x_3) = \frac{(x_1, x_2, x_3) + e_i}{2}$.

The g_i are strict contractions with factor $\frac{1}{2}$, so the operator $H(\Delta) \to H(\Delta)$ given by $X \mapsto \bigcup_{i \in \mathscr{A}} g_i(X)$ has a unique fixed point, which is called the Sierpiński gasket; see Fig. 7. It will be denoted by $\mathbf{S}$.

Let G be the map defined on $\cup_{i \in \mathscr{A}} \Delta_i$ by $G(x_1, x_2, x_3) = 2(x_1, x_2, x_3) - e_i$ if $(x_1, x_2, x_3) \in \Delta_i$; it is 3-to-1, and its reciprocal branches are the contracting maps g_i. The Sierpiński gasket is the set on which G can be iterated infinitely, and by the dynamical system of the Sierpiński gasket we understand the dynamical system $(\mathbf{S}, G)$.

Fig. 7 The Sierpiński gasket

4.3 A Generalization of the Minkowski Question Mark Function

We can define a coding for the Sierpiński gasket, as for the Rauzy gasket, by $\nu_G(x) = (\nu(G^n(x))_{n\in\mathbb{N}}$. It is easy to prove that this coding is well defined, except for points with dyadic coordinates, where there are two possible codings, and that the map is continuous except for these points with dyadic coordinates; the reverse map π_G associates to any symbolic sequence (i_n) of elements of $\mathscr{A}$ the unique point in $\cap_{n\in\mathbb{N}} g_{i_0} g_{i_1} \cdots g_{i_n}(\Delta)$. It is continuous.

We can now define a generalization of the Minkowski question mark function.

Proposition 4. *The map $\Theta = \pi_G \circ \nu_F : \mathbf{R} \to \mathbf{S}$ is well defined and continuous.*

Proof. The map is clearly well defined, except for rational points, which may have two codings. A direct study shows that the point $(\frac{1}{2}, \frac{1}{2}, 0)$ admits the codings 21^∞ and 12^∞ and that these two codings have the same image (again $(\frac{1}{2}, \frac{1}{2}, 0)$) by π_G, so the image does not depend on the choice of the coding and is well defined. This property easily extends to all rational points.

Continuity is clear for the irrational points, since ν_F is continuous in these points. A local study shows that symbolic coding of points close to $(\frac{1}{2}, \frac{1}{2}, 0)$ must have a long prefix common with one of the two possible codings for this point; hence their images by Θ are close, which proves the continuity at this point. A similar proof works for any rational point. □

Proposition 5. *The dynamical systems* $(\mathbf{S}, G)$ *and* $(\mathbf{R}, F)$ *are conjugate by* Θ.

Proof. This is an immediate consequence of the fact that π_F and π_G conjugate, respectively, except for a countable set, $(\mathbf{S}, G)$ and $(\mathbf{R}, F)$ to the shift on $\mathscr{A}^\infty$. □

Proposition 6. *The restriction of* Θ *to the segment of the boundary of* Δ *joining* e_1 *and* e_2 *is the Minkowski ? function.*

Proof. It suffices to remark that the restriction of F to this segment is exactly the function ϕ and the restriction of G is the function γ, so the conjugacy must be the question mark function. □

Remark 4. Another higher dimensional generalization of the question mark function has been described in [12].

5 The Apollonian Gasket

The Apollonian gasket $\mathbf{A}$ can be described as follows: consider three pairwise tangent circles, which define a curvilinear triangle in the complex plane. Remove from this triangle the unique disk which is tangent to the three circles; we obtain three smaller triangles, each delimited by three pairwise tangent circles, and we can iterate the procedure. The limit set is the Apollonian gasket.

Although it might seem to depend on the initial configuration of circles, there is only one Apollonian gasket up to conjugacy by a Möbius transformation. Indeed, the triangle of tangency points completely determines the centers of the three circles, which are on the tangents to the circumscribed circle at the tangency points; but the group of Möbius transformations acts transitively on the set of triangles; since Möbius transformations preserve circles, this action extends to the family of Apollonian gaskets.

It will be convenient to take as tangency points $0, 1$, and $\frac{1+i}{2}$, so the circles are the circles C_1, C_2 with radius $\frac{1}{2}$ and respective centers $1 + \frac{i}{2}$ and $\frac{i}{2}$ and the horizontal axis, which is the generalized circle C_3 with infinite radius (see Fig. 8). We will call $\mathbf{A}$ the subset of $\mathscr{C}$ defined by this gasket.

The tangent circle to these three circles in the bounded region is the circle C of center $\frac{1}{2} + \frac{i}{8}$ and radius $\frac{1}{8}$. Note that C, C_j, C_k also define a new version of the Apollonian gasket, which we will denote by $\mathbf{A}_i$. We can find a Möbius transformation h_i which preserves C_j and C_k and sends C_i to C.

We denote by h_i the corresponding matrix in $SL(2, \mathscr{C})$. Computation shows that

$$h_1 = \begin{pmatrix} 1 & 0 \\ 1 & 1 \end{pmatrix}, \qquad h_2 = \begin{pmatrix} 0 & 1 \\ -1 & 2 \end{pmatrix}, \qquad h_3 = \begin{pmatrix} i & 1 \\ 2i & 2-i \end{pmatrix}.$$

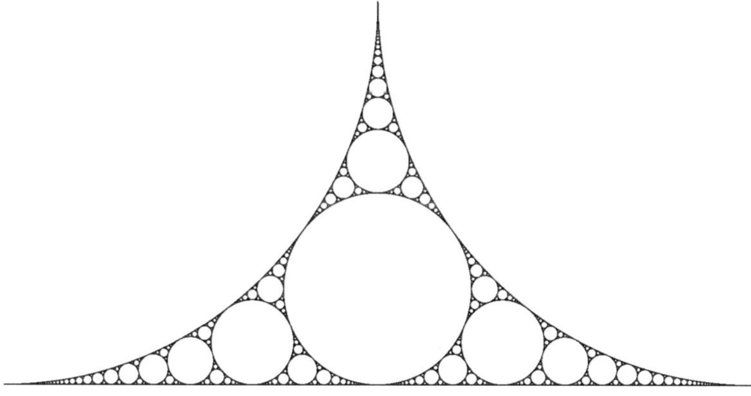

Fig. 8 The Apollonian gasket

Since the map h_i sends $\mathbf{A}$ to $\mathbf{A}_i$, the Apollonian gasket satisfies $\mathbf{A} = \cup_{i \in \mathscr{A}} h_i \mathbf{A}$; it is the solution of a conformal parabolic IFS. It has been thoroughly investigated (see for example [9, 10, 15]). Its exact Hausdorff dimension is not known, but it has been proved that its Hausdorff measure is finite.

By taking the inverse of the h_i, one can define a map $H : \mathbf{A} \to \mathbf{A}$ which is 3-to-1 and a coding map ν_H. Exactly as in the previous section, one can prove the following proposition:

Proposition 7. *There exists a homeomorphism* $\mathbf{R} \to \mathbf{A}$ *which conjugates the dynamical system* $(\mathbf{R}, F)$ *to* $(\mathbf{A}, H)$.

This map is certainly not a diffeomorphism, since it takes an equilateral triangle to a curvilinear triangle with angles 0; for the same reason, it cannot be a diffeomorphism in any rational point of the Rauzy gasket. It is however more regular than the conjugacy defined in the previous section.

Proposition 8. *The restriction of the conjugacy to the lower boundary of the Rauzy gasket in the identity.*

Proof. We already remarked in the proof of Proposition 6 that the restriction of F to this lower boundary was the function ϕ defined by $\phi(x) = \frac{x}{1-x}$ if $x < \frac{1}{2}$ and $\phi(x) = 2 - \frac{1}{x}$ if $x \geq \frac{1}{2}$; but the formulas above show that the restriction of H to the segment $[0, 1]$ is given by the same formula, hence the result. □

It follows immediately that the restriction of the map to the boundary of any triangle in the complement of the Rauzy gasket is smooth; that is, the restriction to the irrational points of the complement of $\mathbf{R}_{\text{AR}}$ is smooth.

6 Relation with the Fully Subtractive Algorithm

6.1 The Fully Subtractive Algorithm

The fully subtractive algorithm has been treated for instance in [8,11]. We first recall its definition and some results.

The fully subtractive algorithm is defined on the positive cone $\mathbb{R}^3_{\geq 0}$; it subtracts the smallest number from the two others, i.e., it is given by the map $S : \mathbb{R}^3_{\geq 0} \mapsto \mathbb{R}^3_{\geq 0}$ defined by

$$\tilde{S} : (x_1, x_2, x_3) \mapsto \begin{cases} (x_1, x_2 - x_1, x_3 - x_1) & \text{if} \quad x_1 \leq x_2, x_1 \leq x_3, \\ (x_1 - x_2, x_2, x_3 - x_2) & \text{if} \quad x_2 \leq x_1, x_2 \leq x_3, \\ (x_1 - x_3, x_2 - x_3, x_3) & \text{if} \quad x_3 \leq x_1, x_3 \leq x_2, \end{cases}$$

Note that the definition is again not completely consistent; for the set of vectors having two coordinates equal to each other we make an arbitrary choice. Since the algorithm is clearly equivariant under permutations of coordinates, the algorithm is often defined on the quotient space given by $x_1 \leq x_2 \leq x_3$, with a reordering of the image vector; this removes the problem, but makes the geometry less clear.

By considering the action of $\tilde{S}$ on projective space, we can define a map $S : \Delta \to \Delta$, with barycentric coordinates. If one coordinate is 0, the point is fixed. Thus, the set of fixed points of S is the boundary of Δ. The map S is 3-to-1; its restriction to the set Γ_i defined by $x_i \leq \inf(x_j, x_k)$ is a homeomorphism from Γ_i to Δ.

Computation shows that the segment $x_i = \frac{1}{2}$ is invariant by S. Indeed, if $z < y < \frac{1}{2}$ and $y + z = \frac{1}{2}$, we have $\tilde{S}(\frac{1}{2}, y, z) = (\frac{1}{2} - z, y - z, z)$, so that after renormalization $S(\frac{1}{2}, y, z) = (\frac{1}{2}, \frac{y-z}{1-2z}, \frac{z}{1-2z})$. It follows that the restriction of S to Δ_i preserves this triangle.

The restriction of S to the segment $(\frac{1}{2}, \frac{1}{2} - z, z)$ is conjugate by $(\frac{1}{2}, \frac{1}{2} - z, z) \mapsto 2z$ to the map ϕ of Sect. 4.1. The points $(\frac{1}{2}, \frac{1}{2} - z, z)$, with $z \in \mathbb{Q}$ of this segment, are sent by a power of S first to $(\frac{1}{2}, \frac{1}{4}, \frac{1}{4})$ and then, after an arbitrary choice (since S is ill-defined in this point) to one of the endpoints $(\frac{1}{2}, \frac{1}{2}, 0)$ or $(\frac{1}{2}, 0, \frac{1}{2})$ which are fixed by S. The other points have their orbit contained in the interior of the segment. By linearity, the map S sends the segment joining the fixed point e_i to a point P on $x_i = \frac{1}{2}$ to the segment joining e_i to $S(P)$. Hence the orbit of any point in the interior of Δ_i either ends in a finite number of steps on the boundary (if it is on the segment joining e_i to a rational point) or tends to the vertex e_i, since computation shows that in that case the coordinate x_i tends to a limit which can only be 1; this last set has obviously full measure in Δ_i.

Hence there are three attractors of the system (Δ, S)—the vertices of the standard simplex. Figure 9 shows the action of S on Γ_3, with the preimages by S of the four triangles Δ_T and Δ_i, for $i \in \mathcal{A}$. Figure 10 shows the three basins of attraction, distinguished by different colors.

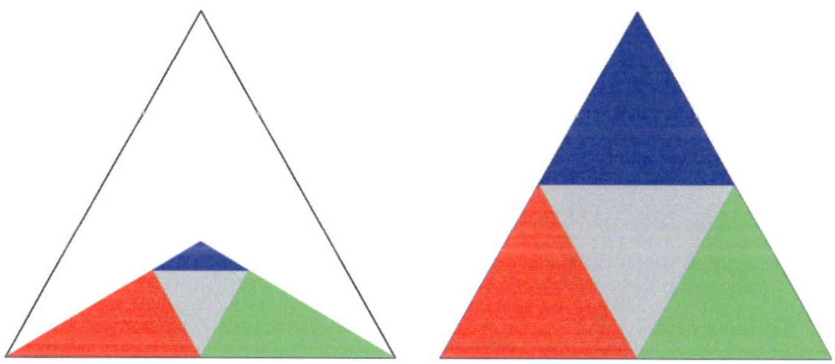

Fig. 9 The action of S on Γ_3

Fig. 10 Basins of attraction of the dynamical system (Δ, S). (More precisely, the set $S^{-8}(\Delta_i)$ is depicted for all i and colored by a different color)

6.2 *The Fully Subtractive Algorithm as an Extension of the Rauzy Gasket*

The map F associated with the Rauzy gasket was not defined on the central triangle; we will now extend it on all of Δ, by enlarging the set of definition to include points with negative coordinates. Let Δ' denote the convex span of $\{(-1,1,1),(1,-1,1),(1,1,-1)\}$, and let $\Delta'_i = \{(x_1,x_2,x_3) \in \Delta' \mid x_i \geq x_j, x_k\}$. Note that, with the notations of Sect. 3, Δ'_i contains Δ_i; we extend F to a map F' on Δ' by extending to Δ'_i the formula on Δ_i. It is the projective map associated with the piecewise linear map $\tilde{F}'$:

$$\tilde{F}' : (x_1,x_2,x_3) \mapsto \begin{cases} (x_1 - x_2 - x_3, x_2, x_3) & \text{if} \quad x_1 \geq x_2, x_1 \geq x_3, \\ (x_1, x_2 - x_1 - x_3, x_3) & \text{if} \quad x_2 \geq x_1, x_2 \geq x_3, \\ (x_1, x_2, x_3 - x_1 - x_2) & \text{if} \quad x_3 \geq x_1, x_3 \geq x_2. \end{cases}$$

Proposition 9. *(Δ', F') is conjugate to (Δ, S).*

Proof. It is enough to prove it for the piecewise linear maps $\tilde{F}'$ and $\tilde{S}$. Let P be the matrix that sends the canonical basis to the vertices of Δ'; let A (resp. B) be the matrix of $\tilde{F}'$ on the cone on Δ'_1 (resp. the matrix of $\tilde{S}$ on the cone on Γ_1). We have

$$\tilde{P} = \begin{pmatrix} -1 & 1 & 1 \\ 1 & -1 & 1 \\ 1 & 1 & -1 \end{pmatrix}, \qquad A = \begin{pmatrix} 1 & -1 & -1 \\ 0 & 1 & 0 \\ 0 & 0 & 1 \end{pmatrix}, \qquad B = \begin{pmatrix} 1 & 0 & 0 \\ -1 & 1 & 0 \\ -1 & 0 & 1 \end{pmatrix}.$$

Computation shows that $P(\Gamma_i) = \Delta'_i$ and that $B = P^{-1}AP$. □

In Fig. 10, the Rauzy gasket appears as the complement of the three basins of attraction; the dynamical system of the Rauzy gasket is the chaotic part of the fully subtractive algorithm.

6.3 Two Properties of the Rauzy Gasket

It is known that the fully subtractive algorithm is, in continued fraction terms, not convergent: for almost every point, the symbolic dynamics does not define the point. More precisely, we have the following:

Theorem 3 ([8, 11], Theorem 1). *For almost all $x \in \mathbb{R}^3_{\geq 0}$ we have*

$$\lim_{j \to +\infty} \tilde{S}^j(x) \neq (0,0,0).$$

We can restate the theorem in terms of the projective map on the simplex:

Corollary 2. *Almost any point of Δ tends to one of the three vertices under the action of S.*

Since the map P sends the Rauzy gasket to the complement of the attraction basins, we obtain the following:

Corollary 3. *The set* **R** *has zero Lebesgue measure.*

Remark 5. We could give a direct proof of this corollary along the line of the proof of [11]. We consider the restriction of F' on Δ, and we want to prove that, almost surely, it cannot be iterated infinitely inside Δ. Since $F'(\mathring{\Delta}_T)$ is disjoint from Δ, it suffices to consider the restriction to one of the Δ_i, say Δ_1. The main problem is the indifferent fixed point in e_1; to avoid this, we define, for any $x \in \Delta_1$, the integer n_x which is the smallest integer such that $F'^{n_x}(x)$ is not in Δ_1, and we consider the map $x \to F'^{n_x}(x)$. One computes explicitly the continuity domain of this map and its reciprocal branches; one then shows that the branches are uniformly contracting and that the Jacobian has a bounded distortion property. We can then consider the cylinders defined by the symbolic dynamics associated with this map and prove that the proportion of any cylinder which goes to Δ_T and leaves the simplex under the next iteration is bounded from below, which implies the corollary.

Proposition 10. *For all $y \in P^{-1}\mathbf{R}$, $\delta > 0$ and $i \in \mathscr{A}$, there exists $y_i \in \mathring{\Delta}$, $|y - y_i| < \delta$ such that*

$$\lim_{j \to +\infty} S^j(y_i) = e_i.$$

In other words, any uncolored point in $\mathring{\Delta}$ in Fig. 10 has all three colors in any of its neighborhood.

Proof. Let us first denote A_i the basin of attraction to the attractor e_i:

$$A_i = \left\{ x \in \mathring{\Delta} \;\middle|\; \lim_{j \to +\infty} S^j(x) = e_i \right\}.$$

Using Remark 3 one can see that $\mathbf{R} = \lim_{j \to +\infty}(V_j)$ where V_j is the set of antecedents of order j by F of the three vertices. Thus it suffices to show that for all j that every point $x \in P^{-1}(V_j)$ lies on all three boundaries of the sets A_i for all i. The proof is by induction on j.

The point $P^{-1}e_1 = (0, \frac{1}{2}, \frac{1}{2})$ is on the boundary of Δ_2 and Δ_3; it is also the limit of the points $\left(\frac{1}{2n+1}, \frac{n}{2n+1}, \frac{n}{2n+1}\right)$ which are preimages of e_1 under S; hence $P^{-1}V_0$ lies on the boundary of the sets A_i.

Suppose it is true for $j < N$. We apply S^{-1} to both sides of the equation $P^{-1}(V_{N-1}) \subset \bigcap_{i \in \mathscr{A}} \overline{A_i}$. On the left-hand side we have $S^{-1}P^{-1}(V_{N-1}) = P^{-1}F'^{-1}(V_{N-1}) = P^{-1}(V_N)$. And on the right-hand side we have $S^{-1} \bigcap_{i \in \mathscr{A}} \overline{A_i} = \bigcap_{i \in \mathscr{A}} \overline{A_i}$.

□

7 Final Remarks

We have restricted ourselves to ternary Arnoux–Rauzy words. However, the definition of episturmian words immediately extends to any finite alphabet, with the same renormalization procedure related to the fully subtractive algorithm, and we can define a Rauzy gasket in dimension d. Since the result for the fully subtractive algorithm in [8] is valid for any dimension, we can use it to prove that the Rauzy gasket in dimension d has Lebesgue measure 0.

The Rauzy gasket can be seen as a generalized Julia set for the dynamical system associated to the subtractive algorithm, and it shares some properties of a Julia set. One would like to know more about the Hausdorff dimension of **R** and the invariant measure of the underlying dynamical system; a first step should be to understand better the conjugacy with the Apollonian gasket: can we extend the regularity found on the boundary? Does it preserve Hausdorff dimension and measure? This would not completely solve the problem, since the Hausdorff dimension of the Apollonian gasket is not exactly known, but it is known (see [9, 15]) that its Hausdorff measure is finite.

It is a curious fact that the map S is dual (in the linear algebra sense) of F; this can be used to give a natural extension of the dynamical system of the Rauzy gasket as the skew product:

$$\overline{F}: \mathbf{R} \times P^{-1}\mathbf{R} \to \mathbf{R} \times P^{-1}\mathbf{R} \qquad (x,y) \mapsto (f_i^{-1}(x), {}^t f_i(y)) \text{ if } x \in \Delta_i,$$

where f_i is branch of F^{-1} such that $f_i(\Delta) = \Delta_i$. This remark might be useful to study the invariant measures for F.

Acknowledgments The second author acknowledges financial support by the Czech Science Foundation grant GAČR 201/09/0584, by the grants MSM6840770039 and LC06002 of the Ministry of Education, Youth, and Sports of the Czech Republic, and by the grant of the Grant Agency of the Czech Technical University in Prague grant No. SGS11/162/OHK4/3T/14.

References

1. Arnoux, P., Rauzy, G.: Représentation géométrique de suites de complexité $2n+1$. Bull. Soc. Math. France **119**, 199–215 (1991)
2. Boshernitzan, M.: A unique ergodicity of minimal symbolic flows with linear block growth. J. Anal. Math. **46**, 196–213 (1993)
3. Cassaigne, J., Ferenczi, S., Zamboni, L.Q.: Imbalances in Arnoux-Rauzy sequences. Ann. Inst. Fourier **50**, 1265–1276 (2000)
4. Droubay, X., Justin, J., Pirillo, G.: Episturmian words and some constructions of de Luca and Rauzy. Theoret. Comput. Sci. **255**(1-2), 539–553 (2001)
5. Fogg, N.P.: Substitutions in Arithmetics, Dynamics and Combinatorics, 1st edn. Springer, New York (2002)
6. Glen, A., Justin, J.: Episturmian words: a survey. Theoret. Inf. Appl. **49**(3), 403–442 (2009)
7. Justin, J., Pirillo, G.: Episturmian words and episturmian morphisms. Theoret. Comput. Sci. **276**(1-2), 281–313 (2002). doi:http://dx.doi.org/10.1016/S0304-3975(01)00207-9
8. Kraaikamp, C., Meester, R.: Ergodic properties of a dynamical system arising from percolation theory. Ergod. Theor. Dyn. Syst. **15**, 653–661 (1995)
9. Mauldin, R.D., Urbański, M.: Dimension and measures for a curvilinear Sierpiński gasket or Apollonian packing. Adv. Math. **136**(1), 26–38 (1998)

10. McMullen, C.T.: Hausdorff dimension and conformal dynamics III: computation of dimension. Am. J. Math. **120**(4), 691–72 (1998)
11. Meester, R.W.J., Nowicki, T.: Infinite clusters and critical values in two dimensional circle percolation. Isr. J. Math. **68**(1), 63–81 (1989)
12. Panti, G.: Multidimensional continued fractions and a Minkowski function. Monatsh. Math. **154**, 247–264 (2008)
13. Rauzy, G.: Nombres algébriques et substitutions. Bull. Soc. Math. France **110**, 147–178 (1982)
14. Rauzy, G.: Suites à termes dans un alphabet fini. Séminaire de Théorie des Nombres de Bordeaux **Année 1982–1983**(exposé 25) (1983)
15. Sullivan, D.: Entropy, Hausdorff measure old and new, and limit sets of geometrically finite Kleinian groups. Acta Math. **153**, 259–277 (1984)
16. Wozny, N., Zamboni, L.Q.: Frequencies of factors in Arnoux-Rauzy sequences. Acta Arith. **96**, 261–278 (2001)

On the Hausdorff Dimension of Graphs of Prevalent Continuous Functions on Compact Sets

Frédéric Bayart and Yanick Heurteaux

Abstract Let K be a compact set in $\mathbb{R}^d$ with positive Hausdorff dimension. Using a fractional Brownian motion, we prove that in a prevalent set of continuous functions on K, the Hausdorff dimension of the graph is equal to $\dim_{\mathscr{H}}(K)+1$. This is the largest possible value. This result generalizes a previous work due to J.M. Fraser and J.T. Hyde which was exposed in the conference *Fractals and Related Fields II*. The case of α-Hölderian functions is also discussed.

1 Introduction

Let $d \geq 1$ and let K be a compact subset in $\mathbb{R}^d$. Denote by $\mathscr{C}(K)$ the set of continuous functions on K with real values. This is a Banach space when equipped with the supremum norm, $\|f\|_\infty = \sup_{x\in K}|f(x)|$. The graph of a function $f \in \mathscr{C}(K)$ is the set

$$\Gamma_f^K = \{(x, f(x))\ ;\ x \in K\} \subset \mathbb{R}^{d+1}.$$

It is often difficult to obtain the exact value of the Hausdorff dimension of the graph Γ_f^K of a given continuous function f. For example, a famous conjecture says that the Hausdorff dimension of the graph of the Weierstrass function

$$f(x) = \sum_{k=0}^{+\infty} 2^{-k\alpha}\cos(2^k x),$$

F. Bayart (✉) • Y. Heurteaux
Laboratoire de Mathématiques, Clermont Université, Université Blaise Pascal, BP 10448, F-63000 Clermont-Ferrand, France

Laboratoire de Mathématiques, CNRS, Complexe Scientifique des Cézeaux, UMR 6620, F-63177 Aubiere Cedex, France
e-mail: Frederic.Bayart@math.univ-bpclermont.fr; Yanick.Heurteaux@math.univ-bpclermont.fr

J. Barral and S. Seuret (eds.), *Further Developments in Fractals and Related Fields*, Trends in Mathematics, DOI 10.1007/978-0-8176-8400-6_2,

where $0 < \alpha < 1$, satisfies

$$\dim_{\mathscr{H}}\left(\Gamma_f^{[0,2\pi]}\right) = 2 - \alpha.$$

This is the natural expected value, but, to our knowledge, this conjecture is not yet solved.

If we add some randomness, the problem becomes much easier and Hunt proved in [12] that the Hausdorff dimension of the graph of the random Weierstrass function

$$f(x) = \sum_{k=0}^{+\infty} 2^{-k\alpha}\cos(2^k x + \theta_k)$$

where $(\theta_k)_{k\geq 0}$ is a sequence of independent uniform random variables is almost surely equal to the expected value $2-\alpha$.

In the same spirit we can hope to have a generic answer to the following question:

"What is the Hausdorff dimension of the graph of a continuous function?"

Curiously, the answer to this question depends on the type of genericity we consider. If genericity is relative to the Baire category theorem, Mauldin and Williams proved at the end of the 1980s the following result:

Theorem 1 ([14]). *For quasi-all functions $f \in \mathscr{C}([0,1])$, we have*

$$\dim_{\mathscr{H}}\left(\Gamma_f^{[0,1]}\right) = 1.$$

This theorem was recently generalized to the case of a metric compact set K. In that situation, the Hausdorff dimension of the graph of quasi-all functions $f \in \mathscr{C}(K)$ is equal to $\dim_{\mathscr{H}}(K)$ (see [1]).

This statement on the Hausdorff dimension of the graph is very surprising because it seems to say that a generic continuous function is quite regular. Indeed it is convenient to think that there is a deep correlation between strong irregularity properties of a function and large values of the Hausdorff dimension of its graph.

This curious result seems to indicate that genericity in the sense of the Baire category theorem is not "the good notion of genericity" for this question. In fact, when genericity is related to the notion of prevalence (see Sect. 2 for a precise definition), Fraser and Hyde recently obtained the following result.

Theorem 2 ([7]). *Let $d \in \mathbb{N}^*$. The set*

$$\left\{f \in \mathscr{C}([0,1]^d)\ ;\ \dim_{\mathscr{H}}\left(\Gamma_f^{[0,1]^d}\right) = d+1\right\}$$

is a prevalent subset of $\mathscr{C}([0,1]^d)$.

This result says that the Hausdorff dimension of the graph of a generic continuous function is as large as possible and is much more in accordance with the idea that a generic continuous function is strongly irregular.

The main tool in the proof of Theorem 2 is the construction of a fat Cantor set in the interval $[0,1]$ and a stochastic process on $[0,1]$ whose graph has almost surely Hausdorff dimension 2. This construction is difficult to generalize to a compact set $K \neq [0,1]$. Nevertheless, there are in the literature stochastic processes whose almost sure Hausdorff dimension of their graph is well known. The most famous example is the fractional Brownian motion. Using such a process, we are able to prove the following generalization of Theorem 2.

Theorem 3. *Let $d \geq 1$ and let $K \subset \mathbb{R}^d$ be a compact set such that $\dim_{\mathscr{H}}(K) > 0$. The set*

$$\left\{f \in \mathscr{C}(K)\ ;\ \dim_{\mathscr{H}}\left(\Gamma_f^K\right) = \dim_{\mathscr{H}}(K) + 1\right\}$$

is a prevalent subset of $\mathscr{C}(K)$.

In this chapter, we have decided to focus to the notion of Hausdorff dimension of graphs. Nevertheless, we can mention that there are also many papers that deal with the generic value of the dimension of graphs when the notion of dimension is for example the lower box dimension (see [6, 10, 13, 17]) or the packing dimension (see [11, 15]).

The chapter is devoted to the proof of Theorem 3 and is organized as follows. In Sect. 2 we recall the basic facts on prevalence. In particular we explain how to use a stochastic process in order to prove prevalence in functional vector spaces. In Sect. 3, we prove an auxiliary result on fractional Brownian motion which will be the key of the main theorem. We finish the proof of Theorem 3 in Sect. 4. Finally, in a last section, we deal with the case of α-Hölderian functions.

2 Prevalence

Prevalence is a notion of genericity which generalizes to infinite-dimensional vector spaces, the notion of "almost everywhere with respect to Lebesgue measure." This notion has been introduced by Christensen in [3] and has been widely studied since then. In fractal and multifractal analysis, some properties which are true on a dense G_δ-set are also prevalent (see for instance [8, 9] or [2]), whereas some are not (see for instance [9] or [16]).

Definition 1. Let E be a complete metric vector space. A Borel set $A \subset E$ is called *Haar-null* if there exists a compactly supported probability measure μ such that, for any $x \in E$, $\mu(x+A) = 0$. If this property holds, the measure μ is said to be *transverse* to A.

A subset of E is called *Haar-null* if it is contained in a Haar-null Borel set. The complement of a Haar-null set is called a *prevalent* set.

The following results enumerate important properties of prevalence and show that this notion supplies a natural generalization of "almost every" in infinite-dimensional spaces:

- If A is Haar-null, then $x+A$ is Haar-null for every $x \in E$.
- If $\dim(E) < +\infty$, A is Haar-null if and only if it is negligible with respect to the Lebesgue measure.
- Prevalent sets are dense.
- The intersection of a countable collection of prevalent sets is prevalent.
- If $\dim(E) = +\infty$, compacts subsets of E are Haar-null.

In the context of a functional vector space E, a usual way to prove that a set $A \subset E$ is prevalent is to use a stochastic process. More precisely, suppose that W is a stochastic process defined on a probability space $(\Omega, \mathscr{F}, \mathbb{P})$ with values in E and satisfies.

$$\forall f \in E, \quad f+W \in A \quad \text{almost surely.}$$

Replacing f by $-f$, we get that the law μ of the stochastic process W is such that

$$\forall f \in E, \quad \mu(f+A) = 1.$$

In general, the measure μ is not compactly supported. Nevertheless, if we suppose that the vector space E is also a Polish space (that is if we add the hypothesis that E is separable), then we can find a compact set $Q \subset E$ such that $\mu(Q) > 0$. It follows that the compactly supported probability measure $\nu = (\mu(Q))^{-1}\mu_{|Q}$ is transverse to $E \setminus A$.

3 On the Graph of a Perturbed Fractional Brownian Motion

In this section, we prove an auxiliary result which will be the key of the proof of Theorem 3. For the definition and the main properties of the fractional Brownian motion, we refer to [5, Chap. 16].

Theorem 4. *Let K be a compact set in $\mathbb{R}^d$ such that $\dim_{\mathscr{H}}(K) > 0$ and $\alpha \in (0,1)$. Define the stochastic process in $\mathbb{R}^d$*

$$W(x) = W^1(x_1) + \cdots + W^d(x_d) \tag{1}$$

where $W^1, \ldots, W^d$ are independent fractional Brownian motions starting from 0 with Hurst parameter equal to α. Then, for any function $f \in \mathscr{C}(K)$,

$$\dim_{\mathscr{H}}\left(\Gamma^K_{f+W}\right) \geq \min\left(\frac{\dim_{\mathscr{H}}(K)}{\alpha}, \dim_{\mathscr{H}}(K) + 1 - \alpha\right) \quad \textit{almost surely.}$$

Let us remark that the conclusion of Theorem 4 is sharp. More precisely, suppose that $f = 0$ and let $\varepsilon > 0$. It is well known that the fractional Brownian motion is almost surely uniformly $(\alpha - \varepsilon)$-Hölderian. It follows that the stochastic process W is also uniformly $(\alpha - \varepsilon)$-Hölderian on K. It is then straightforward that the graph Γ_W^K satisfies

$$\dim_{\mathscr{H}}\left(\Gamma_W^K\right) \leq \dim_{\mathscr{H}}(K) + 1 - (\alpha - \varepsilon) \quad \text{a.s..}$$

On the other hand, the function

$$\Phi : x \in K \longmapsto (x, W(x)) \in \mathbb{R}^{d+1}$$

is almost surely $(\alpha - \varepsilon)$-Hölderian. It follows that

$$\dim_{\mathscr{H}}\left(\Gamma_W^K\right) \leq \frac{\dim_{\mathscr{H}}(K)}{\alpha - \varepsilon} \quad \text{a.s..}$$

The proof of Theorem 4 is based on the following lemma.

Lemma 1. *Let $s > 0$, $\alpha \in (0,1)$ and W be the process defined as in Eq. (1). There exists a constant $C := C(s) > 0$ such that for any $\lambda \in \mathbb{R}$, for any $x, y \in \mathbb{R}^d$,*

$$\mathbb{E}\left[\frac{1}{(\|x-y\|^2 + (\lambda + W(x) - W(y))^2)^{s/2}}\right] \leq \begin{cases} C\,\|x-y\|^{1-s-\alpha} & \text{provided } s > 1 \\ C\,\|x-y\|^{-\alpha s} & \text{provided } s < 1. \end{cases}$$

Proof. Observe that $W(x) - W(y)$ is a centered gaussian variable with variance

$$\sigma^2 = h_1^{2\alpha} + \cdots + h_d^{2\alpha}$$

where $h = (h_1, \ldots, h_d) = x - y$. Hölder's inequality yields

$$\|h\|^{2\alpha} \leq \sigma^2 \leq d^{1-\alpha}\|h\|^{2\alpha}.$$

Now,

$$\mathbb{E}\left[\frac{1}{(\|x-y\|^2 + (\lambda + W(x) - W(y))^2)^{s/2}}\right] = \int \frac{e^{-u^2/(2\sigma^2)}}{(\|h\|^2 + (\lambda + u)^2)^{s/2}} \frac{\mathrm{d}u}{\sigma\sqrt{2\pi}}.$$

Suppose that $s > 1$. We get

$$\begin{aligned} \mathbb{E}\left[\frac{1}{(\|x-y\|^2 + (\lambda + W(x) - W(y))^2)^{s/2}}\right] &\leq \int \frac{\mathrm{d}u}{(\|h\|^2 + (\lambda + u)^2)^{s/2}\sigma\sqrt{2\pi}} \\ &= \int \frac{\|h\|\,\mathrm{d}v}{(\|h\|^2 + (\|h\|v)^2)^{s/2}\sigma\sqrt{2\pi}} \end{aligned}$$

$$\leq \|h\|^{1-s-\alpha} \frac{1}{\sqrt{2\pi}} \int \frac{\mathrm{d}v}{(1+v^2)^{s/2}}$$
$$:= C\,\|x-y\|^{1-s-\alpha}.$$

In the case when $0 < s < 1$, we write

$$\mathbb{E}\left[\frac{1}{(\|x-y\|^2+(\lambda+W(x)-W(y))^2)^{s/2}}\right] \leq \int \frac{e^{-v^2/2}}{(\lambda+\sigma v)^s}\frac{\mathrm{d}v}{\sqrt{2\pi}}$$
$$\leq \|h\|^{-\alpha s}\int \frac{e^{-v^2/2}}{(\gamma+v)^s}\frac{\mathrm{d}v}{\sqrt{2\pi}}$$

where $\gamma = \lambda\sigma^{-1}$. On the other hand,

$$\int \frac{e^{-v^2/2}\,\mathrm{d}v}{(\gamma+v)^s} = \int \frac{e^{-(v-\gamma)^2/2}\,\mathrm{d}v}{v^s} \leq \int_{-1}^{1}\frac{\mathrm{d}v}{v^s} + \int_{\mathbb{R}\setminus[-1,1]}\frac{e^{-(v-\gamma)^2/2}\,\mathrm{d}v}{v^s}$$
$$\leq \int_{-1}^{1}\frac{\mathrm{d}v}{v^s} + \int_{\mathbb{R}} e^{-x^2/2}\,\mathrm{d}x$$

which is a constant C independent of γ and α. □

We are now able to finish the proof of Theorem 4. We use the potential theoretic approach (for more details on the potential theoretic approach of the calculus of the Hausdorff dimension, we can refer to [5, Chap. 4]). Suppose first that $\dim_{\mathscr{H}}(K) > \alpha$ and let δ be a real number such that

$$\alpha < \delta < \dim_{\mathscr{H}}(K).$$

There exists a probability measure m on K whose δ-energy $I_\delta(m)$, defined by

$$I_\delta(m) = \iint_{K\times K}\frac{\mathrm{d}m(x)\,\mathrm{d}m(y)}{\|x-y\|^\delta}$$

is finite. Conversely, to prove that the Hausdorff dimension of the graph $\Gamma^K_{f+W_\omega}$ is at least $\dim_{\mathscr{H}}(K)+1-\alpha$, it suffices to find, for any $s < \dim_{\mathscr{H}}(K)+1-\alpha$, a measure m_ω on $\Gamma^K_{f+W_\omega}$ with finite s-energy.

Let $(\Omega,\mathscr{F},\mathbb{P})$ be the probability space where the fractional Brownian motions $W^1,\dots,W^d$ are defined. For any $\omega\in\Omega$, define m_ω as the image of the measure m on the graph $\Gamma^K_{f+W_\omega}$ via the natural projection

$$x\in K \longmapsto (x, f(x)+W_\omega(x)).$$

Set $s = \delta+1-\alpha$ which is greater than 1. The s-energy of m_ω is equal to

$$
\begin{aligned}
I_s(m_\omega) &= \iint_{\Gamma^K_{f+W_\omega}\times\Gamma^K_{f+W_\omega}} \frac{\mathrm{d}m_\omega(X)\,\mathrm{d}m_\omega(Y)}{\|X-Y\|^s} \\
&= \iint_{K\times K} \frac{\mathrm{d}m(x)\,\mathrm{d}m(y)}{\left(\|x-y\|^2+\left(f(x)+W_\omega(x)-(f(y)+W_\omega(y))\right)^2\right)^{s/2}}.
\end{aligned}
$$

Fubini's theorem and Lemma 1 ensure that

$$
\begin{aligned}
\mathbb{E}\left[I_s(m_\omega)\right] &= \iint_{K\times K} \mathbb{E}\left[\frac{1}{\left(\|x-y\|^2+\left((f(x)-f(y))+(W(x)-W(y))\right)^2\right)^{s/2}}\right] \mathrm{d}m(x)\,\mathrm{d}m(y) \\
&\le C\iint_{K\times K} \|x-y\|^{1-s-\alpha}\mathrm{d}m(x)\,\mathrm{d}m(y) \\
&= CI_\delta(m) \\
&< +\infty.
\end{aligned}
$$

We deduce that for $\mathbb{P}$-almost all $\omega\in\Omega$, the energy $I_s(m_\omega)$ is finite. Since s can be chosen arbitrary close to $\dim_{\mathscr{H}}(K)+1-\alpha$, we get

$$
\dim_{\mathscr{H}}\left(\Gamma^K_{f+W_\omega}\right) \ge \dim_{\mathscr{H}}(K)+1-\alpha \quad \text{almost surely.}
$$

In the case where $\dim_{\mathscr{H}}(K)\le\alpha$, we proceed exactly the same way, except that we take any $\delta<\dim_{\mathscr{H}}(K)$ and we set $s=\frac{\delta}{\alpha}$ which is smaller than 1. We then get

$$
\dim_{\mathscr{H}}\left(\Gamma^K_{f+W}\right) \ge \frac{\dim_{\mathscr{H}}(K)}{\alpha} \qquad \text{almost surely.}
$$

4 Proof of Theorem 3

We can now prove Theorem 3. Let K be a compact set in $\mathbb{R}^d$ satisfying $\dim_{\mathscr{H}}(K)>0$. Remark first that for any function $f\in\mathscr{C}(K)$, the graph Γ^K_f is included in $K\times\mathbb{R}$. It follows that

$$
\dim_{\mathscr{H}}\left(\Gamma^K_f\right) \le \dim_{\mathscr{H}}(K\times\mathbb{R}) = \dim_{\mathscr{H}}(K)+1.
$$

Define

$$
G=\left\{f\in\mathscr{C}(K);\ \dim_{\mathscr{H}}\left(\Gamma^K_f\right)=\dim_{\mathscr{H}}(K)+1\right\}.
$$

Theorem 4 says that for any α such that $0<\alpha<\min(1,\dim_{\mathscr{H}}(K))$, the set G_α of all continuous functions $f\in\mathscr{C}(K)$ satisfying $\dim_{\mathscr{H}}\left(\Gamma^K_f\right)\ge\dim_{\mathscr{H}}(K)+1-\alpha$ is prevalent in $\mathscr{C}(K)$. Finally, we can write

$$
G=\bigcap_{n\ge 0} G_{\alpha_n}
$$

where $(\alpha_n)_{n\geq 0}$ is a sequence decreasing to 0 and we obtain that G is prevalent in $\mathscr{C}(K)$.

Remark 1. It is an easy consequence of Ascoli's theorem that the law of the process W is compactly supported in $\mathscr{C}(K)$ (remember that W is almost surely $(\alpha-\varepsilon)$-Hölderian). Then, we do not need to use that $\mathscr{C}(K)$ is a Polish space to obtain Theorem 3.

Remark 2. Let $K=[0,1]$ and $f\in\mathscr{C}([0,1])$. Theorem 3 implies that the set $G\bigcap(f+G)$ is prevalent. We can then write

$$f=f_1-f_2 \quad\text{with}\quad \dim_{\mathscr{H}}\left(\Gamma_{f_1}^{[0,1]}\right)=2 \quad\text{and}\quad \dim_{\mathscr{H}}\left(\Gamma_{f_2}^{[0,1]}\right)=2$$

where f_1 and f_2 are continuous functions.

On the other hand, it was recalled in Theorem 1 that the set

$$\tilde{G}=\left\{f\in\mathscr{C}([0,1])\ ;\ \dim_{\mathscr{H}}\left(\Gamma_f^{[0,1]}\right)=1\right\}$$

contains a dense G_δ-set of $\mathscr{C}([0,1])$. It follows that any continuous function $f\in\mathscr{C}([0,1])$ can be written

$$f=f_1-f_2 \quad\text{with}\quad \dim_{\mathscr{H}}\left(\Gamma_{f_1}^{[0,1]}\right)=1 \quad\text{and}\quad \dim_{\mathscr{H}}\left(\Gamma_{f_2}^{[0,1]}\right)=1$$

where f_1 and f_2 are continuous functions.

We can then ask the following question: given a real number $\beta\in(1,2)$ can we write an arbitrary continuous function $f\in\mathscr{C}([0,1])$ in the following way:

$$f=f_1-f_2 \quad\text{with}\quad \dim_{\mathscr{H}}\left(\Gamma_{f_1}^{[0,1]}\right)=\beta \quad\text{and}\quad \dim_{\mathscr{H}}\left(\Gamma_{f_2}^{[0,1]}\right)=\beta$$

where f_1 and f_2 are continuous functions?

We do not know the answer to this question.

5 The Case of α-Hölderian Functions

Let $0<\alpha<1$ and let $\mathscr{C}^\alpha(K)$ be the set of α-Hölderian functions in K endowed with the standard norm

$$\|f\|_\alpha=\sup_{x\in K}|f(x)|+\sup_{(x,y)\in K^2}\frac{|f(x)-f(y)|}{\|x-y\|^\alpha}.$$

It is well known that the Hausdorff dimension of the graph Γ_f^K of a function $f\in\mathscr{C}^\alpha(K)$ satisfies

$$\dim_{\mathscr{H}}\left(\Gamma_f^K\right) \leq \min\left(\frac{\dim_{\mathscr{H}}(K)}{\alpha}, \dim_{\mathscr{H}}(K)+1-\alpha\right) \tag{2}$$

(see, e.g., the remark following the statement of Theorem 4). It is then natural to ask if inequality (2) is an equality in a prevalent set of $\mathscr{C}^{\alpha}(K)$. This is indeed the case as said in the following result.

Theorem 5. *Let $d \geq 1$, $0 < \alpha < 1$ and $K \subset \mathbb{R}^d$ be a compact set with strictly positive Hausdorff dimension. The set*

$$\left\{f \in \mathscr{C}^{\alpha}(K)\ ;\ \dim_{\mathscr{H}}\left(\Gamma_f^K\right) = \min\left(\frac{\dim_{\mathscr{H}}(K)}{\alpha}, \dim_{\mathscr{H}}(K)+1-\alpha\right)\right\}$$

is a prevalent subset of $\mathscr{C}^{\alpha}(K)$.

This result generalizes to arbitrary compact subsets of positive dimension in $\mathbb{R}^d$ a previous work of Clausel and Nicolay (see [4, Theorem 2]).

Proof. Let $\alpha < \alpha' < 1$ and let W be the stochastic process defined in Theorem 4 with Hurst parameter α' instead of α. The stochastic process $W_{|K}$ takes values in $\mathscr{C}^{\alpha}(K)$. Moreover, if $\alpha < \alpha'' < \alpha'$, the injection

$$f \in \mathscr{C}^{\alpha''}(K) \longmapsto f \in \mathscr{C}^{\alpha}(K)$$

is compact. It follows that the law of the stochastic process $W_{|K}$ is compactly supported in $\mathscr{C}^{\alpha}(K)$ (W is α''-Hölderian). Then, Theorem 4 ensures that the set

$$\left\{f \in \mathscr{C}^{\alpha}(K)\ ;\ \dim_{\mathscr{H}}\left(\Gamma_f^K\right) \geq \min\left(\frac{\dim_{\mathscr{H}}(K)}{\alpha'}, \dim_{\mathscr{H}}(K)+1-\alpha'\right)\right\}$$

is prevalent in $\mathscr{C}^{\alpha}(K)$. Using a sequence $(\alpha_n)_{n\geq 0}$ decreasing to α, we get the conclusion of Theorem 5. □

References

1. Balka, R., Buczolich, Z., Elekes, M.: Topological Hausdorff dimension and level sets of generic continuous functions on fractals. arXiv:1108.5578 (2011)
2. Bayart, F., Heurteaux, Y.: Multifractal analysis of the divergence of Fourier series: the extreme cases. arXiv:1110:5478, submitted (2011)
3. Christensen, J.P.R.: On sets of Haar measure zero in Abelian Polish groups. Israel J. Math. **13**, 255–260 (1972)
4. Clausel, M., Nicolay, S.: Some prevalent results about strongly monoHölder functions. Nonlinearity **23**, 2101–2116 (2010)
5. Falconer, K.J.: Fractal geometry: Mathematical foundations and applications. Wiley, Hoboken (2003)

6. Falconer, K.J., Fraser, J.M.: The horizon problem for prevalent surfaces, Math. Proc. Cambridge Philos. Soc. **151**, 355–372 (2011)
7. Fraser, J.M., Hyde, J.T.: The Hausdorff dimension of graphs of prevalent continuous functions. arXiv:1104.2206 Real Anal. Exchange **37**, 333–352 (2012)
8. Fraysse, A, Jaffard S.: How smooth is almost every function in a Sobolev space?. Rev. Mat. Iboamericana **22**, 663–682 (2006)
9. Fraysse, A., Jaffard, S., Kahane, J.P.: Quelques propriétés génériques en analyse. (French) [Some generic properties in analysis]. C. R. Math. Acad. Sci. Paris **340**, 645–651 (2005)
10. Gruslys, V., Jonušas, J., Mijovič, V., Ng, O., Olsen, L., Petrykiewicz I.: Dimensions of prevalent continuous functions. Monash. Math. **166**, 153–180 (2012)
11. Humke, P.D., Petruska, G.: The packing dimension of a typical continuous function is 2. Bull. Am. Math. Soc. (N.S.) **27**, 345–358 (1988–89)
12. Hunt, B.R.: The Hausdorff dimension of graphs of Weierstrass functions. Proc. Am. Math. Soc. **126**, 791–800 (1998)
13. Hyde, J.T., Laschos, V., Olsen, L., Petrykiewicz, I., Shaw, A.: On the box dimensions of graphs of typical functions. J. Math. Anal. Appl. **391**, 567–581 (2012)
14. Mauldin, R.D., Williams, S.C.: On the Hausdorff dimension of some graphs. Trans. Am. Math. Soc. **298**, 793–803 (1986)
15. McClure, M.: The prevalent dimension of graphs, Real Anal. Exchange **23**, 241–246 (1997)
16. Olsen, L.: Fractal and multifractal dimensions of prevalent measures. Indiana Univ. Math. J. **59**, 661–690 (2010)
17. Shaw, A.: Prevalence, M. Math Dissertation, University of St. Andrews (2010)

Hausdorff Dimension and Diophantine Approximation

Yann Bugeaud

Abstract In this survey chapter, we explain how the theory of Hausdorff dimension and Hausdorff measure is used to answer certain questions in Diophantine approximation. The final section is devoted to a discussion around the Diophantine properties of the points lying in the middle third Cantor set.

1 Introduction

The main goal of this survey chapter is to point out how the theory of Hausdorff dimension and Hausdorff measure can be used to solve various questions in Diophantine approximation. We also point out several open problems, which hopefully will motivate further research.

Throughout this text we denote by dim the Hausdorff dimension.

Let ξ be an irrational real number. By the theory of continued fractions (or by Dirichlet's *Schubfachprinzip*), there exist infinitely many rational numbers p/q such that

$$|\xi - p/q| < q^{-2}.$$

For any given $\varepsilon > 0$, a covering argument (easy half of the Borel–Cantelli lemma) shows that, for almost all real numbers ξ (throughout Sects. 1 and 2, "almost all" always refers to the Lebesgue measure), there are only finitely many rational numbers p/q such that

$$|\xi - p/q| < q^{-2-\varepsilon}. \tag{1}$$

Y. Bugeaud (✉)
Institut de Recherche Mathématique Avancée, Mathématiques, Université de Strasbourg, 7, rue René Descartes, F-67084 Strasbourg Cedex, France
e-mail: bugeaud@math.unistra.fr

J. Barral and S. Seuret (eds.), *Further Developments in Fractals and Related Fields*, Trends in Mathematics, DOI 10.1007/978-0-8176-8400-6_3,

However, for certain irrational real numbers ξ, inequality (1) has infinitely many solutions.

Definition 1. The irrationality exponent of an irrational real number ξ, denoted by $\mu(\xi)$, is the supremum of the real numbers μ such that

$$|\xi - p/q| < q^{-\mu}$$

has infinitely many solutions in rational numbers p/q.

By means of the theory of continued fractions, for every real number $\mu \geq 2$, it is easy to construct real numbers ξ such that $\mu(\xi) = \mu$; see, e.g., [17] or [12]. Moreover, Jarník [16] proved in 1929 that

$$\dim\{\xi \in [0,1] : \mu(\xi) \geq \mu\} = \frac{2}{\mu}, \tag{2}$$

a result established independently by Besicovitch [8] a few years later. In 1931 Jarník [17] refined his result from [16] by using generalized Hausdorff measures. Although this is not explicitly written in [17], he showed that

$$\dim\{\xi \in [0,1] : \mu(\xi) = \mu\} = \frac{2}{\mu}. \tag{3}$$

The irrationality exponent μ introduced in Definition 1 is an example of an *exponent of approximation*, that is, of a function defined on the set of real irrational numbers by means of consideration from Diophantine approximation.

Definition 2. The spectrum of an exponent of approximation is the set of values taken by this exponent.

For an exponent of approximation w, there are two natural questions:

* **(Q1)** To determine the spectrum of w
* **(Q2)** For a real number w_0, to determine the Hausdorff dimension of the set at level w_0, that is, of the set of real numbers ξ such that $w(\xi) = w_0$

In the case of the irrationality exponent, we have seen above that both questions have been answered. We introduce in Sect. 2 three (classical) families of exponents of approximation and discuss both questions for these exponents. We will see that, in certain cases, to answer **(Q2)** is the only known way to solve the apparently simpler **(Q1)**.

Section 3 is devoted to a survey of recent results on Diophantine approximation of elements of the middle third Cantor set.

2 Three Families of Exponents of Approximation

We define two families of exponents of approximation which generalize the irrationality exponent. A third family of exponents gives certain information on the expansions of real numbers to integer bases. Further classical families of exponents are discussed in [13].

For $n \geq 1$, the accuracy with which real numbers are approximated by algebraic numbers of degree at most n is measured by means of the exponents w_n^*, introduced in 1939 by Koksma [19]. Recall that the height $H(P)$ of an integer polynomial $P(X)$ is the maximum of the moduli of its coefficients, and the height $H(\alpha)$ of an algebraic number α is the height of its minimal polynomial over $\mathbf{Z}$.

Definition 3. Let $n \geq 1$ be an integer and let ξ be an irrational real number. We denote by $w_n^*(\xi)$ the supremum of the real numbers w^* for which the inequality

$$|\xi - \alpha| \leq H(\alpha)^{-w^*-1} \tag{4}$$

is satisfied for infinitely many algebraic numbers α of degree at most n.

Clearly, every irrational real number ξ satisfies

$$\mu(\xi) = w_1^*(\xi) + 1.$$

This shows that the exponents w_n^* with $n \geq 2$ extend in a natural way the irrationality exponent μ.

The introduction of the exponent -1 in Eq. (4) is explained on p.48 of [10]. The reader is directed to this monograph for known results on the exponents w_n^*. We only mention here that $w_n^*(\xi) = \min\{n, d-1\}$ for every real algebraic number ξ of degree $d \geq 2$ and that almost all real numbers ξ satisfy $w_n^*(\xi) = n$ for $n \geq 1$. Wirsing [25] proved that $w_n^*(\xi) \geq (n+1)/2$ for every transcendental real number ξ and every $n \geq 1$. With the exception of a result of Davenport and Schmidt [14], showing that $w_2^*(\xi) \geq 2$ for every real number ξ not of degree at most 2, there has been no significant improvement of Wirsing's statement during the last fifty years.

Problem 1. Are there an integer $n \geq 3$ and a transcendental real number ξ such that $w_n^*(\xi) < n$?

Other natural exponents of approximation, which also extend the irrationality exponent μ, take into account the accuracy with which the first integral powers of a given real number are simultaneously approximated by rational numbers with the same denominator. Throughout this text, $\|.\|$ denotes the distance to the nearest integer.

Definition 4. Let $n \geq 1$ be an integer and let ξ be an irrational real number. We denote by $\lambda_n(\xi)$ the supremum of the real numbers λ for which the inequality

$$\max_{1 \leq m \leq n} \|q\xi^m\| \leq q^{-\lambda}$$

has infinitely many solutions q in $\mathbf{Z}_{\geq 1}$.

We have $\lambda_n(\xi) = \max\{1/n, 1/(d-1)\}$ for $n \geq 1$ for every real algebraic number ξ of degree $d \geq 2$. Furthermore, $\lambda_n(\xi) \geq 1/n$ for every irrational real number ξ and every $n \geq 1$; see [12] for further results.

We end this list of definitions with the exponents v_b which were introduced in [1]. They provide information on the lengths of blocks of digits 0 (or of digits $b-1$) occurring in the expansion of ξ to base b.

Definition 5. Let ξ be an irrational real number. Let b be an integer with $b \geq 2$. We denote by $v_b(\xi)$ the supremum of the real numbers v for which the inequality

$$||b^n \xi|| < (b^n)^{-v}$$

has infinitely many solutions n in $\mathbf{Z}_{\geq 1}$.

Let $b \geq 2$ be an integer. Clearly, we have

$$w_1^*(\xi) \geq v_b(\xi), \quad \mu(\xi) \geq v_b(\xi) + 1, \tag{5}$$

for all irrational real numbers ξ. These inequalities are rarely sharp since almost all real numbers ξ satisfy $v_b(\xi) = 0$.

For every $b \geq 2$ and for every v with $0 < v < +\infty$ the real number

$$\xi_{b,v} := \sum_{j \geq 1} b^{-\lfloor (v+1)^j \rfloor}, \tag{6}$$

where $\lfloor . \rfloor$ denotes the integer part, satisfies $v_b(\xi_{b,v}) = v$. Since

$$v_b\left(\sum_{j \geq 1} b^{-j^2}\right) = 0 \quad \text{and} \quad v_b\left(\sum_{j \geq 1} b^{-j!}\right) = +\infty,$$

this shows that the spectrum of v_b is equal to $[0, +\infty]$. Thus, we have given the answer to **(Q1)** for the exponent v_b.

Now, we explain how to show that, for every real number $v > 0$, we have

$$\dim\{\xi \in [0,1] : v_b(\xi) \geq v\} = \dim\{\xi \in [0,1] : v_b(\xi) = v\} = \frac{1}{v+1}. \tag{7}$$

To bound this Hausdorff dimension from above by $1/(v+1)$ is an immediate application of a covering argument (easy half of the Hausdorff–Cantelli lemma). To prove that this dimension is at least equal to $1/(v+1)$ is much more interesting. We construct inductively a large Cantor-type set contained in $\{\xi \in [0,1] : v_b(\xi) \geq v\}$.

Let $(n_k)_{k\geq 1}$ be a rapidly increasing sequence of integers with $n_1 = 1$ and $n_2 > 2/v$. Set $\mathscr{E}_1 = [0,1]$. For $p = 1,\ldots,b^{n_2}-1$, set

$$E_{2,p} = [p/b^{n_2} - 1/b^{(v+1)n_2}, p/b^{n_2} + 1/b^{(v+1)n_2}]$$

and put $\mathscr{E}_2 = E_{2,1} \cup \ldots \cup E_{2,b^{n_2}-1}$. Assume that, for some $k \geq 2$, the set $\mathscr{E}_k$ has been constructed and is equal to a finite union of intervals $E_{k,1},\ldots,E_{k,t_k}$ of length $2b^{-(v+1)n_k}$ and centered at rational numbers of denominator b^{n_k}. Let $E_{k,p}$ be such an interval. Let denote by $\mathscr{E}_{k+1,p}$ the set of intervals of the form

$$[a/b^{n_{k+1}} - 1/b^{(v+1)n_{k+1}}, a/b^{n_{k+1}} + 1/b^{(v+1)n_{k+1}}],$$

with a an integer, which are contained in $E_{k,p}$. There are at least

$$m_{k+1} := b^{n_{k+1}}(2b^{-(v+1)n_k}) - 2$$

such intervals and the distance between any two distinct such intervals always exceeds

$$\varepsilon_{k+1} := b^{-n_{k+1}}/2.$$

Putting $\mathscr{E}_{k+1} = \mathscr{E}_{k+1,1} \cup \ldots \cup \mathscr{E}_{k+1,t_k}$, we have completed the inductive step of the construction. Set

$$\mathscr{K} := \bigcap_{k\geq 1} \mathscr{E}_k.$$

By construction, every element ξ in $\mathscr{K}$ satisfies $v_b(\xi) \geq v$. The mass distribution principle (as stated, e.g., on p. 59 of [15] or on p. 97 of [10]) shows that

$$\dim \mathscr{K} \geq \liminf_{k\to+\infty} \frac{\log(m_1 \ldots m_k)}{-\log(m_{k+1}\varepsilon_{k+1})}.$$

In our situation, if the sequence $(n_k)_{k\geq 1}$ grows sufficiently rapidly, we deduce that

$$\dim\{\xi \in [0,1] : v_b(\xi) \geq v\} \geq \frac{1}{v+1}.$$

To get the same lower bound for the smaller set

$$\dim\{\xi \in [0,1] : v_b(\xi) = v\}$$

we need to use refined Hausdorff measures, but there is no additional difficulty.

In the above proof, we have used that, if I is an interval of positive length $|I|$ contained in $[0,1]$, then, for n large enough in terms of $|I|$, there are around $b^n|I|$ rational points of the form a/b^n in I, and these points are regularly spaced.

The same strategy was used by Jarník to establish Eqs. (2) and (3). He proved that the set of rational numbers p/q in $[0,1]$ is evenly distributed in the following sense:

For I as above, if Q is large enough in terms of $|I|$, then there are $\gg Q^2$ rational numbers p/q in I with $Q \ll q \ll Q$ and such that the distance between any two of them is $\gg Q^{-2}$ (here and below, the constants implicit in $\ll$ and $\gg$ are numerical constants). This is proved on p.99 of [10]; see also on p. 142 of [15] for a weaker result, however sufficient to derive Eqs. (2) and (3).

To formalize the properties of distribution needed to apply the method described above, Baker and Schmidt [2] have introduced the notion of *regular systems*; see Sect. 5.4 of [10]. Several authors prefer to use ubiquitous systems [6], which give more flexibility.

For $n \geq 1$, explicit examples of real numbers have been constructed in [12] in order to show that the spectrum of w_n^* includes the interval $[2n-1, +\infty)$. However, the next question remains open.

Problem 2. Let $n \geq 2$ be an integer. To construct explicitly a real number w with $n < w < 2n-1$ and a real number ξ such that $w_n^*(\xi) = w$.

Apparently, there is no suitable multidimensional generalization of the theory of continued fraction which can be used to solve (at least partially) Problem 2.

At present, for $n \geq 2$, the only known way to show that the spectrum of w_n^* includes the interval $[n, +\infty)$ is by means of the next theorem, proved in 1970 by Baker and Schmidt [2].

Theorem 1. *Let $n \geq 1$ be an integer. For every real number $w^* \geq n$, we have*

$$\dim\{\xi \in [0,1] : w_n^*(\xi) = w^*\} = \frac{n+1}{w^*+1}. \tag{8}$$

We check that Eqs. (3) and (8) for $n = 1$ coincide. Baker and Schmidt used the same strategy as explained above. The difficult point is to prove that algebraic numbers of bounded degree are evenly distributed. Note that, since an algebraic number of degree n and height H is a root of an integer polynomial of degree n and with all coefficients bounded in absolute value by H, their number does not exceed $(2H+1)^{n+1}$. With I as above, it is proved in [2] that, if H is large enough in terms of $|I|$, then there are $\gg H^{n+1}|I|$ algebraic numbers α of degree n in I with $H(\alpha) \ll H(\log H)^{c(n)}$ and such that the difference between any two of them if $\gg H^{-n-1}$. Here, $c(n)$ is a constant depending only on n. A deep result of Beresnevich [4] from 1999 shows that the above statement remains true with $c(n) = 0$.

Thus, we have seen that **(Q1)** and **(Q2)** are answered for the exponents of approximation v_b and w_n^*. The situation is different and much more complicated for the exponents λ_n.

For $n \geq 1$, Bugeaud [12] constructed explicit examples of real numbers in order to show that the spectrum of λ_n includes the interval $[1, +\infty)$. However, the next question remains open.

Problem 3. Let $n \geq 2$ be an integer. To construct explicitly a real number λ with $1/n < \lambda < 1$ and a transcendental real number ξ such that $\lambda_n(\xi) = \lambda$.

At present, the only known way to show that the spectrum of λ_2 is the whole interval $[1/2,+\infty)$ is the proof by Beresnevich, Dickinson, Vaughan and Velani [7, 23] that, for every λ in $[1/2,1]$, we have

$$\dim\{\xi \in [0,1] : \lambda_2(\xi) = \lambda\} = \frac{2-\lambda}{1+\lambda}.$$

The Jarník–Besicovitch theorem (2) was recently extended by Budarina et al. [9] as follows (see also [12] for an alternative proof).

Theorem 2. *Let $n \geq 2$ be an integer. Let $\lambda \geq n-1$ be a real number. Then, we have*

$$\dim\{\xi \in [0,1] : \lambda_n(\xi) = \lambda\} = \frac{2}{n(1+\lambda)}.$$

The assumption $\lambda \geq n-1$ in Theorem 2 is quite restrictive when $n \geq 3$ but is optimal for $n = 2$.

We end this section with two open questions.

Problem 4. Let $n \geq 3$ be an integer. To determine the spectrum of λ_n.

Problem 5. Let $n \geq 3$ be an integer. Let λ be a real number with $1/n < \lambda < n-1$. To determine

$$\dim\{\xi \in [0,1] : \lambda_n(\xi) = \lambda\}.$$

Partial results towards the resolution of the difficult Problems 4 and 10 have been given by Beresnevich [5]. He proved that, for an integer n and a real number λ satisfying $1/n \leq \lambda < 3/(2n-1)$, we have

$$\dim\{\xi \in [0,1] : \lambda_n(\xi) \geq \lambda\} \geq \frac{n+1}{\lambda+1} - (n-1). \tag{9}$$

He conjectured that equality holds in Eq. (9).

3 Approximation to Points in the Middle Third Cantor Set

Throughout this section, we denote by K the middle third Cantor set. Certain results stated below are valid for more general Cantor-type sets, however. We recall that the Hausdorff dimension of K satisfies

$$\dim K = \frac{\log 2}{\log 3}.$$

The first significant result on Diophantine approximation to elements of K was proved in 2001 by Weiss [24]. He showed that the irrationality exponent of almost all ξ in K (in this section, almost all refers to the standard measure supported by K) is equal to 2.

For $v > 0$, the real number $\xi_{3,v}$ defined in Eq. (6) satisfies

$$v_3(\xi_{3,v}) = v, \quad \text{and thus } \mu(\xi_{3,v}) \ge v + 1,$$

by Eq. (5) or by noticing that

$$\left| \xi_{3,v} - \sum_{j=1}^{J} 3^{-\lfloor (v+1)^j \rfloor} \right| < 2 \cdot 3^{-\lfloor (v+1)^{J+1} \rfloor} \le 6 \cdot (3^{-\lfloor (v+1)^J \rfloor})^{v+1},$$

for every sufficiently large integer J. At first sight, for $v \ge 1$, it could seem that $\xi_{3,v}$ satisfies

$$\mu(\xi_{3,v}) = v_3(\xi_{3,v}) + 1 = v + 1. \tag{10}$$

This is, however, not clear since there may exist very good rational approximants to $\xi_{3,v}$ which are not obtained by truncation of the infinite sum giving $\xi_{3,v}$, that is, which are not of the form $\sum_{j=1}^{J} 3^{-\lfloor (v+1)^j \rfloor}$ for some $J \ge 1$. However, if v is sufficiently large — precisely, if $v \ge (\sqrt{5}+1)/2$ — a simple argument based on triangle inequalities (see, e.g., Sect. 8 of [21]) implies Eq. (10).

In fact, it appears that Eq. (10) holds for every $v \ge 1$, as proved in [11] by means of the following observation: the continued fraction expansion of a suitable rational translate of $-\xi_{3,v}$ can be constructed explicitly. The next result is extracted from [11].

Theorem 3. *Let $\mu \ge 2$ be a real number. The middle third Cantor set contains uncountably many elements ξ with $\mu(\xi) = \mu$.*

Theorem 3 answers **(Q1)** for the restriction to K of the irrationality exponent, but **(Q2)** remains open. In order to attack the latter question, the authors of [21] investigated the analogous problem for the exponent v_3. They proved the following statement.

Theorem 4. *For any positive real number v we have*

$$\begin{aligned} \dim\{\xi \in K : v_3(\xi) = v\} &= \frac{\log 2}{\log 3} \times \frac{1}{v+1} \\ &= (\dim K) \times \dim\{\xi \in [0,1] : v_3(\xi) = v\}. \end{aligned}$$

Things are more easier with the exponent v_3 than with the irrationality exponent μ. Indeed, if for some $v > 1$ and ξ in K we have

$$|\xi - p/3^n| < (3^n)^{-v},$$

for some rational number $p/3^n$, then $p/3^n$ must lie in K.

We display an immediate consequence of Eqs. (3), (7), and Theorem 4.

Corollary 1. *For any real number $\mu > 2$, we have*

$$\dim\{\xi \in K : \mu(\xi) \geq \mu\} \geq \frac{1}{2} \times (\dim K) \times \dim\{\xi \in [0,1] : \mu(\xi) = \mu\}. \tag{11}$$

Apparently, the methods used in [21] do not allow us to replace the $\geq$ sign by the $=$ sign in the left-hand side of Eq. (11). They, however, give the following slight refinement of Eq. (11):

$$\dim\{\xi \in K : \mu \leq \mu(\xi) \leq 2\mu\} \geq \frac{1}{2} \times (\dim K) \times \dim\{\xi \in [0,1] : \mu(\xi) = \mu\}.$$

Regarding the upper bound, Pollington and Velani [22] (see also [20]) used a simple covering argument to establish that

$$\dim\{\xi \in K : \mu(\xi) \geq \mu\} \leq (\dim K) \times \dim\{\xi \in [0,1] : \mu(\xi) \geq \mu\},$$

for every $\mu \geq 2$.

Let $\mu > 2$ be a real number. The authors of [21] speculate at the end of their paper that we have

$$\dim\{\xi \in K : \mu(\xi) \geq \mu\} = (\dim K) \times \dim\{\xi \in [0,1] : \mu(\xi) \geq \mu\}.$$

Problem 6. Let $\mu > 2$ be a real number. To determine

$$\dim\{\xi \in K : \mu(\xi) = \mu\}.$$

A related question involving rational approximation and asymptotic frequencies of digits in a fixed integer base has been investigated in [3].

Very few is known on the expansions of a given irrational real number to two multiplicatively independent bases. The study of the set of values taken by the exponent v_2 at the points in K would give some information on the binary expansions of the elements of K.

A straightforward adaptation of the arguments of Weiss [24] and Kristensen [20] allows us to prove the next result.

Theorem 5. *For any positive real number v we have*

$$\dim\{\xi \in K : v_2(\xi) \geq v\} \leq (\dim K) \times \dim\{\xi \in [0,1] : v_2(\xi) \geq v\}. \tag{12}$$

However, it does not seem to be easy to give a non-trivial lower bound for the left-hand side of Eq. (12).

Problem 7. Let v be a positive real number. To determine

$$\dim\{\xi \in K : v_2(\xi) = v\}.$$

Weiss' result mentioned at the beginning of this section has been extended in Proposition 7.10 from [18] to the exponents w_n^*.

Theorem 6. *Almost all points ξ in the middle third Cantor set satisfy*

$$w_n^*(\xi) = n, \quad \textit{for every } n \geq 1$$

and

$$\lambda_n(\xi) = 1/n, \quad \textit{for every } n \geq 1.$$

The last part of Theorem 6 follows by combining Proposition 7.10 from [18] with a classical transference principle. We omit the details.

We conclude this text by a last open problem; see Sect. 6 of [12] for a small contribution towards its resolution.

Problem 8. Let $n \geq 1$ be an integer and $w \geq n$ and $\lambda \geq 1/n$ be real numbers. To determine the Hausdorff dimension of the sets

$$\{\xi \in K : w_n^*(\xi) = w\}$$

and

$$\{\xi \in K : \lambda_n(\xi) = \lambda\}.$$

For $n \geq 2$, to determine the spectra of the restriction to K of w_n^* and of λ_n.

References

1. Amou, M., Bugeaud, Y.: Expansions in integer bases and exponents of Diophantine approximation. J. London Math. Soc. **81**, 297–316 (2010)
2. Baker, A., Schmidt, W.M.: Diophantine approximation and Hausdorff dimension. Proc. London Math. Soc. **21**, 1–11 (1970)
3. Barral, J., Seuret, S.: Ubiquity and large intersections properties under digit frequencies constraints. Math. Proc. Cambridge Philos. Soc. **145**, 527–548 (2008)
4. Beresnevich, V.: On approximation of real numbers by real algebraic numbers. Acta Arith. **90**, 97–112 (1999)
5. Beresnevich, V.: Rational points near manifolds and metric Diophantine approximation. Ann. Math. **175**, 187–235 (2012)
6. Beresnevich, V., Dickinson, D., Velani, S.L.: Measure theoretic laws for limsup sets. Mem. Amer. Math. Soc. **179**(846), 1–91 (2006)

7. Beresnevich, V., Dickinson, D., Velani, S.L.: Diophantine approximation on planar curves and the distribution of rational points, with an appendix by R.C. Vaughan: 'Sums of two squares near perfect squares'. Ann. Math. **166**, 367–426 (2007)
8. Besicovitch, A.S.: Sets of fractional dimension (IV): on rational approximation to real numbers. J. London Math. Soc. **9**, 126–131 (1934)
9. Budarina, N., Dickinson, D., Levesley, J.: Simultaneous Diophantine approximation on polynomial curves. Mathematika **56**, 77–85 (2010)
10. Bugeaud, Y.: Approximation by algebraic numbers, Cambridge Tracts in Mathematics. Cambridge University Press, Cambridge (2004)
11. Bugeaud, Y.: Diophantine approximation and Cantor sets. Math. Ann. **341**, 677–684 (2008)
12. Bugeaud, Y.: On simultaneous rational approximation to a real number and its integral powers. Ann. Inst. Fourier (Grenoble) **60**, 2165–2182 (2010)
13. Bugeaud, Y., Laurent, M.: Exponents of Diophantine approximation. In: Diophantine Geometry Proceedings, Scuola Normale Superiore Pisa, Ser. CRM, vol. 4, pp. 101–121 (2007)
14. Davenport, H., Schmidt, W.M.: Approximation to real numbers by quadratic irrationals. Acta Arith. **13**, 169–176 (1967)
15. Falconer, K.: Fractal Geometry: Mathematical Foundations and Applications. Wiley, Chichester (1990)
16. Jarník, V.: Diophantischen Approximationen und Hausdorffsches Mass. Mat. Sbornik **36**, 371–382 (1929)
17. Jarník, V.: Über die simultanen Diophantische Approximationen. Math. Z. **33**, 505–543 (1931)
18. Kleinbock, D., Lindenstrauss, E., Weiss, B.: On fractal measures and Diophantine approximation. Selecta Math. **10**, 479–523 (2004)
19. Koksma, J.F. : Über die Mahlersche Klasseneinteilung der transzendenten Zahlen und die Approximation komplexer Zahlen durch algebraische Zahlen. Monatsh. Math. Phys. **48**, 176–189 (1939)
20. Kristensen, S.: Approximating numbers with missing digits by algebraic numbers. Proc. Edinb. Math. Soc. **49**, 657–666 (2006)
21. Levesley, J., Salp, C., Velani, S.L.: On a problem of K. Mahler: Diophantine approximation and Cantor sets. Math. Ann. **338**, 97–118 (2007)
22. Pollington, A.D., Velani, S.: Metric Diophantine approximation and 'absolutely friendly' measures. Selecta Math. **11**, 297–307 (2005)
23. Vaughan, R.C., Velani, S.: Diophantine approximation on planar curves: the convergence theory. Invent. Math. **166**, 103–124 (2006)
24. Weiss, B.: Almost no points on a Cantor set are very well approximable. R. Soc. Lond. Proc. Ser. A Math. Phys. Eng. Sci. **457**, 949–952 (2001)
25. Wirsing, E.: Approximation mit algebraischen Zahlen beschränkten Grades. J. Reine Angew. Math. **206**, 67–77 (1961)

Singular Integrals on Self-similar Subsets of Metric Groups

Vasilis Chousionis and Pertti Mattila

Abstract In this chapter we study singular integrals on small (i.e., measure zero and lower than full dimensional) subsets of metric groups. The main examples of the groups we have in mind are Euclidean spaces and Heisenberg groups. We shall pay particular attention to the behaviour of singular integral operators on self-similar subsets.

1 Introduction

The general question we are interested in here is as follows: how is the L^2-boundedness of singular integral operators related to geometric properties of the underlying sets and measures? A little more precisely, in some space, say d-dimensional space in terms of Hausdorff dimension, we study singular integral operators on s-dimensional subsets with $s < d$. The spaces we are mainly interested in, are Euclidean spaces and Heisenberg groups, but we shall say something also in more general metric groups. Such questions in Euclidean spaces have been studied systematically for more than 20 years; the book [9] of David and Semmes is a good source for background information. This survey focuses mostly to our recent progress in Heisenberg groups in [5, 6]. The general setting is the following:

We assume that (G,d) is a complete separable metric group with the following properties:

V. Chousionis (✉)
Department of Mathematics, University of Illinois-Urbana Champaign,
1409 W. Green St, IL 61801 Urbana, USA
e-mail: vchous@illinois.edu

P. Mattila
Department of Mathematics and Statistics, University of Helsinki,
Helsinki P.O. Box 68, FI-00014, Finland
e-mail: pertti.mattila@helsinki.fi

J. Barral and S. Seuret (eds.), *Further Developments in Fractals and Related Fields*,
Trends in Mathematics, DOI 10.1007/978-0-8176-8400-6_4,

1. The left translations $\tau_q : G \to G$,

$$\tau_q(x) = q \cdot x, x \in G,$$

are isometries for all $q \in G$.

2. There exist dilations $\delta_r : G \to G, r > 0$, which are continuous group homomorphisms for which:

   (a) $\delta_1 =$ identity
   (b) $d(\delta_r(x), \delta_r(y)) = rd(x,y)$ for $x, y \in G, r > 0$
   (c) $\delta_{rs} = \delta_r \circ \delta_s$

   It follows that for all $r > 0$, δ_r is a group isomorphism with $\delta_r^{-1} = \delta_{\frac{1}{r}}$.

Euclidean spaces, Heisenberg groups and the more general Carnot groups are the main examples of such groups.

Let μ be a finite Borel measure on G and let $K : G \times G \setminus \{(x,y) : x = y\} \to \mathbf{R}$ be a Borel measurable kernel which is bounded away from the diagonal, that is K is bounded in $\{(x,y) : d(x,y) > \delta\}$ for all $\delta > 0$. The truncated singular integral operators associated to μ and K are defined for $f \in L^1(\mu)$ and $\varepsilon > 0$ as

$$T_\varepsilon(f)(y) = \int_{G \setminus B(x,\varepsilon)} K(x,y) f(y) \mathrm{d}\mu y,$$

and the maximal singular integral operator is defined as usual:

$$T_K^*(f)(x) = \sup_{\varepsilon > 0} |T_\varepsilon(f)(x)|.$$

For a vector-valued kernel $K = (K_1, \dots, K_l)$ we define

$$T_K^*(f)(x) = \max_{1 \le j \le l} \{T_{K_j}^*(f)(x)\}.$$

Then T_K^* bounded in $L^2(\mu)$ means that

$$\int T_K^*(f)^2 \mathrm{d}\mu \le C \int |f|^2 \mathrm{d}\mu \quad \text{for all } f \in L^2(\mu).$$

We are particularly interested in the following class of kernels.

Definition 1. For $s > 0$ the s-homogeneous kernels are of the form,

$$K_\Omega(x,y) = \frac{\Omega(x^{-1} \cdot y)}{d(x,y)^s}, \ x, y \in G \setminus \{(x,y) : x = y\},$$

where $\Omega : G \to \mathbf{R}$ is a continuous and homogeneous function of degree zero, that is,

$$\Omega(\delta_r(x)) = \Omega(x) \quad \text{for all } x \in G, r > 0.$$

We shall discuss results saying that such maximal singular integral operators are often unbounded on fractal type sets. We shall mostly restrict to s-dimensional Ahlfors–David regular, briefly s-regular, and Borel measures μ, which means that for some positive and finite constant C,

$$r^s/C \leq \mu(B(x,r)) \leq Cr^s \quad \text{for all} x \in spt\mu, 0 < r < d(spt\mu).$$

Here $B(x,r)$ is the closed ball with centre x and radius r, and $d(E)$ denotes the diameter of E. A closed set E is called s-regular if the s-dimensional Hausdorff measure $\mathscr{H}^s\lfloor E$ restricted to E is s-regular.

First we shall review briefly some of the Euclidean results. Recent surveys are [15, 24].

2 The One-Dimensional Case

We start with the following result from [16] for one-dimensional sets. It characterizes geometrically the 1-regular measures on which the singular integral operator related to the one-dimensional Riesz kernel

$$R_1(x) = x/|x|^2, x \in \mathbf{R}^n$$

is bounded in $L^2(\mu)$. Note that in the complex plane this kernel is essentially the Cauchy kernel $1/z = \bar{z}/|z|^2$.

Theorem 1. *Let μ be a 1-regular measure in $\mathbf{R}^n$. The following two conditions are equivalent:*

1. *$T^*_{R_1}$ is bounded in $L^2(\mu)$.*
2. *$spt\mu \subset \Gamma$ where Γ is a curve with $\mathscr{H}^1(\Gamma \cap B(x,r)) \leq Cr$ for all $x \in \mathbf{R}^n$ and for all $r > 0$.*

The key for the proof was the following identity found by Melnikov in [18] for $z_1, z_2, z_3 \in \mathbf{C}$:

$$c(z_1,z_2,z_3)^2 = \sum_\sigma \frac{1}{(z_{\sigma(1)} - z_{\sigma(3)})\overline{(z_{\sigma(2)} - z_{\sigma(3)})}}, \tag{1}$$

where σ runs through all six permutations of 1,2 and 3, and $c(z_1,z_2,z_3)$ is the reciprocal of the radius of the circle passing through z_1, z_2 and z_3. It is called the Menger curvature of this triple. It vanishes exactly when the three points lie on the same line. In general it measures how far they are from being collinear. Melnikov and Verdera used this identity to give a new proof for the boundedness of the Cauchy singular integral operator on Lipschitz graphs in [19]. Integrating the above identity with respect to all three variables and using Fubini's theorem, one can prove Theorem 1 by proving that the conditions (1) and (2) are both equivalent to

$$\int_B\int_B\int_B c(x,y,z)^2 \mathrm{d}\mu x \mathrm{d}\mu y \mathrm{d}\mu z \leq C d(B)$$

for all balls $B \subset \mathbf{R}^n$ and for all $r > 0$.

The identity (1) connects the sum over permutations, which is a kind of symmetrization over the three variables, to a nice geometric object. But already the fact that this sum is non-negative is unexpected and useful. The proof of the identity is an exercise.

In the plane, Theorem 1 remains valid if the kernel R_1 is replaced by any of its coordinate parts $x_1/|x|$ or $x_2/|x|, x = (x_1,x_2) \in \mathbf{R}^2$, because the symmetrization method described earlier works in this case as well. Recently, in [4], Theorem 1 was extended to all kernels $k_n(x) = x_1^{2n-1}/|x|^{2n}$, $n \in \mathbf{N}$. It should be noted that the proof in [4] also depends on some good symmetrization properties of the kernels k_n.

Based on earlier work of many people Theorem 1 gives the following corollary:

Corollary 1. *Let E be a compact 1-regular subset of the complex plane. The following three conditions are equivalent:*

1. *E is removable for bounded analytic functions.*
2. *E is removable for Lipschitz harmonic functions.*
3. *E is purely unrectifiable.*

Here the pure unrectifiablity of E means that E meets every rectifiable curve in zero length. The removability of E for bounded analytic functions means that if E is contained in an open set U, any bounded analytic function in $U \setminus E$ can be extended analytically to U. The removability for Lipschitz harmonic functions is analogous, but since Lipschitz functions on $U \setminus E$ can be uniquely extended as Lipschitz functions, (2) means that any Lipschitz function in U which is harmonic in $U \setminus E$ is harmonic in U.

David showed later in [7] that instead of AD-regularity it is enough to assume that E has finite one-dimensional Hausdorff measure. Still later Tolsa gave in [23] a characterization of removability for all compact subsets of the complex plane in terms of Menger curvature. A consequence of this is that (1) and (2) in the above corollary are equivalent for any compact set E. An amusing feature is that nobody knows how to prove this without going through the Menger curvature characterization. For a survey, see [24] or [20]. Tolsa's result is

Theorem 2. *Let E be a compact subset of the complex plane. The following three conditions are equivalent:*

1. *E is not removable for bounded analytic functions.*
2. *E is not removable for Lipschitz harmonic functions.*
3. *There is a finite Borel measure μ supported in E such that $\mu(E) > 0$, $\mu(B) \leq d(B)$ for all discs B and*

$$\int\int\int c(x,y,z)^2 \mathrm{d}\mu x \mathrm{d}\mu y \mathrm{d}\mu z < \infty.$$

3 The Higher-Dimensional Case

The higher dimensional analogues of the above results are unknown. Let R_m be the vector-valued m-dimensional Riesz kernel:

$$R_m(x) = x/|x|^{m+1}, x \in \mathbf{R}^n.$$

Let μ be an m-regular measure and E an m-regular set in $\mathbf{R}^n$. The natural questions are when m is an integer, is it true that:

(a) $T^*_{R_m}$ is bounded in $L^2(\mu)$ if and only if $spt\mu$ is uniformly rectifiable?
(b) When $m = n-1$, E is removable for Lipschitz harmonic functions if and only if E is purely unrectifiable?

The reason that the Riesz kernel $|x|^{-n}x$ appears in connection of removable sets of Lipschitz harmonic functions is that it is essentially the gradient of the fundamental solution of the Laplacian.

The m-dimensional pure unrectifiability can be defined, for example, as the property that the set intersects every m-dimensional C^1 surface in a set of zero m-dimensional measure. The uniform rectifiability is a quantitative concept of rectifiability due to David and Semmes; see [9]. For one-dimensional sets it means exactly the condition (ii) of Theorem 1. It is known that the "if" part in (a) and the "only if" part in (b) are true. Some partial results for the converse can be found in [12, 14, 17]; they are discussed also in the book [13]. The main problem for the converse is to prove that boundedness such as in (a) implies some sort of rectifiability. One characterization of the rectifiability of E is approximation of E with m-dimensional planes almost everywhere at all small scales. The partial results referred to above are in the spirit that the boundedness implies such approximation almost everywhere at some, but maybe not all, small scales. Such partial results hold also in Heisenberg groups and we shall below formulate them more precisely there.

One can also consider the Riesz kernels when m is not an integer. Vihtilä showed in [26] that then $T^*_{R_m}$ is never bounded in $L^2(\mu)$ for m-regular measures μ.

4 Self-similar Sets and Singular Integrals

We shall now return to the general setting of Introduction. Let $\mathscr{S} = \{S_1,\dots,S_N\}, N \geq 2$, be an iterated function system (IFS) of similarities of the form

$$S_i = \tau_{q_i} \circ \delta_{r_i} \tag{2}$$

where $q_i \in G, r_i \in (0,1)$ and $i = 1,\dots,N$. The self-similar set C with respect to $\mathscr{S}$ is the unique non-empty compact set such that

$$C = \bigcup_{i=1}^{N} S_i(C).$$

If this system satisfies the strong separation condition, that is, the sets $S_i(C)$ are pairwise disjoint for $i = 1, \ldots, N$, it follows by a general metric space result of Schief in [21] (which holds also under the open set condition) that

$$0 < \mathscr{H}^s(C) < \infty \text{ for } \sum_{i=1}^{N} r_i^s = 1,$$

and the Hausdorff measure $\mathscr{H}^s \lfloor C$ is d-regular.

The following result was proved in [6]:

Theorem 3. *Let $\mathscr{S} = \{S_1, \ldots, S_N\}$ be an IFS in G satisfying the strong separation condition, let C be the corresponding s-dimensional self-similar set, and let K_Ω be an s-homogeneous kernel. If there exists a fixed point x for some $S_{i_1} \circ \cdots \circ S_{i_k}$; $S_{i_1} \circ \cdots \circ S_{i_k}(x) = x$, such that*

$$\int_{C \setminus S_{i_1} \circ \cdots \circ S_{i_k}(C)} K_\Omega(x, y) \mathrm{d}\mathscr{H}^s y \neq 0,$$

then the maximal operator $T^*_{K_\Omega}$ is unbounded in $L^2(\mathscr{H}^s \lfloor C)$; moreover, $\|T^*_{K_\Omega}(1)\|_{L^\infty(\mathscr{H}^s \lfloor C)} = \infty$.

Remark 1. Since such fixed points are dense in C, we have infinitely many points in a dense set and it suffices to check the condition at any one of them. Even when the ambient space is Euclidean, Theorem 3 provides new information about the behaviour of general homogeneous singular integrals on self-similar sets. For any kernel $K_\Omega(x) = \frac{\Omega(x/|x|)}{|x|^s}, x \in \mathbb{R}^n \setminus \{0\}, s \in (0, n)$, where Ω is continuous, one can easily find Sierpiński-type s-dimensional self-similar sets C_s for which one can check using Theorem 3 that the corresponding operator $T^*_{K_\Omega}$ is unbounded. For example, it follows that the operator associated to the kernel $z^3/|z|^4, z \in \mathbf{C} \setminus \{0\}$, is unbounded on many simple one-dimensional self-similar sets. In the case of the Sierpiński gasket this is immediate while in the case of the $1/4$-Cantor set it requires more computational work and it was checked after compiling a computer program. In [11], Huovinen considered such kernels in the plane and he proved that the a.e. existence of principal values of operators associated to any kernel $\frac{z^{2n-1}}{|z|^{2n}}$, for $n \geq 1$ implies rectifiability.

5 Self-similar Sets in Heisenberg Groups

For an introduction to Heisenberg groups and some of the facts mentioned below, see, for example, [2] or [1]. Below we state the basic facts needed in this survey.

The Heisenberg group $\mathbb{H}^n$, identified with $\mathbf{R}^{2n+1}$, is a non-abelian group where the group operation is given by

$$p \cdot q = (p_1 + q_1, \ldots, p_{2n} + q_{2n}, p_{2n+1} + q_{2n+1} + A(p,q)),$$

where

$$A(p,q) = -2 \sum_{i=1}^{n} (p_i q_{i+n} - p_{i+n} q_i).$$

We will denote points $p \in \mathbb{H}^n$ by $p = (p', p_{2n+1}), p' \in \mathbf{R}^{2n}, p_{2n+1} \in \mathbf{R}$. For any $q \in \mathbb{H}^n$ and $r > 0$, let again $\tau_q : \mathbb{H}^n \to \mathbb{H}^n$ be the left translation

$$\tau_q(p) = q \cdot p,$$

and define the dilation $\delta_r : \mathbb{H}^n \to \mathbb{H}^n$ by

$$\delta_r(p) = (rp_1, \ldots, rp_{2n}, r^2 p_{2n+1}).$$

A natural metric d on $\mathbb{H}^n$ is defined by

$$d(p,q) = \|p^{-1} \cdot q\|$$

where

$$\|p\| = (\|(p_1, \ldots, p_{2n})\|^4_{\mathbf{R}^{2n}} + p^2_{2n+1})^{\frac{1}{4}}.$$

The metric is left invariant, that is, $d(q \cdot p_1, q \cdot p_2) = d(p_1, p_2)$, and the dilations satisfy $d(\delta_r(p_1), \delta_r(p_2)) = rd(p_1, p_2)$. All the conditions of the general setting of Introduction are satisfied.

A subgroup G of $\mathbb{H}^n$ is called homogeneous if it is closed and invariant under the dilations; $\delta_r(G) = G$ for all $r > 0$. Every homogeneous subgroup G is a linear subspace of $\mathbf{R}^{2n+1}$. We call G a k-subgroup if its linear dimension is k. The homogeneous subgroups fall into two categories, vertical and horizontal: the vertical homogeneous k-subgroups are the linear subspaces of $\mathbf{R}^{2n+1}$ of the form $V \times \mathscr{T}$ where V is a $(k-1)$-dimensional linear subspace of $\mathbf{R}^n$ and $\mathscr{T}$ is the t-, that is, p_{2n+1}-axis. Their Hausdorff dimension is $k+1$. The horizontal homogeneous k-subgroups are those k-dimensional linear subspaces of $\mathbf{R}^{2n}$ on which A vanishes identically. Their Hausdorff dimension is k. The Haar measure on a k-subgroup is just the k-dimensional Lebesgue measure on it. We denote the set of these measures by $\mathscr{H}(n,k)$.

In this section we consider certain families of self-similar sets in $\mathbb{H}^n$ and we discuss their relations with Riesz-type transforms.

Definition 2. Let $Q=[0,1]^{2n}\subset\mathbf{R}^{2n}$ and $r\in(0,\frac{1}{2})$. Let $z_j\in\mathbf{R}^{2n}, j=1,\dots,2^{2n}$, be distinct points such that $z_{j,i}\in\{0,1-r\}$ for all $j=1,\dots,2^{2n}$ and $i=1,\dots,2n$. We consider the following 2^{2n+2} similitudes depending on the parameter r:

$$S_j=\tau_{(z_j,0)}\delta_r,\quad \text{for } j=1,\dots,2^{2n},$$

$$S_j=\tau_{(z_{\lfloor j\rfloor_{2^{2n}}},\frac{1}{4})}\delta_r,\quad \text{for } j=2^{2n}+1,\dots,2\cdot 2^{2n},$$

$$S_j=\tau_{(z_{\lfloor j\rfloor_{2\cdot 2^{2n}}},\frac{1}{2})}\delta_r,\quad \text{for } j=2\cdot 2^{2n}+1,\dots,3\cdot 2^{2n},$$

$$S_j=\tau_{(z_{\lfloor j\rfloor_{3\cdot 2^{2n}}},\frac{3}{4})}\delta_r,\quad \text{for } j=3\cdot 2^{2n}+1,\dots,2^{2n+2},$$

where $\lfloor j\rfloor_m:=j \mod m$ and $1\le\lfloor j\rfloor_m\le m$.

Theorem 4. *Let $r\in(0,\frac{1}{2})$ and $\mathscr{S}_r=\{S_1,\dots,S_{2^{2n+2}}\}$ where the S_j's are the similitudes of Definition 2. Let K_r be the self-similar set defined by*

$$K_r=\bigcup_{j=1}^{2^{2n+2}}S_j(K_r).$$

Then the sets $S_j(K_r)$ are disjoint for $j=1,\dots,2^{2n+2}$, and

$$0<\mathscr{H}^s(K_r)<\infty\quad \textit{with}\quad s=\frac{(2n+2)\log(2)}{\log(\frac{1}{r})}.$$

We give a sketch of the proof. It is similar to the one given by Strichartz in [22] in the case $r=1/2$. He obtains then a fractal tiling of $\mathbb{H}^n$. It is enough to find some set $R\supset K$ such that for all $j=1,\dots,2^{2n+2}$:

1. $S_j(R)\subset R$.
2. The sets $S_j(R)$ are disjoint.

This is established by finding a continuous function $\varphi:Q\to\mathbf{R}$ such that the set

$$R=\{q\in\mathbb{H}^n:q'\in Q\quad\text{and}\quad\varphi(q')\le q_{2n+1}\le\varphi(q')+1\}$$

satisfies (1) and (2).

This will follow immediately if we find some continuous $\varphi:Q\to\mathbf{R}$ which satisfies for all $j=1,\dots,2^{2n}$,

$$\tau_{(z_j,0)}\delta_r(R)=\{q\in\mathbb{H}^n:q'\in Q_j\quad\text{and}\quad\varphi(q')\le q_{2n+1}\le\varphi(q')+r^2\},\quad (3)$$

where $Q_j=\tau_{(z_j,0)}(\delta_r(Q))$. Since

$$\tau_{(z_j,0)}\delta_r(R) = \{p \in \mathbb{H}^n : p' \in Q_j \quad \text{and} \quad r^2\varphi(\frac{p'-z_j}{r}) - 2\sum_{i=1}^{n}(z_{j,i}p_{i+n} - z_{j,i+n}p_i)$$
$$\leq p_{2n+1} \leq r^2\varphi(\frac{p'-z_j}{r}) - 2\sum_{i=1}^{n}(z_{j,i}p_{i+n} - z_{j,i+n}p_i) + r^2\},$$

proving Eq. (3) amounts to showing that

$$\varphi(w) = r^2\varphi(\frac{w-z_j}{r}) - 2\sum_{i=1}^{n}(z_{j,i}w_{i+n} - z_{j,i+n}w_i) \quad \text{for } w \in Q_j,\ j = 1,\dots,2^{2n}. \quad (4)$$

Such a function φ is found with an application of the Banach fixed point theorem to a contraction T satisfying

$$T(f)(w) = r^2 f(\frac{w-z_j}{r}) - 2\sum_{i=1}^{n}(z_{j,i}w_{i+n} - z_{j,i+n}w_i) \quad \text{for } w \in Q_j.$$

6 Riesz-Type Kernels in Heisenberg Groups

Definition 3. The s-Riesz kernels in $\mathbb{H}^n$, $s \in (0, 2n+2)$, are defined as

$$R_s(p) = (R_{s,1}(p), \dots, R_{s,2n+1}(p))$$

where

$$R_{s,i}(p) = \frac{p_i}{\|p\|^{s+1}} \quad \text{for } i = 1,\dots,2n$$

and

$$R_{s,2n+1}(p) = \frac{p_{2n+1}}{\|p\|^{s+2}}.$$

Notice that these kernels are antisymmetric,

$$R_s(p^{-1}) = (R_s(p))^{-1},$$

and s-homogeneous,

$$R_s(\delta_r(p)) = \frac{1}{r^s}R_s(p).$$

Let μ be a finite Borel measure in $\mathbb{H}^n$. The image $f_{\#}\mu$ under a map $f : \mathbb{H}^n \to \mathbb{H}^n$ is the measure on $\mathbb{H}^n$ defined by

$$f_{\#}\mu(A) = \mu\left(f^{-1}(A)\right) \quad \text{for all } A \subset \mathbb{H}^n.$$

For $a \in \mathbb{H}^n$ and $r > 0$, $T_{a,r} : \mathbb{H}^n \to \mathbb{H}^n$ is defined for all $p \in \mathbb{H}^n$ by

$$T_{a,r}(p) = \delta_{1/r}(a^{-1} \cdot p).$$

Definition 4. We say that ν is a *tangent measure* of μ at $a \in \mathbb{H}^n$ if ν is a Radon measure on $\mathbb{H}^n$ with $\nu(\mathbb{H}^n) > 0$ and there are positive numbers c_i and r_i, $i = 1, 2, \dots$, such that $r_i \to 0$ and

$$c_i T_{a,r_i\#}\mu \to \nu \text{ weakly as } i \to \infty.$$

We denote by $Tan(\mu, a)$ the set of all tangent measures of μ at a.

The numbers c_i are normalization constants which are needed to keep ν non-trivial and locally finite. Often one can use $c_i = \mu(B(a, r_i))^{-1}$.

The following result was proved in [5] (recall that $\mathscr{H}(n,k)$ denotes the set of the Haar measures of the k-subgroups):

Theorem 5. *Let $s \in (0, 2n+2)$ and let μ be an s-regular measure in $\mathbb{H}^n$. If $T^*_{R_s}$ is bounded in $L^2(\mu)$, then*

1. *s is an integer in $[1, 2n+1]$*
2. *For μ-a.e. $a \in \mathbb{H}^n$, the set of tangent measures of μ at a,* $\mathrm{Tan}(\mu, a)$, *contains measures in $\mathscr{H}(n,s)$*

One can show that the s-dimensional self-similar sets of Theorem 4 do not have tangent measures in $\mathscr{H}(n,s)$; they are too spread at all scales for that. This leads to

Corollary 2. *The maximal operators $T^*_{R_s}$ are unbounded in $L^2(\mathscr{H}^s\lfloor C)$ for the s-dimensional self-similar sets of Theorem 4.*

Theorem 5 corresponds to what is known in $\mathbf{R}^n$ for s-regular sets and Riesz kernels in this respect (in other respects much more is known by results of Tolsa, Volberg and others; see, e.g., [10, 25]). The disadvantage here is that the kernels are not natural in the same way as Riesz kernels in $\mathbf{R}^n$; they do not seem to relate to any function classes. Analogues of harmonic functions lead to other kernels which we look at now.

7 Δ_{h}-Removability and Singular Integrals

The Lie algebra of left invariant vector fields in $\mathbb{H}^n$ is generated by

$$X_i := \partial_i + 2x_{i+n}\partial_{2n+1}, \quad Y_i := \partial_{i+n} - 2x_i\partial_{2n+1}, \quad T := \partial_{2n+1},$$

for $i = 1, \dots, n$. In fact, these vector fields generate the whole group and metric structure of $\mathbb{H}^n$.

If f is a real function defined on an open set of $\mathbb{H}^n$ its h-gradient is given by

$$\nabla_{\mathrm{h}} f = (X_1 f, \dots, X_n f, Y_1 f, \dots, Y_n f).$$

The h-divergence of a function $\phi = (\phi_1, \dots, \phi_{2n}) : \mathbb{H}^n \to \mathbf{R}^{2n}$ is defined as

$$div_{\mathrm{h}}\phi = \sum_{i=1}^{n}(X_i\phi_i + Y_i\phi_{i+n}).$$

The sub-Laplacian in $\mathbb{H}^n$ is given by

$$\Delta_{\mathrm{h}} = \sum_{i=1}^{n}(X_i^2 + Y_i^2)$$

or equivalently

$$\Delta_{\mathrm{h}} = div_{\mathrm{h}}\nabla_{\mathrm{h}}.$$

Definition 5. Let $U \subset \mathbb{H}^n$ be an open set. A real-valued function f on U is called Δ_{h}-harmonic, or simply harmonic, on U if $\Delta_{\mathrm{h}} f = 0$ on U.

We shall consider removable sets for Lipschitz solutions of the sub-Laplacian:

Definition 6. A compact set $C \subset \mathbb{H}^n$ will be called removable, or Δ_{h}-removable for Lipschitz Δ_{h}-harmonic functions, if for every open set U with $C \subset U$ and every Lipschitz function $f : U \to \mathbf{R}$,

$$\Delta_{\mathrm{h}} f = 0 \text{ in } U \setminus C \text{ implies } \Delta_{\mathrm{h}} f = 0 \text{ in } U.$$

Fundamental solutions for sub-Laplacians in homogeneous Carnot groups are defined in accordance with the classical Euclidean setting. In particular, in the case of the sub-Laplacian in $\mathbb{H}^n$:

Definition 7 (Fundamental solutions). A function $\Gamma : \mathbf{R}^{2n+1} \setminus \{0\} \to \mathbf{R}$ is a fundamental solution for Δ_{h} if:

1. $\Gamma \in C^\infty(\mathbf{R}^{2n+1} \setminus \{0\})$
2. $\Gamma \in L^1_{\mathrm{loc}}(\mathbf{R}^{2n+1})$ and $\lim_{\|p\| \to \infty} \Gamma(p) \to 0$
3. For all $\varphi \in C_0^\infty(\mathbf{R}^{2n+1})$

$$\int_{\mathbf{R}^{2n+1}} \Gamma(p)\Delta_{\mathrm{h}}\varphi(p)\mathrm{d}p = -\varphi(0).$$

It also follows easily that for every $p \in \mathbb{H}^n$,

$$\Gamma * \Delta_{\mathrm{h}}\varphi(p) = -\varphi(p) \quad \text{for all } \varphi \in C_0^\infty(\mathbf{R}^{2n+1}). \tag{5}$$

Convolutions are defined as usual by

$$f * g(p) = \int f(q^{-1} \cdot p)g(q)\mathrm{d}q$$

for $f, g \in L^1$ and $p \in \mathbb{H}^n$.

The fundamental solution Γ of Δ_{h} is given by

$$\Gamma(p) = C_d \|p\|^{2-d} \quad \text{for} \quad p \in \mathbb{H}^n \setminus \{0\}$$

where $d = 2n+2$ is the Hausdorff dimension of $\mathbb{H}^n$.

Let $K = \nabla_{\mathrm{h}} \Gamma$, then $K = (K_1, \dots, K_{2n}) : \mathbb{H}^n \to \mathbf{R}^{2n}$ where

$$K_i(p) = c_d \frac{p_i |p'|^2 + p_{i+n} p_{2n+1}}{\|p\|^{d+2}} \quad \text{and} \quad K_{i+n}(p) = c_d \frac{p_{i+n} |p'|^2 - p_i p_{2n+1}}{\|p\|^{d+2}}, \tag{6}$$

for $i = 1, \dots, n$, $p \in \mathbb{H}^n \setminus \{0\}$ and $c_d = (2-d)C_d$. We will also use the following notation:

$$\Omega_i(p) = c_d \frac{(p_i |p'|^2 + p_{i+n} p_{2n+1})}{\|p\|^3} \quad \text{and} \quad \Omega_{i+n}(p) = c_d \frac{(p_{i+n} |p'|^2 - p_i p_{2n+1})}{\|p\|^3}, \tag{7}$$

for $i = 1, \dots, n$ and $p \in \mathbb{H}^n \setminus \{0\}$. Hence,

$$K_i(p) = \frac{\Omega_i(p)}{\|p\|^{d-1}} \quad \text{and} \quad K(p) = \frac{\Omega(p)}{\|p\|^{d-1}}, \tag{8}$$

for $i = 1, \dots, 2n, \Omega = (\Omega_1, \dots, \Omega_{2n})$ and $p \in \mathbb{H}^n \setminus \{0\}$. The functions Ω_i are homogeneous and hence, recalling Definition 1, the kernels K_i are $(d-1)$-homogeneous.

The following proposition asserts that K is a standard kernel.

Proposition 1. *For all $i = 1, \dots, 2n$:*

1. $|K_i(p)| \lesssim \|p\|^{1-d}$ *for* $p \in \mathbb{H}^n \setminus \{0\}$
2. $|\nabla_{\mathrm{h}} K_i(p)| \lesssim \|p\|^{-d}$ *for* $p \in \mathbb{H}^n \setminus \{0\}$
3. $|K_i(p^{-1} \cdot q_1) - K_i(p^{-1} \cdot q_2)| \lesssim \max \left\{ \frac{d(q_1, q_2)}{d(p, q_1)^d}, \frac{d(q_1, q_2)}{d(p, q_2)^d} \right\}$ *for* $q_1, q_2 \neq p \in \mathbb{H}^n$

The following theorem, which makes use of Proposition 1, was proved in [6]. With d replaced by n, it is also valid for Lipschitz harmonic functions in $\mathbb{R}^n$, as it was shown in [17].

Theorem 6. *Let C be a compact subset of $\mathbb{H}^n$.*

1. *If $\mathscr{H}^{d-1}(C) = 0$, C is removable.*
2. *If $\dim C > d-1$, C is not removable.*

8 Δ_{h}-Removable Self-similar Cantor Sets in $\mathbb{H}^n$

In this section we consider a modified class of the self-similar Cantor sets C in $\mathbb{H}^n$ which were introduced in Sect. 3. Notice that there is one piece $S_0(C_{r,N})$ of $C_{r,N}$

below, which is well separated from the others. This is in order to make the condition of Theorem 3 easily checkable. It is very probable that also the more symmetric self-similar sets of Sect. 3 would satisfy that condition, but the calculation would become much more complicated.

Let $Q=[0,1]^{2n}\subset \mathbf{R}^{2n}, r>0, N\in 2\mathbf{N}$, be such that $r<\frac{1}{N}<\frac{1}{2}$. Let $z_j\in\mathbf{R}^{2n}, j=1,\dots,N^{2n}$, be distinct points such that $z_{j,i}\in\{\frac{l}{N}: l=0,1,\dots,N-1\}$ for all $j=1,\dots,N^{2n}$ and $i=1,\dots,2n$.

The similarities $\mathscr{S}_{r,N}=\{S_0,\dots,S_{\frac{1}{2}N^{2n+2}}\}$, depending on the parameters r and N, are defined as follows:

$$S_0=\delta_r,$$

$$S_j=\tau_{(z_{\lfloor j\rfloor_{N^{2n}}},\frac{1}{2}+\frac{i}{N^2})}\circ\delta_r,\ \text{ for } i=0,\dots,\frac{N^2}{2}-1 \quad\text{and}\quad j=iN^{2n}+1,\dots,(i+1)N^{2n},$$

where $\lfloor j\rfloor_m := j \mod m$.

Let $C_{r,N}$ be the self-similar set defined by

$$C_{r,N}=\bigcup_{j=0}^{\frac{1}{2}N^{2n+2}} S_j(C_{r,N}).$$

Then

$$0<\mathscr{H}^s(C_{r,N})<\infty \quad\text{with}\quad s=\frac{\log(\frac{1}{2}N^{2n+2}+1)}{\log(\frac{1}{r})}.$$

Denote by C_{d-1} the set C_{r_{d-1},N_0} for which

$$0<\mathscr{H}^{2n+1}(C_{r_{d-1},N_0})<\infty.$$

Theorem 7. *The Cantor set C_{d-1} satisfies $0<\mathscr{H}^{d-1}(C_{d-1})<\infty$ and is removable.*

The proof of Theorem 7 can be found in [6] and to prove it one verifies the condition of the general Theorem 3.

9 Concluding Comments

As discussed above, the question for what kind of 1-regular measures the singular integral operators based on the one-dimensional Riesz kernel are L^2-bounded is solved. So are the corresponding removability questions, both even much more generally than for regular measures and sets. For other integral dimensional Riesz kernels in $\mathbf{R}^n$ and Riesz-type kernels in $\mathbb{H}^n$ we have partial results for general regular measures and sets. For other kernels, such as the gradient of the fundamental solution of the sub-Laplacian, we only know results for some special self-similar sets. A natural direction would be to proceed further with self-similar sets, studying more systematically their properties and defining conditions in relation with kernels and L^2-boundedness. The L^2-boundedness on some particular self-similar sets for kernels adapted to them was shown in [8], by David, and in [3].

Acknowledgements P.M and V.C were supported by the Academy of Finland.

References

1. Bonfiglioli, A., Lanconelli, E., Uguzzoni, F.: Stratified Lie Groups and Potential Theory for Their Sub-Laplacians. Springer Monographs in Mathematics, Berlin (2007)
2. Capogna, L., Danielli, D., Pauls, S.D., Tyson, J.T.: An Introduction to the Heisenberg Group and the Sub-Riemannian Isoperimetric Problem. Birkhäuser, Basel (2007)
3. Chousionis, V.: Singular integrals on Sierpinski gaskets. Publ. Mat. **53**(1), 245–256 (2009)
4. Chousionis, V., Mateu, J., Prat, L., Tolsa, X.: Calderón-Zygmund kernels and rectifiability in the plane. Adv. Math. **231**(1), 535–568 (2012)
5. Chousionis, V., Mattila, P.: Singular integrals on Ahlfors-David subsets of the Heisenberg group. J. Geom. Anal. **21**(1), 56–77 (2011)
6. Chousionis, V., Mattila, P.: Singular integrals on self-similar sets and removability for Lipschitz harmonic functions in Heisenberg groups, to appear in J. Reme. Angew. Math
7. David, G.: Unrectifiable 1-sets have vanishing analytic capacity. Rev. Math. Iberoam. **14**, 269–479 (1998)
8. David, G.: Des intégrales singulières bornées sur un ensemble de Cantor. C. R. Acad. Sci. Paris Sr. I Math. **332**(5), 391–396 (2001)
9. David, G., Semmes, S.: Analysis of and on uniformly rectifiable sets. Mathematical Surveys and Monographs, vol. 38. American Mathematical Society, Providence (1993)
10. Eiderman, V. Nazarov, F., Volberg, A.: Vector-valued Riesz potentials: Cartan-type estimates and related capacities. Proc. London Math. Soc. **101**(3), 727–758 (2010)
11. Huovinen, P.: Singular integrals and rectifiability of measures in the plane. Ann. Acad. Sci. Fenn. Math. Diss. **109**, 1–63 (1997)
12. Lorent, A.: A generalised conical density theorem for unrectifiable sets. Ann. Acad. Sci. Fenn. Math. **28**(2), 415–431 (2003)
13. Mattila, P.: Geometry of Sets and Measures in Euclidean Spaces. Cambridge University Press, Cambridge (1995)
14. Mattila, P.: Singular integrals and rectifiability. In: Proceedings of the 6th International Conference on Harmonic Analysis and Partial Differential Equations (El Escorial, 2000), vol. Extra, pp. 199–208. Publ. Mat. (2002)
15. Mattila, P.: Removability, singular integrals and rectifiability. Rev. Roumaine Math. Pures Appl. **54**(5–6), 483–491 (2009)
16. Mattila, P., Melnikov, M.S., Verdera, J.: The Cauchy integral, analytic capacity, and uniform rectifiability. Ann. of Math. **144**(1–2), 127–136 (1996)
17. Mattila, P., Paramonov, P.V.: On geometric properties of harmonic Lip1-capacity. Pacific J. Math. **171**(2), 469–490 (1995)
18. Melnikov, M.S.: Analytic capacity: discrete approach and curvature of a measure. Sbornik: Mathematics **186**(6), 827–846 (1995)
19. Melnikov, M.S., Verdera, J.: A geometric proof of the L^2 boundedness of the Cauchy integral on Lipschitz graphs. Internat. Math. Res. Notices **7**, 325–331 (1995)
20. Pajot, H.: Analytic capacity, rectifiability, Menger curvature and the Cauchy integral. Lecture Notes in Mathematics, vol. 1799. Springer, Berlin (2002)

21. Schief, A.: Self-similar sets in complete metric spaces. Proc. Amer. Math. Soc. **124**, 481–490 (1996)
22. Strichartz, R.: Self-similarity on nilpotent Lie groups. Contemp. Math. **140**, 123–157 (1992)
23. Tolsa, X.: Painlevé's problem and the semiadditivity of analytic capacity. Acta Math. **190**(1), 105–149 (2003)
24. Tolsa, X.: Analytic capacity, rectifiability, and the Cauchy integral. Contemp. International Congress of Mathematicians, vol. II, pp. 1505–1527. Eur. Math. Soc., Zürich (2006)
25. Tolsa, X.: Calderon-Zygmund capacities and Wolff potentials on Cantor sets. J. Geom. Anal. **21**(1), 195–223 (2011)
26. Vihtilä, M.: The boundedness of Riesz s-transforms of measures in $\mathbb{R}^n$. Proc. Amer. Math. Soc. **124**, 481–490 (1996)

Multivariate Davenport Series

Arnaud Durand and Stéphane Jaffard

Abstract We consider series of the form $\sum a_n\{n \cdot x\}$, where $n \in \mathbf{Z}^d$ and $\{x\}$ is the sawtooth function. They are the natural multivariate extension of Davenport series. Their global (Sobolev) and pointwise regularity are studied and their multifractal properties are derived. Finally, we list some open problems which concern the study of these series.

1 Introduction

Let $\lfloor \cdot \rfloor$ denote integer part and let $\{\cdot\}$ be the *centered sawtooth function* defined by

$$\{x\} = \begin{cases} x - \lfloor x \rfloor - \frac{1}{2} & \text{if } x \notin \mathbf{Z}, \\ 0 & \text{else.} \end{cases} \tag{1}$$

The purpose of this chapter is to investigate regularity properties of the multivariate functions which are defined by

$$\forall x \in \mathbf{R}^d \qquad f(x) = \sum_{n \in \mathbf{Z}^d_*} a_n \{n \cdot x\}, \tag{2}$$

A. Durand (✉)
Laboratoire de Mathématiques, UMR 8628, Université Paris-Sud, 91405 Orsay Cedex, France
e-mail: arnaud.durand@math.u-psud.fr

S. Jaffard
Laboratoire d'Analyse et de Mathématiques Appliquées UMR 8050, Université Paris-Est - Créteil Val-de-Marne, 61 avenue du Général de Gaulle, 94010, Créteil Cedex, France
e-mail: jaffard@u-pec.fr

J. Barral and S. Seuret (eds.), *Further Developments in Fractals and Related Fields*, Trends in Mathematics, DOI 10.1007/978-0-8176-8400-6_5,

where $n \cdot x$ denotes the standard inner product between the vectors n and x, and $(a_n)_{n \in \mathbf{Z}_*^d}$ is a real-valued sequence indexed by the set $\mathbf{Z}_*^d = \mathbf{Z}^d \setminus \{0\}$. With a slight abuse, the vectors n for which a_n is nonvanishing will be referred to as the *frequencies* of the series.

In the one-variable case, examples of such functions can be traced back to the *Habilitationsschrift* of Riemann, see [40, 43]; they were later considered by Hecke [28], and also Hardy, who studied the series

$$\mathfrak{H}_\beta(x) = \sum_{n=1}^{\infty} \frac{\{nx\}}{n^\beta}. \tag{3}$$

It seems however that the general one-dimensional case was first considered only in 1937 by Davenport in [19, 20]. The first of these papers starts with the following remarkable identity, which establishes in all generality the connection with Fourier series:

$$\sum_{n=1}^{\infty} a_n \{nx\} = \sum_{m=1}^{\infty} c_m \sin(2\pi m x) \quad \text{with} \quad c_m = -\frac{1}{\pi m} \sum_{\substack{n \in \mathbf{N} \\ n | m}} n a_n. \tag{4}$$

One of the fascinating aspects of these expansions is that their study lies at the crossroad between several areas of mathematics. They appear naturally in several problems related with analytic number theory; this actually was the motivation of Davenport for studying them, see also the recent studies by de la Bretèche and Tenenbaum (such as [17] for instance). They were later considered in connection with harmonic analysis, see, e.g., [35] and references therein where a function space point of view is developed, and it is shown in which sense an arbitrary one-periodic odd function can be expanded on this system. Convergence properties of these series at particular points are related with the Diophantine approximation properties of these points, see [17, 35]. Recently, Brémont studied the L^2 and almost-sure convergence of these series, see [15]. The multifractal analysis of these functions shows connections between their pointwise regularity properties and geometric measure theory, see [35] and also [47] for an extension of Davenport series with translated phases. Note also that examples of Davenport series valued in $\mathbf{R}^2$ were proposed by H. Lebesgue as space-filling functions; this study was developed in [38, 39], where the connections between Davenport series and space-filling functions are examined.

In this chapter, we shall investigate the multivariate case, which has not been considered up to now. Our main motivation is that multivariate Davenport series are natural examples of multifractal fields. The recent increase of interest in such fields is motivated by the relevance of multifractal analysis techniques in image classification, see [1, 2]. Indeed, the validation of 2D multifractal analysis algorithms requires the introduction and the mathematical study of collections of multifractal fields of various kinds. However, very few multivariate multifractal models have been studied up to now (see however the Ph.D. thesis of Oppenheim [41] for an early analysis of a multifractal function of several variables where Diophantine approximation properties are involved and [4] for fields generated by random wavelet series). Other cases of random fields which have recently been studied are Lévy fields, which are a natural extension of Lévy processes to the multivariate setting; their multifractal analysis has recently been performed by the authors, see [24]. The scarcity of existing results is partly due to the fact that the derivation of the multifractal properties of multivariate functions lies on variants of ubiquity methods which can prove much more involved in the multidimensional setting. Therefore, extending the collection of available multivariate models, and elucidating their multifractal properties, is an important issue. An additional motivation of this chapter is to draw a comparison between the multifractal behavior of Davenport series and Lévy fields. Indeed, they are both constructed as superpositions of piecewise linear functions which display jumps along hyperplanes; the main difference being that the locations of these hyperplanes are random in the case of Lévy fields, whereas they are determined by arithmetic conditions in the case of Davenport series. We shall see that the multifractal properties of Davenport series bear similarities with those of Lévy fields so that they can be seen as a kind of deterministic counterpart of these fields.

The chapter is organized as follows. In Sect. 2, we establish the relationships between the Davenport and Fourier coefficients of Davenport series, in the normally convergent case. We shall see that this relationship extends to more general, and actually distributional, settings in Sect. 8. The main purpose of this chapter is the study of pointwise regularity properties of Davenport series. The key step consists in analyzing the locations and magnitudes of the jumps of Davenport series. Preliminary results concerning this study are collected in Sects. 3 and 4. In Sect. 5, an upper bound of the Hölder exponent is derived and, as a consequence, cases where this exponent vanishes everywhere are worked out. A difficult question (which is far from being closed, even in the one-variable case) is to understand when this upper bound is sharp; the purpose of Sect. 6 is to show that this is the case when the frequencies of the Davenport series are sufficiently sparse. Implications for multifractal analysis are stated in Sect. 7. In Sect. 8, we shall consider convergence properties of Davenport series in the Sobolev spaces H^s for $s \in \mathbf{R}$, especially when the sequence of coefficients does not belong to ℓ^1; this study will also be the occasion to draw bridges with arithmetic functions in several variables, a topic which has been barely scratched until now (see however [16, 25] and references therein). Concluding remarks and open problems are collected in Sect. 9. Finally, the proofs of the main results are completed in Sects. 10 and 11. This is the occasion for us to exhibit deep connections with the theory of sets of large intersection and with the Duffin–Schaeffer and Catlin conjectures in the metric theory of Diophantine approximation, see Sect. 11.2.

2 Relationships Between Davenport and Fourier Series

We start by establishing some conventions. First, note that functions such as Eq. (2) are necessarily odd and $\mathbf{Z}^d$-periodic. Since $\{-x\} = -\{x\}$, it follows that the system supplied by the $\{n \cdot x\}$, for $n \in \mathbf{Z}^d$, is redundant. The choice made for one-dimensional Davenport series is to use only these functions for $n \geq 1$, as, e.g., in Eq. (4) above. We shall make a different choice in dimension $d \geq 2$, which will preserve the symmetry of the decomposition. Specifically, we shall keep both functions $\{n \cdot x\}$ and $\{-n \cdot x\}$, and, without loss of generality, we shall assume that the sequence $(a_n)_{n \in \mathbf{Z}^d}$ of Davenport coefficients is an odd sequence indexed by $\mathbf{Z}^d$, which implies uniqueness of the decomposition.

The function spaces that we shall consider are composed of $\mathbf{Z}^d$-periodic odd functions, and the sequence spaces that we shall consider are composed of odd sequences. Therefore, we shall use the following conventions concerning spaces: With a slight abuse of notations, ℓ^p will denote the space of odd sequences which belong to $\ell^p(\mathbf{Z}^d)$, L^2 is the space of odd locally square-integrable functions which are $\mathbf{Z}^d$-periodic, and, more generally, if E is a space of functions defined on $\mathbf{R}^d$, we shall also denote by E the space of odd functions that belong locally to E and are $\mathbf{Z}^d$-periodic.

Another convention concerns divisibility in several dimensions. Let $n, m \in \mathbf{Z}^d_*$. If $m = ln$ for some $l \in \mathbf{Z}^*$, we say that l and n are divisors of m. The fact that the term "divisor" applies without distinction to elements of $\mathbf{Z}^d_*$ and of $\mathbf{Z}^*$ will not create confusions because the context will always be clear. If $l = \pm 1$ are the only integer divisors of m, we say that m is *irreducible*; this means that its components are coprime. Throughout the chapter, $\mathbf{N}$ denotes the set of positive integers.

Finally, the *support* of a sequence $a = (a_n)_{n \in \mathbf{Z}^d}$ is

$$\operatorname{supp}(a) = \{n \in \mathbf{Z}^d \mid a_n \neq 0\}.$$

Let us now investigate the relationship between Davenport and Fourier series. To this end, let us assume that the series (2) converges normally, that is, that the sequence $(a_n)_{n \in \mathbf{Z}^d}$ belongs to ℓ^1 (convergence properties in different functional settings will be investigated in Sect. 8). Then, f belongs to L^∞, hence to L^2 and the Fourier series expansion of f converges in L^2. Since f is odd, it may be written in the form

$$f(x) = \sum_{m \in \mathbf{Z}^d} c_m \sin(2\pi m \cdot x) \quad \text{with} \quad c_m = \int_{[0,1)^d} f(x) \sin(2\pi m \cdot x)\, \mathrm{d}x.$$

Here, we adopt the same convention for Fourier series as for Davenport series, that is, we assume that the expansion is taken on all frequencies of $\mathbf{Z}^d$, but that the sequence $(c_m)_{m \in \mathbf{Z}^d}$ is odd. Since Eq. (2) is normally convergent,

$$c_m = \sum_{n \in \mathbf{Z}^d_*} a_n \int_{[0,1)^d} \{n \cdot x\} \sin(2\pi m \cdot x)\, \mathrm{d}x$$

for all $m \in \mathbf{Z}^d_*$. A straightforward computation shows that the above integral is equal to zero except if n is a divisor of m, in which case there exists an integer $l \in \mathbf{Z}^*$ such that $ln = m$, and the integral is equal to $-1/(2\pi l)$. As a consequence, the Fourier coefficients of f are given by

$$c_m = -\frac{1}{2\pi} \sum_{\substack{(l,n)\in\mathbf{Z}^*\times\mathbf{Z}^d_*\\ ln=m}} \frac{a_n}{l}. \tag{5}$$

Note that without making any assumption on the summability of the sequence $(a_n)_{n\in\mathbf{Z}^d}$, the above formula still enables us to define a sequence $(c_m)_{m\in\mathbf{Z}^d}$. This detour via Fourier series will allow us to study the convergence of the series (2) even when $(a_n)_{n\in\mathbf{Z}^d}$ does not belong to ℓ^1, see Sect. 8. Indeed, we shall see that, in many functional settings, when the associated Fourier series converges, then the partial sums of the series $\sum_n a_n\{n\cdot x\}$ converge to the same limit.

3 Discontinuities of Davenport Series

Let us consider a bounded function $g : \mathbf{R}^d \to \mathbf{R}$. By definition, the magnitude of the jump of g at any fixed point $x_0 \in \mathbf{R}^d$ is

$$\Delta_g(x_0) = \limsup_{x\to x_0} g(x) - \liminf_{x\to x_0} g(x)$$

(which may possibly vanish, in which case g is continuous at x_0). The magnitude of the jumps can also be expressed by means of local oscillations. To be specific, let us recall that the *oscillation* of the function g on a bounded subset Ω of $\mathbf{R}^d$ is defined by

$$\mathrm{Osc}_g(\Omega) = \sup_{x\in\Omega} g(x) - \inf_{x\in\Omega} g(x).$$

Letting $\mathrm{B}(x,r)$ denote the open ball with center x and radius r, it is easy to see that the magnitude of the jump of the function g at the point x_0 satisfies

$$\Delta_g(x_0) = \lim_{r\to 0} \mathrm{Osc}_g(\mathrm{B}(x_0,r)).$$

We shall now determine the set of points at which the Davenport series f defined by Eq. (2) has a discontinuity, and we shall study the magnitude of the corresponding jump. We assume in what follows that the sequence $(a_n)_{n\in\mathbf{Z}^d}$ belongs to ℓ^1.

Given a vector with integer coordinates $q \in \mathbf{Z}^d_*$ and an integer $p \in \mathbf{Z}$, let $H_{p,q}$ denote the hyperplane

$$H_{p,q} = \{x \in \mathbf{R}^d \mid p = q\cdot x\}. \tag{6}$$

It is clear that multiplying p and the components of q by a common integer value leaves the hyperplane unchanged. In order to ensure the uniqueness of the representation, it is sufficient to assume that p and the components of q are coprime and that q belongs to the subset $\mathbf{Z}^d_+$ of $\mathbf{Z}^d_*$ formed by the vectors whose first nonvanishing coordinate is positive. In fact, one easily checks that any hyperplane $H_{p,q}$ may be indexed in a unique manner by a pair (p,q) that belongs to

$$\mathscr{H}_d = \left\{(p,q) \in \mathbf{Z}\times\mathbf{Z}^d_+ \mid \gcd(p,q) = 1\right\},$$

where $\gcd(p,q)$ is the greatest common divisor of the integer p and the components of the vector q. Furthermore, let $\{\cdot\}_\star$ denote the restriction of the sawtooth function $\{\cdot\}$ to the open interval $(-1/2,1/2)$. Then, $\{\cdot\}_\star$ is continuous everywhere except at the origin: $\{x\}_\star$ makes a jump of size -1 when x crosses zero in the upward direction. In addition, $\{x\}$ is the sum of $\{x-p\}_\star$ over all the integers $p\in\mathbf{Z}$. Along with the fact that the sequence $(a_n)_{n\in\mathbf{Z}^d}$ and the function $\{\cdot\}_\star$ are both odd, this enables us to rewrite the definition (2) of the Davenport series f in the form

$$f(x)=\sum_{(p,q)\in\mathscr{H}_d} f_{p,q}(x) \qquad \text{with} \qquad f_{p,q}(x)=2\sum_{l=1}^{\infty} a_{lq}\{l(q\cdot x-p)\}_\star,$$

where all the series converge normally. This decomposition enlightens the fact that f is the superposition of a family of functions that are continuous everywhere except maybe on a specific hyperplane of the above kind. To be precise, each function $f_{p,q}$ is continuous everywhere except maybe on the hyperplane $H_{p,q}$, and the fact that (p,q) belongs to $\mathscr{H}_d$ implies that these hyperplanes are distinct. Moreover, when a point x crosses $H_{p,q}$, the real points $l(q\cdot x-p)$, for $l\geq 1$, all cross zero, so that $f_{p,q}(x)$ makes a jump of magnitude $|A_q|$, where

$$A_q=2\sum_{l=1}^{\infty} a_{lq}. \tag{7}$$

Note that, in the case where the latter sum vanishes, the function $f_{p,q}$ is actually continuous on the whole space, including the hyperplane $H_{p,q}$.

The analysis of the discontinuities of the Davenport series f begins with a first remark: As the series (2) is normally convergent, its sum f is a function in L^∞ so that the potential discontinuities must have finite magnitude. We shall now show that the set of points at which f is not continuous is exactly

$$\bigcup_{\substack{(p,q)\in\mathscr{H}_d\\ A_q\neq 0}} H_{p,q}. \tag{8}$$

First, note that if a point x_0 does not belong to the latter set, then the above decomposition entails that f is a sum of uniformly convergent series of functions that are continuous at x_0, thereby being continuous at x_0 as well. Conversely, if a point x_0 belongs to a hyperplane $H_{p,q}$ indexed by a pair $(p,q)\in\mathscr{H}_d$ for which A_q does not vanish, and to no other hyperplane of that form (which is the case of Lebesgue-almost every point of $H_{p,q}$), then f has a discontinuity at x_0 of magnitude $|A_q|$, that is,

$$\Delta_f(x_0)=|A_q|>0.$$

More generally, suppose that x_0 belongs to a (possibly infinite) collection of hyperplanes H_{p_i,q_i} indexed by pairs $(p_i,q_i)\in\mathscr{H}_d$ for which A_{q_i} do not vanish. The previous case shows that, for any specific value of i, the function f has a discontinuity of magnitude exactly $|A_{q_i}|$ on a dense set of points of H_{p_i,q_i}. Therefore, one can pick a point y_i arbitrarily close to x_0 at which f has a discontinuity of magnitude exactly $|A_{q_i}|$. It follows that

$$\Delta_f(x_0)\geq \max_i |A_{q_i}|>0,$$

so that f exhibits a discontinuity at x_0.

Given that $|A_q|$ is the magnitude of the jump of the Davenport series f at Lebesgue-almost every point of the hyperplane $H_{p,q}$, we shall call with a slight abuse $|A_q|$ the magnitude of the jump of f on $H_{p,q}$.

We see here a sharp contrast with Fourier series: The series (2) will usually exhibit discontinuities no matter how fast the coefficients a_n decay. The following proposition shows that even more is true: The zero function is the only continuous Davenport series.

Proposition 1. *Let f be a Davenport series with coefficients given by a sequence $a = (a_n)_{n\in\mathbf{Z}^d}$ in ℓ^1. If f is a continuous function, then*

$$\forall n \in \mathbf{Z}^d \qquad a_n = 0.$$

Proof. The continuity of f implies that $A_q = 0$ for all vectors $q \in \mathbf{Z}^d$. Given an irreducible vector q, let $b_l^q = a_{lq}$ for any integer $l \geq 1$. Then, the sequence $(b_l^q)_{l\geq 1}$ is in $\ell^1(\mathbf{N})$ and satisfies

$$\forall l \geq 1 \qquad \sum_{k=1}^{\infty} b_{kl}^q = 0.$$

Haar proved that these conditions imply that $b_l^q = 0$ for all $l \geq 1$, see [42, Chap. 1, no. 129]. This argument holds in all directions q, so the result follows. □

We refer to the next section for more general results that explain how to recover the coefficients a_n from the values A_q.

4 The Jump Operator

In order to study the regularity properties of normally convergent Davenport series, it is useful to consider the linear operator J which maps the sequence of Davenport coefficients to the sequence of jumps and which is defined by

$$\forall (a_n)_{n\in\mathbf{Z}^d} \in \ell^1 \qquad J\left((a_n)_{n\in\mathbf{Z}^d}\right) = (A_q)_{q\in\mathbf{Z}^d} \in \ell^\infty, \tag{9}$$

where the coefficients A_q are given by Eq. (7). The key results concerning this mapping follow from those obtained in [35] in the one-dimensional case; this is due to a remarkable decomposition that we now present.

Let $\mathscr{I}^d$ denote the subset of $\mathbf{Z}_*^d$ formed by the irreducible vectors and let V_m denote the vector space of odd sequences $(a_n)_{n\in\mathbf{Z}^d}$ that are supported by the multiples of such a vector $m \in \mathscr{I}^d$, that is, such that

$$\operatorname{supp}\left((a_n)_{n\in\mathbf{Z}^d}\right) \subseteq \mathbf{Z}m.$$

Any odd sequence indexed by $\mathbf{Z}^d$ may be decomposed as a sum of sequences b^m such that $b^m \in V_m$. However, as the vector subspaces V_m and V_{-m} coincide, in order to ensure the uniqueness of the decomposition, we shall privilege the irreducible vectors whose first nonvanishing coordinate is positive. The set of those vectors is therefore $\mathscr{I}_+^d = \mathscr{I}^d \cap \mathbf{Z}_+^d$. As a consequence, we obtain the following unconditional Schauder decomposition:

$$\ell^1 = \bigoplus_{m \in \mathscr{I}_+^d} \left(V_m \cap \ell^1\right), \tag{10}$$

meaning that any sequence in ℓ^1 may be written in a unique manner as the sum in the ℓ^1 sense of an unconditionally summable family indexed by $m \in \mathscr{I}_+^d$ of sequences in $V_m \cap \ell^1$. Moreover, the operator J maps the subspace $V_m \cap \ell^1$ to $V_m \cap \ell^\infty$. It follows that, in order to study J, it suffices to analyze its restriction J_m to the subspace $V_m \cap \ell^1$, for any fixed vector $m \in \mathscr{I}_+^d$.

On top of that, let S_m denote the operator of subsampling with step m, which is defined by

$$S_m((a_n)_{n \in \mathbf{Z}^d}) = (a_{lm})_{l \geq 1}$$

for any odd sequence $(a_n)_{n \in \mathbf{Z}^d}$. As we consider odd sequences only, it is clear that the restriction of S_m to V_m is one-to-one. In fact, we even see that S_m maps $V_m \cap \ell^p$ onto $\ell^p(\mathbf{N})$.

In the one-dimensional case, as already mentioned above and illustrated by Eq. (4), one assumes that the sequence of Davenport coefficients is supported on $\mathbf{N}$, instead of supposing that they form an odd sequence indexed by $\mathbf{Z}$. Thus, the operator J has a simpler counterpart which has already been considered in [35]; this is the *jump operator* $\mathscr{J}$ defined by

$$\mathscr{J}\left((b_n)_{n \geq 1}\right) = \left(\sum_{l=1}^{\infty} b_{lq}\right)_{q \geq 1} \in \ell^\infty(\mathbf{N}), \tag{11}$$

for any sequence $(b_n)_{n \geq 1} \in \ell^1(\mathbf{N})$. The following straightforward lemma shows that all the mappings J_m essentially reduce to $\mathscr{J}$.

Lemma 1. *Let us consider a vector $m \in \mathscr{I}_+^d$. Then, for any sequence $a = (a_n)_{n \in \mathbf{Z}^d}$ in $V_m \cap \ell^1$,*

$$S_m(J_m(a)) = 2\mathscr{J}(S_m(a)).$$

Therefore, the following diagram is commutative:

$$\begin{array}{ccc} V_m \cap \ell^1 & \xrightarrow{J_m} & V_m \cap \ell^\infty \\ \sim \big\downarrow S_m & & \sim \big\downarrow S_m \\ \ell^1(\mathbf{N}) & \xrightarrow{2\mathscr{J}} & \ell^\infty(\mathbf{N}) \end{array}$$

It follows from Lemma 1 that the mapping J can be inverted on each subspace $V_m \cap \ell^1$ by means of an inversion formula for the one-dimensional jump operator $\mathscr{J}$. This formula has been obtained in [35] and is recalled in the statement of Proposition 2 below. It makes use of the *Möbius function* μ, which is defined on the positive integers by $\mu(n) = 0$ if n is not square-free, and by $\mu(n) = (-1)^k$ if n is square-free and admits exactly k prime divisors. The inversion formula holds on the subspace $\mathrm{T}(\mathbf{N})$ of $\ell^1(\mathbf{N})$ that is formed by the sequences $(b_n)_{n\geq 1}$ for which the series $\sum_n \tau(n)|b_n|$ converges, where $\tau(n)$ denotes the number of divisors of n; this merely means that the restriction of $\mathscr{J}$ to that subspace is one-to-one. It is well-known that the sequence $\tau(n)$ grows slower than any positive power of n, in the sense that $\tau(n) = \mathrm{o}(n^\varepsilon)$ as n goes to infinity, for all $\varepsilon > 0$. This is a plain consequence of the fact that

$$\limsup_{n\to\infty} \frac{\log\log n}{\log n} \log \tau(n) = \log 2, \tag{12}$$

see, e.g., [3, Theorem 13.12]. This implies in particular that, for any real γ larger than one, $\mathrm{T}(\mathbf{N})$ contains the space $\mathscr{F}^\gamma(\mathbf{N})$ of all the sequences $b = (b_n)_{n\geq 1}$ such that

$$|b|_{\mathscr{F}^\gamma(\mathbf{N})} = \sup_{n\geq 1} n^\gamma |b_n| < \infty.$$

Thus, the restriction of $\mathscr{J}$ to each $\mathscr{F}^\gamma(\mathbf{N})$ is one-to-one. The next result even shows that $\mathscr{J}$ is a bicontinuous automorphism of $\mathscr{F}^\gamma(\mathbf{N})$. We refer to [35] for its proof.

Proposition 2. *The operator* $\mathscr{J}$ *induces a one-to-one mapping from* $\mathrm{T}(\mathbf{N})$ *into* $\ell^1(\mathbf{N})$. *More specifically, for any sequence* $B = (B_q)_{q\geq 1}$ *in the image set* $\mathscr{J}(\mathrm{T}(\mathbf{N}))$, *the equation*

$$B = \mathscr{J}(b)$$

admits exactly one solution $b = (b_n)_{n\geq 1}$ *in* $\mathrm{T}(\mathbf{N})$, *namely, the sequence defined by*

$$\forall n \geq 1 \qquad b_n = \sum_{l=1}^{\infty} \mu(l) B_{ln}. \tag{13}$$

Moreover, for any real $\gamma > 1$, *the operator* $\mathscr{J}$ *induces a bicontinuous automorphism of the space* $\mathscr{F}^\gamma(\mathbf{N})$ *whose inverse is given by Eq. (13).*

Thanks to Lemma 1, Proposition 2 naturally extends to the multivariate setting. In fact, we now establish that the higher-dimensional jump operator J induces a bicontinuous automorphism on the space $\mathscr{F}^\gamma$ defined as follows.

Definition 1. The space $\mathscr{F}^\gamma$ is the vector space composed of the odd sequences $a = (a_n)_{n\in\mathbf{Z}^d}$ satisfying

$$|a|_{\mathscr{F}^\gamma} = \sup_{n\in\mathbf{Z}^d_*} |n|^\gamma |a_n| < \infty.$$

If γ is larger than the dimension d of the ambient space, it is clear that $\mathscr{F}^\gamma$ may be seen as a vector subspace of ℓ^1, so that the operator J is well defined on $\mathscr{F}^\gamma$.

In the opposite case, $\mathscr{F}^\gamma$ is not necessarily included in ℓ^1. However, if $\gamma > 1$, the formula (9) still has a meaning, because all the series (7) converge, which enables us to define the operator J on $\mathscr{F}^\gamma$ as well. This may also be seen as a consequence of Lemma 1, along with the fact that $S_m(\mathscr{F}^\gamma)$ is contained in $\ell^1(\mathbf{N})$.

Proposition 3. *For any $\gamma > 1$, the jump operator J is a bicontinuous automorphism of $\mathscr{F}^\gamma$, and its inverse is given by*

$$\forall A = (A_q)_{q \in \mathbf{Z}^d} \in \mathscr{F}^\gamma \qquad J^{-1}(A) = \left(\frac{1}{2} \sum_{l=1}^{\infty} \mu(l) A_{ln} \right)_{n \in \mathbf{Z}^d}. \tag{14}$$

Proof. Given $\gamma > 1$, let $a = (a_n)_{n \in \mathbf{Z}^d}$ be a sequence in $\mathscr{F}^\gamma$, and let $A = (A_q)_{q \in \mathbf{Z}^d}$ denote its image under J, that is, $A = J(a)$. Then, for each vector $q \in \mathbf{Z}^d_*$,

$$|A_q| \leq 2 \sum_{l=1}^{\infty} |a_{lq}| \leq 2 \sum_{l=1}^{\infty} \frac{|a|_{\mathscr{F}^\gamma}}{|lq|^\gamma} \leq \frac{2\zeta(\gamma)}{|q|^\gamma} |a|_{\mathscr{F}^\gamma},$$

where ζ is the Riemann zeta function. Therefore, J is continuous on $\mathscr{F}^\gamma$ with operator norm at most $2\zeta(\gamma)$.

In order to study the invertibility of the operator J on $\mathscr{F}^\gamma$, let us begin by observing that, in a way similar to Eq. (10), the latter space may be decomposed as the direct sum over $m \in \mathscr{I}^d_+$ of the subspaces $V_m \cap \mathscr{F}^\gamma$. Moreover, the subsampling operator S_m induces a one-to-one mapping from $V_m \cap \mathscr{F}^\gamma$, which is included in $V_m \cap \ell^1$, onto $\mathscr{F}^\gamma(\mathbf{N})$. Then, letting π_m denote the projection onto V_m, we deduce from Lemma 1 and Proposition 2 that the following diagram is commutative:

$$\begin{array}{ccc}
\mathscr{F}^\gamma & \xrightarrow{\quad J \quad} & \mathscr{F}^\gamma \\
\downarrow \pi_m & & \downarrow \pi_m \\
V_m \cap \mathscr{F}^\gamma & \xrightarrow{\quad J_m \quad} & V_m \cap \mathscr{F}^\gamma \\
\sim \downarrow S_m & & \sim \downarrow S_m \\
\mathscr{F}^\gamma(\mathbf{N}) & \xrightarrow[\sim]{\quad 2\mathscr{J} \quad} & \mathscr{F}^\gamma(\mathbf{N})
\end{array}$$

This implies the invertibility of the operator J on the space $\mathscr{F}^\gamma$. In order to obtain an explicit formula for its inverse J^{-1}, let us make use of the diagram to infer that the equation $A = J(a)$ implies

$$S_m(\pi_m(a)) = \frac{1}{2} \mathscr{J}^{-1}(S_m(\pi_m(A)))$$

for all $m \in \mathscr{I}_{+}^{d}$, where $\mathscr{J}^{-1}$ denotes the inverse of the one-dimensional jump operator $\mathscr{J}$ on $\mathscr{F}^{\gamma}(\mathbf{N})$. By means of Eq. (13), we deduce that

$$S_m(a) = \left(\frac{1}{2} \sum_{l=1}^{\infty} \mu(l) A_{lkm} \right)_{k \geq 1}$$

and Eq. (14) follows. Finally, the method that we used above in order to show the continuity of the operator J also applies to its inverse because the Möbius function is at most one in absolute value; proceeding in this way, we deduce that J^{-1} is continuous with operator norm at most $\zeta(\gamma)/2$. □

We shall show in Sect. 6 below that the statement of Proposition 3 may be extended to the case where $0 < \gamma \leq 1$, up to replacing $\mathscr{F}^{\gamma}$ by a subspace formed by sequences whose support satisfies an additional sparsity assumption.

5 Pointwise Hölder Regularity

The section contains general results on the pointwise regularity of multivariate functions with a dense set of discontinuities, a class in which the typical Davenport series fall. We begin by recalling the appropriate definitions.

Definition 2. Let $f : \mathbf{R}^d \to \mathbf{R}$ be a locally bounded function, $x_0 \in \mathbf{R}^d$ and $\alpha \geq 0$. The function f belongs to $C^{\alpha}(x_0)$ if there exist $C > 0$ and a polynomial P_{x_0} of degree less than α such that for all x in a neighborhood of x_0,

$$|f(x) - P_{x_0}(x)| \leq C\,|x - x_0|^{\alpha}. \tag{15}$$

The *Hölder exponent* of f at x_0 is then defined by

$$h_f(x_0) = \sup\{\alpha \geq 0 \,|\, f \in C^{\alpha}(x_0)\}.$$

Note that h_f takes values in $[0, \infty]$. The following lemma yields an upper bound on the pointwise Hölder exponent of functions that have a dense set of discontinuities and will be applied to Davenport series in the following. It is a direct extension to the multivariate setting of Lemma 1 in [33].

Lemma 2. *Let $f : \mathbf{R}^d \to \mathbf{R}$ be a locally bounded function and let $x_0 \in \mathbf{R}^d$. Then,*

$$h_f(x_0) \leq \liminf_{s \to x_0} \frac{\log \Delta_f(s)}{\log |s - x_0|}. \tag{16}$$

Proof. We may obviously assume that $h_f(x_0)$ is positive, in which case f is continuous at x_0. Then, given a positive real α less than $h_f(x_0)$, there exists $\delta, C > 0$ and a polynomial P_{x_0} such that Eq. (15) holds for any x in the open ball $\mathrm{B}(x_0, \delta)$.

Now, let s be a discontinuity point of f, which thus necessarily differs from x_0. The magnitude $\Delta_f(s)$ of the jump of f at s is positive, as well as $\varepsilon = \Delta_f(s)/6$. Owing to the definition of $\Delta_f(s)$ and the continuity of the polynomial P_{x_0}, there exist two points x_1 and x_2 such that

$$|f(x_1) - f(x_2)| \geq \Delta_f(s) - \varepsilon \qquad \text{and} \qquad |P_{x_0}(x_1) - P_{x_0}(x_2)| \leq \varepsilon.$$

These two points may be chosen arbitrarily close to s, for instance within range $|s - x_0|/2$ from that point. Therefore, at least one of them, denoted by $x(s)$, satisfies

$$|x(s) - s| \leq \frac{|s - x_0|}{2} \qquad \text{and} \qquad |f(x(s)) - P_{x_0}(x(s))| \geq \frac{\Delta_f(s)}{3}. \tag{17}$$

Let L denote the right-hand side of Eq. (16), which we may obviously assume to be finite. Thus, there exists a sequence $(s_n)_{n\geq 1}$ of discontinuity points of f which realizes the lower limit L; the points s_n necessarily differ from x_0 but they converge to that point. For each integer $n \geq 1$, the above procedure yields a point $x(s_n)$ for which Eq. (17) holds. The resulting sequence $(x(s_n))_{n\geq 1}$ thus satisfies

$$\limsup_{n\to\infty} \frac{\log |f(x(s_n)) - P_{x_0}(x(s_n))|}{\log |x(s_n) - x_0|} \leq L.$$

In the meantime, Eq. (15) implies that

$$\liminf_{n\to\infty} \frac{\log |f(x(s_n)) - P_{x_0}(x(s_n))|}{\log |x(s_n) - x_0|} \geq \alpha.$$

We deduce that $\alpha \leq L$, and the result follows from the fact that α may be chosen arbitrarily close to the Hölder exponent $h_f(x_0)$. □

Let us apply Lemma 2 to multivariate Davenport series. For each vector q in $\mathbf{Z}^d_*$, let $\mathscr{P}_q$ denote the set of all integers $p \in \mathbf{Z}$ such that $\gcd(p, q) = 1$. We remark that a vector q is irreducible, that is, belongs to $\mathscr{I}^d$, if and only if $\mathscr{P}_q$ contains zero. Moreover, the set $\mathscr{H}_d$ which indexes the hyperplanes $H_{p,q}$ in a unique manner is actually formed by the pairs (p, q) such that $p \in \mathscr{P}_q$ and $q \in \mathbf{Z}^d_+$. Now, given a point $x_0 \in \mathbf{R}^d$, let $\delta^{\mathscr{P}}_q(x_0)$ denote the distance between x_0 and the hyperplanes $H_{p,q}$ with $p \in \mathscr{P}_q$, that is,

$$\delta^{\mathscr{P}}_q(x_0) = \mathrm{dist}\left(x_0, \bigcup_{p \in \mathscr{P}_q} H_{p,q} \right) = \frac{1}{|q|} \inf_{p \in \mathscr{P}_q} |q \cdot x_0 - p|, \tag{18}$$

where $|q|$ is the Euclidean norm of the vector q. It is obvious that $\delta^{\mathscr{P}}_{-q}(x_0)$ coincides with $\delta^{\mathscr{P}}_{q}(x_0)$, so there is no loss of information in assuming, whenever necessary, that q is in $\mathbf{Z}^d_+$. Also, it is easy to see that the set $\mathscr{P}_q$ is invariant under the translations of the form $p \mapsto p + k \cdot q$ with $k \in \mathbf{Z}^d$, which makes it clear that the function $\delta^{\mathscr{P}}_{q}$ is $\mathbf{Z}^d$-periodic.

The analysis of the discontinuities of Davenport series that we led in Sect. 3 above ensures that every hyperplane $H_{p,q}$ indexed by $(p,q) \in \mathscr{H}_d$ contains a dense set of points at which the Davenport series has a discontinuity of magnitude $|A_q|$, with the proviso that the sum A_q defined by Eq. (7) does not vanish. These observations then yield the following corollary to Lemma 2.

Corollary 1. *Let f be a Davenport series with $a = (a_n)_{n \in \mathbf{Z}^d} \in \ell^1$. Then,*

$$\forall x_0 \in \mathbf{R}^d \qquad h_f(x_0) \leq \liminf_{\substack{q \to \infty \\ q \in \operatorname{supp}(A)}} \frac{\log |A_q|}{\log \delta^{\mathscr{P}}_{q}(x_0)},$$

where the sequence $A = (A_q)_{q \in \mathbf{Z}^d}$ of jump sizes is the image of the sequence a under the jump operator J defined by Eq. (9).

In the above statement, we adopt the usual convention according to which the lower limit is infinite if the index set $\operatorname{supp}(A)$ is finite, in which case the bound is trivial. We now illustrate Corollary 1 by pointing out a class of Davenport series whose Hölder exponent vanishes everywhere, as a direct consequence of the previous upper bound. These series are characterized by the fact that the magnitude $|A_q|$ of the jumps does not become too small as q goes to infinity along a subsequence satisfying particular arithmetical properties.

In order to specify these properties, let us begin by observing that $\delta^{\mathscr{P}}_{q}(x_0)$ may sometimes be bounded above by $1/|q|$ infinitely often, up to a logarithmic factor; also, in view of the statement of Corollary 1, we may restrict our attention to the vectors q for which A_q does not vanish. The situation described above then occurs precisely when the support of the sequence A of jump sizes is regular in the sense of the next definition, which makes use of the function κ defined as follows: For every point $x_0 \in \mathbf{R}^d$ and every infinite subset Q of $\mathbf{Z}^d$,

$$\kappa(x_0, Q) = \limsup_{\substack{q \to \infty \\ q \in Q}} \frac{\log \left(\inf\limits_{p \in \mathscr{P}_q} |q \cdot x_0 - p| \right)}{\log |q|}.$$

Note that, as a consequence of the periodicity of the function $\delta^{\mathscr{P}}_{q}$, the function $\kappa(\,\cdot\,, Q)$ is $\mathbf{Z}^d$-periodic.

Definition 3. 1. An infinite subset Q of $\mathbf{Z}^d$ is said to be regular if the following condition holds:

$$\forall x_0 \in \mathbf{R}^d \qquad \kappa(x_0, Q) < 1.$$

2. A Davenport series with coefficients given by a sequence $a \in \ell^1$ is regular if $\operatorname{supp}(J(a))$ is a regular subset of $\mathbf{Z}^d$.

It is clear that any infinite subset of a regular set is also regular. Moreover, the fact that a set Q is regular roughly means that the sets $\mathscr{P}_q$, for $q \in Q$, do not have exceptionally long gaps. In order to elaborate on this remark, let us focus on the one-dimensional case and give some heuristic arguments. In that situation, the sets $\mathscr{P}_q$ are $q\mathbf{Z}$-periodic, so that it suffices to analyze their gaps in the interval $\{1,\ldots,q-1\}$, which clearly amounts to examining the difference between two consecutive numbers prime to q. There are $\phi(q)$ such numbers, where ϕ denotes Euler's totient function. Hence, in the absence of exceptionally long gaps, the intervals between two consecutive numbers prime to q would have length of the order of $q/\phi(q)$. Thus, the infimum arising in the definition of $\kappa(x_0,Q)$ would grow at a comparable rate, up to constants, and $\kappa(x_0,Q)$ would actually vanish. This is due to the fact that $q/\phi(q) = \mathrm{O}(\log\log q)$ as $q \to \infty$; indeed, it is known that

$$\liminf_{q\to\infty} \frac{\phi(q)}{q} \log\log q = \mathrm{e}^{-\gamma},$$

where γ denotes the Euler–Mascheroni constant, see, e.g., [3, Theorem 13.14]. In general, though, there may exist exceptional gaps of length much larger or smaller than $q/\phi(q)$ between the numbers prime to q, so the above arguments are not always applicable. The literature seems rather scarce on that difficult topic, apart from a series of papers by Hooley [29–31].

In some cases, the previous heuristic arguments can be turned into a rigorous proof. For instance, let us suppose that Q is the set $l^{\mathbf{N}} = \{l^k,\ k \geq 1\}$ of integer powers of a given prime number $l \geq 2$. For any $k \geq 1$, the set $\mathscr{P}_{l^k}$ is formed by the nonmultiples of l, thereby having gaps of length one only. Hence, the infimum of $|l^k x_0 - p|$ over $p \in \mathscr{P}_{l^k}$ is at most two, so that $\kappa(x_0, l^{\mathbf{N}}) = 0$. A one-dimensional Davenport series whose coefficients $a = (a_n)_{n\geq 1}$ are supported in $l^{\mathbf{N}}$ will be termed as l-adic. The sequence $A = J(a)$ of jump sizes is then also supported in $l^{\mathbf{N}}$. As a result, any l-adic Davenport series is regular.

Note however that in slightly more complicated examples, the above infimum may not easily be bounded by a constant, because the sets $\mathscr{P}_q$ may be chosen in such a way that they have longer gaps than above. As an illustration, still in dimension one, assume that Q is the set of all primorials of prime numbers, that is, the set of all integers $q_k = p_1 \cdots p_k$ for $k \geq 1$, where p_i is the ith prime number. Then, $\mathscr{P}_{q_k}$ is the sequence of integers prime that are not a multiple of any of the primes $p_1,\ldots,p_k$. It follows that $\mathscr{P}_{q_k}$ has gaps of size at least $p_{k+1} - 1$, which tends to infinity as $k \to \infty$.

We now introduce a definition that bears on the asymptotic behavior of the sequence indexed by $\mathbf{Z}^d$; we shall apply it in what follows to the sequence formed by the jump magnitudes $|A_q|$.

Definition 4. Let $b = (b_q)_{q\in\mathbf{Z}^d}$ be a real-valued sequence and let Q be an infinite subset of $\mathbf{Z}^d$. We say that the sequence b has slow decay on Q if

$$\liminf_{\substack{q\to\infty\\ q\in Q}} \frac{-\log|b_q|}{\log|q|} = 0.$$

Note that this condition is more and more restrictive as the subset Q becomes smaller; to be precise, any sequence with slow decay on a given set has slow decay on all its supersets. In addition, the above definition is in stark contradiction with that of the sets $\mathscr{F}^\gamma$ that we introduced in Sect. 4 and on which the jump operator J is a bicontinuous automorphism. In fact, it is easy to see that the sequences belonging to the sets $\mathscr{F}^\gamma$, for $\gamma > 0$, do not have slow decay on $\mathbf{Z}^d$.

Combining the previous two definitions, and calling upon Corollary 1, we readily deduce the following result.

Corollary 2. *Let $(a_n)_{n\in\mathbf{Z}^d}$ be a sequence in ℓ^1, and let $A = (A_q)_{q\in\mathbf{Z}^d}$ be its image under the jump operator J. Let us assume that A has slow decay on a regular subset of $\mathbf{Z}^d$. Then,*

$$\forall x_0 \in \mathbf{R}^d \qquad h_f(x_0) = 0.$$

A typical situation encompassed by the above setting is that of a regular Davenport series for which the sequence A of jump magnitudes has slow decay. In dimension one, this is the case of the Davenport series of the form

$$f_{l,\alpha}(x) = \sum_{k=1}^{\infty} \frac{\{l^k x\}}{k^\alpha},$$

where l is a prime number and α is larger than one. Indeed, $f_{l,\alpha}$ being an l-adic Davenport series, it is regular. Furthermore, let $b = (b_n)_{n\geq 1}$ denote the sequence of its coefficients, namely, b_n is equal to $k_0^{-\alpha}$ if $n = l^{k_0}$ for some integer $k_0 \geq 1$, and vanishes otherwise. Then, one easily checks that the sequence of jump sizes $B - \mathscr{J}(b)$ admits the same expression except that we have to replace $k_0^{-\alpha}$ by the sum $\sum_{k\geq k_0} k^{-\alpha}$ in the first case. It follows that B has slow decay, and the previous result ensures that $h_{f_{l,\alpha}}(x_0)$ vanishes everywhere.

6 Sparse Davenport Series

Recall that Corollary 1 above provides an upper bound on the Hölder exponent of a general Davenport series. In this section, we shall show that, under further assumptions, this bound gives the correct value of the Hölder exponent, which will ultimately enable us to perform the multifractal analysis of the corresponding series, see Sect. 7. Our main assumption implies that there cannot be too many nonvanishing terms in the Davenport expansions and boils down to a sparsity condition on the support of the sequence of Davenport coefficients. Note that, in one variable, the only case where one can determine the spectrum of singularities of Davenport series without additional assumptions on the coefficients is precisely the case where one assumes that the frequencies satisfy a lacunarity assumption, see [39].

6.1 Sparse Sets and Link with Lacunary and Hadamard Sequences

Formally, we define the notion of sparse set in the following manner. Recall that $\mathrm{B}(0,R)$ denotes the open ball of $\mathbf{R}^d$ with center zero and radius R. In addition, we let # stand for cardinality.

Definition 5. Let Q be a nonempty subset of $\mathbf{R}^d$. The set Q is said to be sparse if

$$\lim_{R\to\infty} \frac{\log\#(Q\cap \mathrm{B}(0,R))}{\log R} = 0.$$

In what follows, we say that an $\mathbf{R}^d$-valued sequence $\lambda = (\lambda_n)_{n\geq 1}$ is sparse if the set of its values, namely, $\{\lambda_n, n \geq 1\}$, is sparse in the sense of the above definition. If there is no redundancy in the sequence, that is, if it is injective, then the sparsity condition suggests that its terms do not accumulate excessively but rather escape to infinity quite fast. Indeed, if λ is sparse and injective, it is possible to rearrange its terms so as to assume that the sequence $(|\lambda_n|)_{n\geq 1}$ is nondecreasing. Then, one easily checks that the latter sequence grows faster than any power function at infinity, specifically,

$$\lim_{n\to\infty} \frac{\log|\lambda_n|}{\log n} = \infty.$$

A notable case where the sparsity condition holds is given by the sequences $(\lambda_n)_{n\geq 1}$ that are both *separated*, meaning that

$$\exists C>0 \quad \forall n,m\geq 1 \qquad n\neq m \quad \Longrightarrow \quad |\lambda_n-\lambda_m|\geq C,$$

and *lacunary*, in the sense that

$$\exists C'>0 \quad \forall n,m\geq 1 \qquad n\neq m \quad \Longrightarrow \quad |\lambda_n-\lambda_m|\geq C'(|\lambda_n|+|\lambda_m|);$$

these two notions are standard in the study of nonharmonic Fourier series, see for instance [37, 46]. Indeed, let us assume that the two above conditions hold, and let $\mathcal{N}_j(\lambda)$ collect the indices of the terms of the sequence within distance between 2^{j-1} and 2^j from the origin, that is,

$$\mathcal{N}_j(\lambda) = \{n\geq 1 \,|\, 2^{j-1}\leq |\lambda_n| < 2^j\},$$

where $j\geq 1$. The lacunarity assumption entails that the open balls $\mathrm{B}(\lambda_n, C'|\lambda_n|)$, for $n\geq 1$, are disjoint. Therefore, the balls $\mathrm{B}(\lambda_n, C'2^{j-1})$ indexed by $n\in\mathcal{N}_j(\lambda)$ do not intersect either. Meanwhile, all these balls are included in the open ball centered at the origin with radius $(1+C'/2)2^j$. Comparing the volume of these balls, we infer that $\#\mathcal{N}_j(\lambda)$ is bounded above by $(1+2/C')^d$. In addition, the set

$$\mathcal{N}_0(\lambda) = \{n\geq 1 \,|\, |\lambda_n| < 1\}$$

is necessarily finite; in fact, as a result of the separateness condition, the balls $\mathrm{B}(\lambda_n, C/2)$, for $\mathscr{N}_0(\lambda)$, are disjoint and included in the open ball centered at zero with radius $1+C/2$, and the same volume comparison argument implies that $\#\mathscr{N}_0(\lambda)$ is at most $(1+2/C)^d$. As a consequence, the sequence λ is sparse.

The above approach also enables us to write the sparse set $\{\lambda_n, n \geq 1\}$ as a finite union of sets of the form $\{\lambda_n^{(k)}, n \geq 1\}$, where each sequence $(\lambda_n^{(k)})_{n\geq 1}$ is a *Hadamard sequence*, which means that it is separated and satisfies

$$\exists C'' > 1 \quad \forall n \geq 1 \qquad \frac{|\lambda_{n+1}^{(k)}|}{|\lambda_n^{(k)}|} \geq C''.$$

Each of these sequences is obtained first by considering the even values of j and retaining only one term of the initial sequence λ among those indexed by $\mathscr{N}_j(\lambda)$ and then by handling the odd values of j. Note that a Hadamard sequence is clearly lacunary, and therefore sparse, but the converse need not hold. In addition, a finite union of Hadamard sequences need not be lacunary.

6.2 Decay of Sequences with Sparse Support and Behavior of the Jump Operator

Recall that the spaces $\mathscr{F}^\gamma$ play an important role in the analysis of the jump operator J; in fact, we proved in Sect. 4 that the jump operator J is a bicontinuous automorphism of the spaces $\mathscr{F}^\gamma$, for $\gamma > 1$. We shall now obtain an analogous result in the case where $0 < \gamma \leq 1$, up to a sparsity assumption. To be precise, the set $\mathscr{F}^\gamma$ will be replaced by the subspace $\mathscr{F}^\gamma_{\mathscr{S}}$ formed by the sequences with sparse support, namely,

$$\mathscr{F}^\gamma_{\mathscr{S}} = \mathscr{S} \cap \mathscr{F}^\gamma,$$

where $\mathscr{S}$ denotes the vector space of all odd sequences $a = (a_n)_{n\in\mathbf{Z}^d}$ for which the support $\operatorname{supp}(a)$ is a sparse subset of $\mathbf{Z}^d$.

Before studying the behavior of the jump operator on the spaces $\mathscr{F}^\gamma_{\mathscr{S}}$, let us point out that they may be used to characterize the decay of a sequence $a \in \mathscr{S}$. Specifically, we measure the rate of decay of such a sequence by considering

$$\gamma_a = \liminf_{\substack{n\to\infty \\ n\in\operatorname{supp}(a)}} \frac{-\log|a_n|}{\log|n|}. \tag{19}$$

In particular, a sequence $a \in \mathscr{S}$ has slow decay on its support in the sense of Definition 4 if and only if γ_a vanishes. One then easily checks that

$$\gamma_a = \sup\{\gamma > 0 \,|\, a \in \mathscr{F}^\gamma_{\mathscr{S}}\}.$$

It will also be useful to remark that, due to the sparsity of the support, γ_a may as well be seen as a critical exponent for the convergence of a series. Indeed, the fact that a is in $\mathscr{S}$ implies that

$$\gamma_a = \sup\left\{\gamma > 0 \,\middle|\, \sum_{n\in\mathbf{Z}^d} |n|\,|a_n|^{1/\gamma} < \infty\right\} = \inf\left\{\gamma > 0 \,\middle|\, \sum_{n\in\mathbf{Z}^d} |n|\,|a_n|^{1/\gamma} = \infty\right\}. \quad (20)$$

As we shall show below, the exponent γ_a will play a crucial role in the study of the multifractal properties of the Davenport series with coefficients given by a, see Corollary 3 below as well as the results of Sect. 7.

Let us now describe the action of J on the spaces $\mathscr{F}^{\gamma}_{\mathscr{S}}$. The next result may be seen as a partial extension of Proposition 3 to the case where γ is no more restricted to be larger than one. However, it is weaker because it does not discuss invertibility properties and the target set of the jump operator J is the intersection

$$\mathscr{F}^{\gamma,-} = \bigcap_{\varepsilon>0} \mathscr{F}^{\gamma-\varepsilon} \quad (21)$$

instead of the mere space $\mathscr{F}^{\gamma}$. Note that $\mathscr{F}^{\gamma,-}$ is endowed with the natural Fréchet topology inherited from the norms on the spaces $\mathscr{F}^{\gamma-\varepsilon}$.

Proposition 4. *For any $\gamma > 0$, the jump operator J induces a continuous mapping in the following way:*

$$\mathscr{F}^{\gamma}_{\mathscr{S}} \quad \stackrel{J}{\longrightarrow} \quad \mathscr{F}^{\gamma,-}$$

Proof. Given $\gamma > 0$, let $a = (a_n)_{n\in\mathbf{Z}^d}$ be a sequence in $\mathscr{F}^{\gamma}_{\mathscr{S}}$, and let $A = (A_q)_{q\in\mathbf{Z}^d}$ denote its image under J, that is, $A = J(a)$. Then, for each vector $q \in \mathbf{Z}^d_*$,

$$|A_q| \le 2\sum_{l=1}^{\infty} |a_{lq}| \le 2|a|_{\mathscr{F}^{\gamma}} \sum_{l=1}^{\infty} \frac{\mathbb{1}_{\{lq\in\mathrm{supp}(a)\}}}{|lq|^{\gamma}}.$$

In order to give an upper bound on the last sum, we split the index set into dyadic intervals. For each integer $j \ge 0$, we have

$$\sum_{l=2^j}^{2^{j+1}-1} \frac{\mathbb{1}_{\{lq\in\mathrm{supp}(a)\}}}{|lq|^{\gamma}} \le \frac{2^{-\gamma j}}{|q|^{\gamma}} \#(\mathrm{supp}\,(a)\cap \mathrm{B}(0,|q|2^{j+1})).$$

The support of the sequence a is sparse; thus, for all $\varepsilon > 0$, its intersection with the open ball centered at the origin with radius $|q|2^{j+1}$ has cardinality at most $C_\varepsilon |q|^{\varepsilon} 2^{\varepsilon j}$ for some real $C_\varepsilon > 0$ that depends on neither q nor j. We deduce that

$$|A|_{\mathscr{F}^{\gamma-\varepsilon}} = \sup_{q\in\mathbf{Z}^d} |q|^{\gamma-\varepsilon}|A_q| \le 2|a|_{\mathscr{F}^{\gamma}} C_\varepsilon \sum_{j=0}^{\infty} 2^{(\varepsilon-\gamma)j} = \frac{2C_\varepsilon}{1-2^{\varepsilon-\gamma}} |a|_{\mathscr{F}^{\gamma}},$$

with the proviso that $\varepsilon < \gamma$. The result follows. □

Note that, even when a sequence a has sparse support, the support of the associated sequence $J(a)$ of jump sizes need not be sparse; this is why the target set in the above statement involves $\mathscr{F}^{\gamma-\varepsilon}$, but not $\mathscr{F}^{\gamma-\varepsilon}_{\mathscr{S}}$. Moreover, the jump operator on $\mathscr{F}^{\gamma}_{\mathscr{S}}$ entails a slight loss in the speed of decay in the sense that the target set is not exactly $\mathscr{F}^{\gamma}$, as in Proposition 3, but rather $\mathscr{F}^{\gamma,-}$. Still, a simple adaptation of the above proof shows that $J(a)$ actually belongs to $\mathscr{F}^{\gamma}$ when a is a sequence in $\mathscr{F}^{\gamma}$ for which $\operatorname{supp}(a)$ is composed by the values of a separated and lacunary sequence, which is stronger than assuming that a is in $\mathscr{S}$.

6.3 Pointwise Regularity of Sparse Davenport Series

Now that the notion of sequence with sparse support has been defined, we are in position to introduce the notion of sparse Davenport series.

Definition 6. A Davenport series with coefficients given by a sequence $a \in \ell^1$ is sparse if the support $\operatorname{supp}(a)$ is a sparse set, that is, $a \in \mathscr{S} \cap \ell^1$.

In order to recover the regularity of the Davenport series f at every point, we shall assume, in addition to the sparsity of the support, that there is no cancellation in the sums (7) defining the jump operator, in the sense that A_q is at least of the order of magnitude of its largest term. To be specific, for any sequence $a = (a_n)_{n \in \mathbf{Z}^d}$ in ℓ^1, we may consider the even sequence $\overline{a} = (\overline{a}_q)_{q \in \mathbf{Z}^d}$ in $\ell^\infty(\mathbf{Z}^d)$ given by

$$\forall q \in \mathbf{Z}^d \qquad \overline{a}_q = \sup_{l \geq 1} |a_{lq}|.$$

This enables us to define in the following manner a sublinear operator M on ℓ^1, which we refer to as the maximal operator:

$$\forall (a_n)_{n \in \mathbf{Z}^d} \in \ell^1 \qquad M((a_n)_{n \in \mathbf{Z}^d}) = (\overline{a}_q)_{q \in \mathbf{Z}^d}.$$

The jump operator and the maximal operator both act on the sequences $a = (a_n)_{n \in \mathbf{Z}^d}$ in ℓ^1, and the asymptotic behavior of these actions may be compared by means of

$$\theta_a = \limsup_{\substack{q \to \infty \\ q \in \operatorname{supp}(\overline{a})}} \frac{\log |A_q|}{\log |\overline{a}_q|},$$

where A and $\overline{a}$ denote the sequences $J(a)$ and $M(a)$, respectively. Incidentally, it is useful to remark that

$$\operatorname{supp}(a) \cup \operatorname{supp}(A) \subseteq \operatorname{supp}(\overline{a}), \tag{22}$$

and that $\operatorname{supp}(A)$ and $\operatorname{supp}(\overline{a})$ coincide asymptotically whenever θ_a is finite, in the sense that they differ by a finite number of points only. The aforementioned assumption may now be expressed by means of the following definition.

Definition 7. A Davenport series with coefficients given by a sequence $a \in \ell^1$ is asymptotically jump canceling if $\theta_a > 1$.

More precisely, assuming that there is no cancellation in the sums defining the jumps sizes A_q amounts to supposing that θ_a is bounded above by one, that is, that the Davenport series is not jump canceling. The next result shows that, in that situation, the upper bound given by Corollary 1 becomes an equality and may actually be replaced by an expression that is easier to handle. Specifically, the jump sizes A_q arising in the bound may be replaced by the Davenport coefficients a_n themselves, and the distance $\delta_q^{\mathscr{P}}(x_0)$ may be replaced by

$$\delta_n(x_0) = \operatorname{dist}\left(x_0, \bigcup_{k\in\mathbf{Z}} H_{k,n}\right) = \frac{1}{|n|} \inf_{k\in\mathbf{Z}} |n\cdot x_0 - k|, \tag{23}$$

which means that we may discard the rather complicated coprimeness condition arising in Eq. (18) and discussed in Sect. 5.

Recall that the Hölder exponent vanishes wherever the Davenport series is not continuous, that is, on the set defined by Eq. (8). This set is a union of hyperplanes which may be written in the form $D_{J(a)}$, where for any odd sequence $b = (b_q)_{q\in\mathbf{Z}^d}$,

$$D_b = \bigcup_{\substack{(p,q)\in\mathscr{H}_d \\ q\in\operatorname{supp}(b)}} H_{p,q}. \tag{24}$$

We may therefore restrict our attention to the points at which the series is continuous, that is, outside the set $D_{J(a)}$. Actually, our approach only enables us to recover the Hölder exponent of the Davenport series outside the set $D_{M(a)}$, which may be larger than $D_{J(a)}$ in view of Eq. (22). Yet, this slight restriction will not prevent us from performing the multifractal analysis of the Davenport series that are not jump canceling, because the two sets then differ by a finite number of hyperplanes only. In the next statement, as before, $A = (A_q)_{q\in\mathbf{Z}^d}$ denotes the image of a under the jump operator J.

Theorem 1. *Let f be a Davenport series with coefficients given by a sequence $a\in\ell^1$. Let us assume that the series is sparse and not asymptotically jump canceling, that is,*

$$a \in \mathscr{S} \qquad \text{and} \qquad \theta_a \leq 1.$$

Then, the Hölder exponent of f at any fixed point $x_0 \in \mathbf{R}^d$ satisfies

$$h_f(x_0) \leq \liminf_{\substack{q\to\infty \\ q\in\operatorname{supp}(A)}} \frac{\log|A_q|}{\log\delta_q^{\mathscr{P}}(x_0)} \leq \liminf_{\substack{n\to\infty \\ n\in\operatorname{supp}(a)}} \frac{\log|a_n|}{\log\delta_n(x_0)}. \tag{25}$$

Moreover, if x_0 does not belong to $D_{M(a)}$, the above quantities coincide and the Hölder exponent may be computed using either of the following two formulae:

$$h_f(x_0) = \liminf_{\substack{q \to \infty \\ q \in \mathrm{supp}(A)}} \frac{\log |A_q|}{\log \delta_q^{\mathscr{P}}(x_0)} \qquad \textit{and} \qquad h_f(x_0) = \liminf_{\substack{n \to \infty \\ n \in \mathrm{supp}(a)}} \frac{\log |a_n|}{\log \delta_n(x_0)}.$$

The proof of Theorem 1 is postponed to Sect. 10 for the sake of clarity. In the course of the proof, we obtain a uniform bound on the Hölder exponent, which may be seen as a consequence of Theorem 1, and which we state now as a separate result for future reference.

Corollary 3. *Let f be a Davenport series with coefficients given by a sequence $a = (a_n)_{n \in \mathbf{Z}^d}$ in ℓ^1. If the series is sparse and not asymptotically jump canceling, then*

$$\forall x_0 \in \mathbf{R}^d \qquad h_f(x_0) \leq \gamma_a,$$

where γ_a is defined by Eq. (19). In particular, the Hölder exponent of f vanishes everywhere when the sequence a has slow decay.

7 Implications for Multifractal Analysis

The preceding results will allow us to perform the multifractal analysis of some multivariate Davenport series with coefficients $a \in \ell^1$. From now on, *we assume that the series is sparse and not asymptotically jump canceling*. We begin by describing the size properties of the iso-Hölder sets, which are formed of the points where f has Hölder exponent equal to a given h, specifically,

$$E_f(h) = \{x \in \mathbf{R}^d \mid h_f(x) = h\}, \tag{26}$$

for $h \in [0, \infty]$. To be precise, we compute the local spectrum of singularities of the series f, that is, the mapping

$$d_f(h, W) = \dim_{\mathrm{H}}(E_f(h) \cap W), \tag{27}$$

where W is a nonempty open subset of $\mathbf{R}^d$. In the previous formula, $\dim_{\mathrm{H}}$ denotes Hausdorff dimension, whose definition is recalled in Sect. 11.2. The spectrum is actually governed by the parameter γ_a which controls the decay of the sequence a and is defined by Eq. (19).

In view of Corollary 3, the Davenport series f has Hölder exponent at most γ_a everywhere. Thus, all the iso-Hölder sets $E_f(h)$, for $h > \gamma_a$, are empty. As a consequence, the spectrum of singularities is supported on $[0, \gamma_a]$, and we may restrict our attention to that interval in what follows. In addition, if γ_a vanishes, that is, when the sequence a has slow decay, then the Hölder exponent of f vanishes everywhere, so that all the iso-Hölder sets are empty, except $E_f(0)$ which is equal to the whole space $\mathbf{R}^d$. That situation being trivial, we may assume from now on that *γ_a is positive.*

The analysis below does not cover the case where γ_a is infinite. Note that this case includes that in which the sequence a has finite support. In that situation, the Davenport series is a finite sum of piecewise linear functions, thereby being smooth except on a locally finite union of hyperplanes where its Hölder exponent vanishes. If γ_a is infinite and the support of a has infinite cardinality, the arguments below only imply that the Hölder exponent of the Davenport series is infinite Lebesgue-almost everywhere in $\mathbf{R}^d$ and that the iso-Hölder sets associated with finite values of the exponent all have Hausdorff dimension at most $d-1$. It seems plausible, though, that the dimension is exactly $d-1$. In what follows, we shall therefore assume that *γ_a is finite.*

The local spectrum of singularities of f on the interval $[0,\gamma_a]$ is then given by the following statement.

Theorem 2. *Let f be a Davenport series with coefficients given by a sequence $a \in \ell^1$. Let us assume that the series is sparse and not asymptotically jump canceling and that $0 < \gamma_a < \infty$. Then, for any real $h \in [0,\gamma_a]$ and any nonempty open subset W of $\mathbf{R}^d$,*

$$d_f(h,W) = d - 1 + \frac{h}{\gamma_a}.$$

Note that the spectrum of singularities does not depend on the particular region W that is considered, and moreover it is nondegenerate, in the sense that its support is not reduced to a single point. Consequently, following the terminology of [36], the Davenport series falls in the category of *homogeneous multifractal functions*.

We get comparable results for the singularity sets, which are composed by the points where the Davenport series f is continuous and has Hölder exponent at most a given h, that is,

$$E'_f(h) = \{x \in \mathbf{R}^d \setminus D_{J(a)} \,|\, h_f(x) \leq h\},$$

where $D_{J(a)}$ is the set of discontinuities of f, given by Eq. (8) and written using the notation (24). In addition, we prove that the singularity sets belong to the category of sets with large intersection introduced by Falconer [26]. This remarkable property essentially asserts that the sets are so omnipresent and large in a measure theoretic sense that their size properties are not altered by taking countable intersections. As a matter of fact, the intersection of countably many sets with large intersection with Hausdorff dimension at least a given real s still has dimension at least s; this is in stark contradiction with the fact that the codimension of the intersection of two subsets is usually expected to be the sum of their codimension, as is the case for affine subspaces. Formally, the class of sets with large intersection are defined in [26] in the following manner. Recall that a G_δ-set is a set that may be expressed as a countable intersection of open sets.

Definition 8. For any $s \in (0,d]$, the class $\mathscr{G}^s$ of sets with large intersection with dimension at least s is defined as the collection of all G_δ-subsets E of $\mathbf{R}^d$ such that

$$\dim_{\mathrm{H}} \bigcap_{n \geq 1} \varsigma_n(E) \geq s$$

for any sequence $(\varsigma_n)_{n\geq 1}$ of similarity transformations of $\mathbf{R}^d$.

The class $\mathscr{G}^s$ is closed under countable intersections and bi-Lipschitz transformations and is the maximal class of G_δ-sets with Hausdorff dimension at least s that satisfies those properties, see [26, Theorem A] for a precise statement. Moreover, every set of the class $\mathscr{G}^s$ has packing dimension equal to d in every nonempty open set, see [26, Theorem D]. In what follows, packing dimension is denoted by $\dim_{\mathrm{P}}$; we refer for example to [27] for a definition of this notion.

Restricting to G_δ-sets will be quite a constraint for us here, so instead of considering the classes $\mathscr{G}^s$ themselves, we shall work with the extended classes $\overline{\mathscr{G}}^s$ defined by the following condition: For all $E \subseteq \mathbf{R}^d$,

$$E \in \overline{\mathscr{G}}^s \qquad \Longleftrightarrow \qquad \exists E' \in \mathscr{G}^s \quad E' \subseteq E.$$

It is clear that the class $\overline{\mathscr{G}}^s$ contains $\mathscr{G}^s$ and, in view of [26, Theorem C(b)], that the G_δ-sets that belong to $\overline{\mathscr{G}}^s$ actually belong to the original class $\mathscr{G}^s$. Moreover, the extended class $\overline{\mathscr{G}}^s$ naturally inherits from $\mathscr{G}^s$ its remarkable properties: $\overline{\mathscr{G}}^s$ is composed of sets with Hausdorff dimension at least s and packing dimension equal to d and is closed under countable intersections and bi-Lipschitz transformations.

The next result describes the size and large intersection properties of the singularity sets of the Davenport series f.

Theorem 3. *Let f be a Davenport series with coefficients given by a sequence $a = (a_n)_{n\in\mathbf{Z}^d}$ in ℓ^1. Let us assume that the series is sparse and not asymptotically jump canceling and that $0 < \gamma_a < \infty$. Then, for any real $h \in (0, \gamma_a]$,*

$$E'_f(h) \in \overline{\mathscr{G}}^{d-1+h/\gamma_a}$$

and, moreover, for any nonempty open subset W of $\mathbf{R}^d$,

$$\dim_{\mathrm{H}}(E'_f(h) \cap W) = d - 1 + \frac{h}{\gamma_a} \qquad \text{and} \qquad \dim_{\mathrm{P}}(E'_f(h) \cap W) = d.$$

We refer to Sect. 11 for the proof of the two above theorems.

8 Convergence and Global Regularity of Davenport Series

We will now give a few results concerning the convergence of Davenport series, when the sequence of coefficients does not belong to ℓ^1. In that case, the sum does not necessarily belong to L^∞, so that Hölder pointwise regularity may not be a relevant notion anymore. We will mainly consider convergence in Sobolev spaces, with both positive and negative indices, which allows us to consider simultaneously convergence in spaces of functions or, more generally, distributions. Specific additional motivations for this section are supplied in Sect. 9, where we show that the determination of global Sobolev regularity exponents is a preliminary step to either the determination of L^q regularity (which is needed for the study of p-exponents, see Sect. 9.4) or the verification of the multifractal formalism (see the beginning of Sect. 9).

8.1 Preliminaries on Multivariate Arithmetic Functions

An arithmetic function is traditionally a mapping defined on **N** and valued in **R** or sometimes in **C**. The usual multivariate extension deals with functions that are defined on $\mathbf{N}^d$, see [16, 25]. In this chapter, we consider a slightly different setting, with *multivariate arithmetic functions* defined on $\mathbf{Z}^d_*$.

A first simple example is supplied by the natural extension to $\mathbf{Z}^d_*$ of the divisor function already mentioned in Sect. 4; this extension is still denoted by τ for simplicity. To be specific, for any $m \in \mathbf{Z}^d_*$, we define $\tau(m)$ as the number of decompositions

$$m = ln \qquad \text{with} \qquad l \in \mathbf{N} \quad \text{and} \quad n \in \mathbf{Z}^d_*. \tag{28}$$

It is clear that $\tau(m)$ coincides with $\tau(\gcd(m))$, where $\gcd(m)$ denotes the greatest common divisor of the components of the vector m. With the help of Eq. (12), this implies that $\tau(m) = \mathrm{o}(|m|^\varepsilon)$ as m goes to infinity, for any fixed $\varepsilon > 0$. In what follows, we shall write indistinctly $l|m$ and $n|m$ when Eq. (28) holds; with a slight abuse, we shall also write $l = m/n$.

We will also make use of the extensions to the multivariate setting of other arithmetic functions, specifically, the sums of zth powers of the divisors. Given $z \in \mathbf{C}$, recall that the one-dimensional arithmetic function σ_z is defined by

$$\sigma_z(m) = \sum_{n|m} n^z,$$

where the sum bears on the positive divisors of the integer m. In the multivariate case, we have to draw a difference between integer and vector divisors. Therefore, we define two functions of the vectors $m \in \mathbf{Z}^d_*$ by

$$\sigma_z(m) = \sum_{\substack{n \in \mathbf{Z}^d_* \\ n|m}} |n|^z \qquad \text{and} \qquad \widetilde{\sigma}_z(m) = \sum_{\substack{l \in \mathbf{N} \\ l|m}} l^z.$$

It is clear that these two functions coincide on **N** and that $\sigma_0(m) = \widetilde{\sigma}_0(m) = \tau(m)$ for all $m \in \mathbf{Z}^d_*$. Moreover, for any $z \in \mathbf{C}$, one easily checks that $\widetilde{\sigma}_z(m) = \sigma_z(\gcd(m))$ and $\sigma_z(m) = |m|^z \widetilde{\sigma}_{-z}(m) = |m|^z \sigma_{-z}(\gcd(m))$. Given that $\sigma_z(l) = l^z \sigma_{-z}(l)$ for any integer $l \in \mathbf{N}$, we deduce that

$$\forall m \in \mathbf{Z}_*^d \qquad \sigma_z(m) = \left(\frac{|m|}{\gcd(m)}\right)^z \sigma_z(\gcd(m)). \tag{29}$$

Finally, recall that the Dirichlet convolution of two arithmetic functions A and B defined on $\mathbf{N}$ is the arithmetic function $A * B$ given by

$$\forall m \in \mathbf{N} \qquad A * B(m) = \sum_{\substack{(l,n)\in\mathbf{N}\times\mathbf{N} \\ ln=m}} A(n)B(l).$$

Similarly, the convolution of a *multivariate* arithmetic functions A defined on $\mathbf{Z}_*^d$ and a *one-dimensional* arithmetic function B defined on $\mathbf{N}$ is the multivariate arithmetic function given by

$$\forall m \in \mathbf{Z}_*^d \qquad A * B(m) = \sum_{\substack{(l,n)\in\mathbf{N}\times\mathbf{Z}_*^d \\ ln=m}} A(n)B(l).$$

8.2 Davenport Expansions Versus Fourier Expansions

Let us now go back to Davenport series. Without any assumption on the sequence $a = (a_n)_{n\in\mathbf{Z}^d}$ of Davenport coefficients, the right-hand side of Eq. (5) can be inverted using the *multivariate Möbius inversion formula*, which is an easy extension of the one-dimensional case and calls upon the Möbius function μ already used in Sect. 4. However, we start by proving it for the sake of completeness.

Lemma 3. *Let f be a multivariate arithmetic function defined on $\mathbf{Z}_*^d$ and let g be the multivariate arithmetic function given by*

$$\forall m \in \mathbf{Z}_*^d \qquad g(m) = \sum_{\substack{n\in\mathbf{Z}_*^d \\ n|m}} f(n).$$

Then, the function f can be recovered from g by $f = g * \mu$, that is,

$$\forall m \in \mathbf{Z}_*^d \qquad f(m) = \sum_{\substack{n\in\mathbf{Z}_*^d \\ n|m}} g(n)\mu\left(\frac{m}{n}\right).$$

Proof. For any vector $m \in \mathbf{Z}_*^d$, we have

$$\sum_{\substack{n\in\mathbf{Z}_*^d \\ n|m}} g(n)\mu\left(\frac{m}{n}\right) = \sum_{\substack{n\in\mathbf{Z}_*^d \\ n|m}} \mu\left(\frac{m}{n}\right) \sum_{\substack{k\in\mathbf{Z}_*^d \\ k|n}} f(k) = \sum_{\substack{k\in\mathbf{Z}_*^d \\ k|m}} f(k) \sum_{\substack{n\in\mathbf{Z}_*^d \\ k|n|m}} \mu\left(\frac{m}{n}\right).$$

Let us observe that the integer vectors $n \in \mathbf{Z}_*^d$ satisfying $k|n|m$ are merely of the form $n = kl$, where l ranges over the divisors of the positive integer m/k. Thus, the last sum satisfies

$$\sum_{\substack{n\in\mathbf{Z}_*^d \\ k|n|m}} \mu\left(\frac{m}{n}\right) = \sum_{l|(m/k)} \mu\left(\frac{m/k}{l}\right) = \sum_{l|(m/k)} \mu(l) = \mathbb{1}_{\{k=m\}}.$$

The last equality follows from the well-known fact that the sum of the Möbius function over all positive divisors of a given natural number n vanishes except if $n = 1$, where the sum is equal to one. The result follows. □

The next proposition results from applying to Eq. (5) the above inversion formula and will be useful in the determination of the Sobolev regularity of Davenport series. It shows how to recover the Davenport coefficients of a series from the knowledge of its Fourier coefficients.

Proposition 5. *Let f be a Davenport series with coefficients given by a sequence $a = (a_n)_{n\in\mathbf{Z}^d}$ in ℓ^1, and let $(c_m)_{m\in\mathbf{Z}^d}$ denote the sequence of its Fourier coefficients. Then,*

$$\forall n \in \mathbf{Z}_*^d \qquad a_n = -\pi \sum_{\substack{m\in\mathbf{Z}_*^d \\ m|n}} \frac{m}{n} \mu\left(\frac{n}{m}\right) c_m.$$

Proof. A straightforward consequence of Eq. (5) is that for all $m \in \mathbf{Z}_*^d$,

$$-\pi c_m m = \sum_{\substack{n\in\mathbf{Z}_*^d \\ n|m}} a_n n.$$

The result now follows from applying Lemma 3 to the arithmetic functions $f(n) = a_n n$ and $g(m) = -\pi c_m m$. Note that these functions take values in $\mathbf{R}^d$ and not merely in $\mathbf{R}$. However, Lemma 3 obviously extends to this case. □

8.3 Regularity of the Sum of a Davenport Series

Without any assumption on the odd sequence $a = (a_n)_{n\in\mathbf{Z}^d}$ of Davenport coefficients, we may define an odd sequence $(c_m)_{m\in\mathbf{Z}^d}$ with the help of Eq. (5). This detour via Fourier series will allow us to study the convergence of the Davenport series $\sum_n a_n\{n\cdot x\}$, even when a is no longer assumed to belong to ℓ^1. Indeed, we shall see that, in many functional settings, when the associated Fourier series $\sum_m c_m \sin(2\pi m\cdot x)$ converges, then the partial sums of the Davenport series converge to the same limit.

In order to be more precise, let us begin by recalling that the spaces $\mathscr{F}^{\gamma,-}$ are defined in terms of the sequence spaces $\mathscr{F}^{\gamma}$ by means of Eq. (21). Moreover, let F^{γ}

denote the space of distributions whose Fourier coefficients belong to $\mathscr{F}^\gamma$ and by $F^{\gamma,-}$ the space of distributions whose Fourier coefficients belong to $\mathscr{F}^{\gamma,-}$.

In addition, for any odd sequence $a = (a_n)_{n\in\mathbf{Z}^d}$, we denote by f^N the partial sums of the corresponding Davenport series, that is,

$$f^N(x) = \sum_{\substack{n\in\mathbf{Z}^d \\ |n|\le N}} a_n\{n\cdot x\}.$$

The next result discusses the convergence properties of the sequence $(f^N)_{N\ge 1}$ in the spaces F^γ and $F^{\gamma,-}$. A noteworthy consequence lies in the fact that the Davenport series $\sum_n a_n\{n\cdot x\}$ converges in the sense of distributions when the coefficients a_n do not increase faster than any polynomial.

Proposition 6. *Let $\gamma\in\mathbf{R}$, and let $a\in\mathscr{F}^\gamma$:*

- *If $\gamma<0$, then the sequence $(f^N)_{N\ge 1}$ converges in $F^{\gamma,-}$ to a distribution f which belongs to F^γ.*
- *If $0\le\gamma\le 2$, then the sequence $(f^N)_{N\ge 1}$ is convergent in $F^{\min\{1,\gamma\},-}$.*
- *If $\gamma>2$, then the sequence $(f^N)_{N\ge 1}$ converges in $F^{1,-}$ to a distribution f which belongs to F^1.*

Proof. It follows from Eq. (5) that the Fourier coefficients of the partial sum f^N are given by

$$c_m^N = -\frac{1}{2\pi}\sum_{\substack{(l,n)\in\mathbf{Z}^*\times\mathbf{Z}_*^d \\ ln=m,\ |n|\le N}} \frac{a_n}{l}.$$

The condition $|n|\le N$ is necessarily satisfied as soon as N is greater than or equal to $|m|$, so that each sequence $(c_m^N)_{N\ge 1}$ is ultimately constant equal to

$$c_m = -\frac{1}{2\pi}\sum_{\substack{(l,n)\in\mathbf{Z}^*\times\mathbf{Z}_*^d \\ ln=m}} \frac{a_n}{l}.$$

Moreover, given that the sequence a belongs to the space $\mathscr{F}^\gamma$, it is easy to check that for all $N\ge 1$ and $m\in\mathbf{Z}_*^d$,

$$|c_m^N| \le \frac{|a|_{\mathscr{F}^\gamma}}{\pi|m|}\sigma_{1-\gamma}(m),$$

which implies in particular that for all $m\in\mathbf{Z}_*^d$,

$$|c_m| \le \frac{|a|_{\mathscr{F}^\gamma}}{\pi|m|}\sigma_{1-\gamma}(m). \tag{30}$$

The following estimates on the one-variable arithmetic functions $\sigma_z(m)$ for $z\in\mathbf{R}$ may be found in [45]:

$$\begin{cases} z<-1 & \Longrightarrow \sigma_z(m)=O(1), \\ -1\le z<0 & \Longrightarrow \forall\varepsilon>0 \quad \sigma_z(m)=O(m^{\varepsilon}), \\ 0\le z\le 1 & \Longrightarrow \forall\varepsilon>0 \quad \sigma_z(m)=O(m^{z+\varepsilon}), \\ z>1 & \Longrightarrow \sigma_z(m)=O(m^{z}). \end{cases}$$

Thanks to Eq. (29), we easily deduce the following estimates on the corresponding multivariate functions:

$$\begin{cases} z<-1 & \Longrightarrow \sigma_z(m)=O(1), \\ -1\le z<0 & \Longrightarrow \forall\varepsilon>0 \quad \sigma_z(m)=O(|m|^{\varepsilon}), \\ 0\le z\le 1 & \Longrightarrow \forall\varepsilon>0 \quad \sigma_z(m)=O(|m|^{z+\varepsilon}), \\ z>1 & \Longrightarrow \sigma_z(m)=O(|m|^{z}). \end{cases} \tag{31}$$

It now follows from Eq. (31) that the $|c_m^N|$ satisfy the estimates of Proposition 6 uniformly in N, and their limits $|c_m|$ satisfy the same estimates.

Convergence in the corresponding function spaces follows immediately by applying the same approach to the differences $c_m^N - c_m$, starting from the observation that for any fixed $\varepsilon>0$,

$$|c_m^N - c_m| \le \frac{|a|_{\mathscr{F}^\gamma}}{\pi |m| N^{\varepsilon}} \sigma_{1-\gamma+\varepsilon}(m)$$

for all $N\ge 1$ and $m\in \mathbf{Z}_*^d$. □

Our purpose is now to determine in which Sobolev spaces the Davenport series with coefficients in the space $\mathscr{F}^\gamma$ do converge. Let us recall that the Sobolev space H^s is characterized by the following condition on the Fourier coefficients: A $\mathbf{Z}^d$-periodic odd distribution f belongs to H^s if the sequence $(c_m)_{m\in\mathbf{Z}^d}$ of its Fourier coefficients satisfies

$$|f|_{H^s}^2 = \sum_{m\in\mathbf{Z}_*^d} |c_m|^2 |m|^{2s} < \infty.$$

Note that, if $s<0$, this defines a space of distributions. In order to state sharp results, we shall also need the following slight modifications of H^s. Specifically, let H^s_δ be the space of all $\mathbf{Z}^d$-periodic odd distributions f whose Fourier coefficients satisfy

$$|f|_{H^s_\delta}^2 = \sum_{m\in\mathbf{Z}_*^d} |c_m|^2 \frac{|m|^{2s}}{(1+\log|m|)^{\delta}} < \infty,$$

and let $H^s_{\delta,+}$ and $H^{s,-}$ be the spaces defined, respectively, by

$$H^s_{\delta,+} = \bigcap_{\varepsilon>0} H^s_{\delta+\varepsilon} \qquad \text{and} \qquad H^{s,-} = \bigcap_{\varepsilon>0} H^{s-\varepsilon}.$$

Before proceeding, let us begin by observing that the Fourier coefficient indexed by $m = (m_1, \ldots, m_d) \in \mathbf{Z}_*^d$ of the function which maps $x = (x_1, \cdots, x_d)$ to $\{x_1\}$ is equal to $\mathbb{1}_{\{m_2=\ldots=m_d=0\}}/m_1$. Therefore, this function fails to belong $H^{1/2}$ but belongs to $H^{1/2}_{1,+}$. It follows that, no matter how large γ is picked, we cannot expect substantially better results than convergence in $H^{1/2}_{1,+}$. In addition, note that if s is less than $1/2$, then

$$|\{n \cdot x\}|_{H^s} = \frac{|n|^s}{2\pi} (2\zeta(2(1-s)))^{1/2}.$$

Thus, for any odd sequence $a = (a_n)_{n \in \mathbf{Z}^d}$, the Davenport series f defined by Eq. (2) has norm $|f|_{H^s}$ bounded above by $\sum_n |a_n| \, |n|^s$, up to a multiplicative constant. As a consequence,

$$\forall s < 1/2 \qquad \sum_{n \in \mathbf{Z}_*^d} |a_n| \, |n|^s < \infty \quad \Longrightarrow \quad f \in H^s.$$

We will now see how this straightforward result can be improved under the assumption that a belongs to the space $\mathscr{F}^\gamma$. We restrict our attention to the situation where $d \geq 2$, the one-dimensional case being thoroughly studied in [35].

Proposition 7. *Let us assume that $d \geq 2$. Let $\gamma \in \mathbf{R}$, and let $a = (a_n)_{n \in \mathbf{Z}^d}$ be a sequence in $\mathscr{F}^\gamma$. Then, the sequence $(f^N)_{N \geq 1}$ converges in the space*

$$\begin{cases} H^{\gamma-d/2}_{1,+} & \text{if } \gamma \leq 0, \\ H^{\gamma-d/2,-} & \text{if } 0 < \gamma \leq 1, \\ H^{(1+\gamma-d)/2,-} & \text{if } 1 < \gamma \leq 2, \\ H^{(1+\gamma-d)/2}_{1,+} & \text{if } 2 < \gamma < d \text{ and } d \geq 3, \\ H^{1/2}_{2,+} & \text{if } \gamma = d \geq 3, \\ H^{1/2}_{1,+} & \text{if } \gamma > d. \end{cases} \tag{32}$$

Proof. It follows from Proposition 6 that the sequence $(f^N)_{N \geq 1}$ converges to a distribution f. Moreover, the Fourier coefficients of f may be bounded with the help of Eq. (30), so that

$$|f|^2_{H^s_\delta} \leq \frac{|a|^2_{\mathscr{F}^\gamma}}{\pi^2} \sum_{m \in \mathbf{Z}_*^d} \frac{|m|^{2(s-1)}}{(1+\log|m|)^\delta} \sigma_{1-\gamma}(m)^2. \tag{33}$$

In order to estimate the above sum, let us split the index set as a union of dyadic domains. Specifically, the above sum is equal to the sum over all integers $j \geq 0$ of

$$\sum_{\substack{m\in\mathbf{Z}^d\\ 2^j\leq|m|<2^{j+1}}} \frac{|m|^{2(s-1)}}{(1+\log|m|)^\delta}\sigma_{1-\gamma}(m)^2 \leq \frac{2^{2(s-1)(j+1)}}{(1+j)^\delta}\mathfrak{m}_\gamma(2^{j+1})\mathfrak{s}_\gamma(2^{j+1}). \tag{34}$$

In the previous bound, $\mathfrak{m}_\gamma$ and $\mathfrak{s}_\gamma$ are defined, respectively, by

$$\mathfrak{m}_\gamma(x) = \sup_{\substack{m\in\mathbf{Z}^d_*\\ |m|<x}} \sigma_{1-\gamma}(m) \qquad \text{and} \qquad \mathfrak{s}_\gamma(x) = \sum_{\substack{m\in\mathbf{Z}^d_*\\ |m|<x}} \sigma_{1-\gamma}(m),$$

for any real $x>0$. The estimates Eq. (31) readily imply the following bounds on $\mathfrak{m}_\gamma$:

$$\mathfrak{m}_\gamma(x) = \begin{cases} \mathrm{O}(x^{1-\gamma}) & \text{if } \gamma<0,\\ \mathrm{O}(x^{1-\gamma+\varepsilon}) \text{ for all } \varepsilon>0 & \text{if } 0\leq\gamma\leq1,\\ \mathrm{O}(x^{\varepsilon}) \text{ for all } \varepsilon>0 & \text{if } 1<\gamma\leq2,\\ \mathrm{O}(1) & \text{if } \gamma>2. \end{cases}$$

Let us now deal with $\mathfrak{s}_\gamma$. For all $x>0$, we have

$$\mathfrak{s}_\gamma(x) = \sum_{\substack{m\in\mathbf{Z}^d_*\\ |m|<x}} \sum_{\substack{n\in\mathbf{Z}^d_*\\ n|m}} |n|^{1-\gamma} \leq \sum_{\substack{n\in\mathbf{Z}^d_*\\ |n|<x}} \frac{x}{|n|}|n|^{1-\gamma} = x\sum_{\substack{n\in\mathbf{Z}^d_*\\ |n|<x}} |n|^{-\gamma},$$

which readily implies that

$$\mathfrak{s}_\gamma(x) = \begin{cases} \mathrm{O}(x^{d+1-\gamma}) & \text{if } \gamma<d,\\ \mathrm{O}(x\log x) & \text{if } \gamma=d,\\ \mathrm{O}(x) & \text{if } \gamma>d. \end{cases}$$

Combining the above bounds, we deduce that the sum in Eq. (34) is bounded above, up to a multiplicative constant, by

$$\begin{cases} 2^{(d+2(s-\gamma))j}/j^\delta & \text{if } \gamma<0,\\ 2^{(d+2(s-\gamma)+\varepsilon)j} \text{ for all } \varepsilon>0 & \text{if } 0\leq\gamma\leq1,\\ 2^{(d+2s-1-\gamma+\varepsilon)j} \text{ for all } \varepsilon>0 & \text{if } 1<\gamma\leq2,\\ 2^{(d+2s-1-\gamma)j}/j^\delta & \text{if } 2<\gamma<d \text{ and } d\geq3,\\ 2^{(2s-1)j}/j^{\delta-1} & \text{if } \gamma=d\geq3,\\ 2^{(2s-1)j}/j^\delta & \text{if } \gamma>d. \end{cases}$$

These estimates are now sufficient to deduce that the limiting distribution f belongs to the spaces given by Eq. (32).

On top of that, convergence in the corresponding spaces follows immediately by applying the same approach to the differences $f^N - f$. As a matter of fact, for any fixed $\eta > 0$, their Fourier coefficients satisfy

$$|c_m^N - c_m| \leq \frac{|a|_{\mathscr{F}\gamma}}{\pi|m|}\left(\frac{1+\log|m|}{1+\log N}\right)^{\eta/2} \sigma_{1-\gamma}(m)$$

for all $N \geq 1$ and $m \in \mathbf{Z}^d_*$, which implies that Eq. (33) also holds when replacing $|f|^2_{H^s_\delta}$ by $(1+\log N)^\eta |f^N - f|^2_{H^s_{\delta+\eta}}$. □

9 Concluding Remarks and Open Problems

The study of the local regularity of Davenport series remains a largely open field of investigations, with many interesting questions at the crossroad of number theory, harmonic analysis and functional analysis; our purpose in this section is to list a few of them that we believe of particular interest. A first one consists in the verification of the *multifractal formalism*. We shall not describe this question here, because its final and most precise formulation (in terms of *wavelet leaders*) requires the introduction of wavelet methods that go beyond the scope of this chapter; we refer to [36] for a mathematical presentation concerning these issues and to [1] for a recent overview on the applications side. The verification of the multifractal formalism is a completely open problem for series of compensated pure jump functions, whether they be Davenport series (in one or several variables) or Lévy processes and fields. Let us just mention that Sect. 8 can be seen as a preliminary step in this direction; indeed, a part of this verification involves the determination of the Sobolev spaces that contain the function f under consideration.

9.1 Optimality of Lemma 2

The only cases where we have been able to determine the exact pointwise Hölder regularity of the sum of a Davenport series are when the bound given by Lemma 2 is optimal. This is not accidental and actually, in all cases of jump functions for which the Hölder exponent has been determined, it turns out that this bound is optimal: This is the case for Lévy processes without Brownian component and their extension to the multivariate setting [23, 24, 34], for the few cases of Markov processes with nonstationary increments whose multifractal analysis has been performed [10] and for the other cases of Davenport series which can be worked out [35, 38, 39]. We shall however give below a simple example of Davenport series where this is not the case.

We shall restrict the discussion to the one-dimensional setting, which is easier to consider and is sufficient to explain why Lemma 2 is not always sharp, even in the setting of Davenport series. Of course, in all generality, Eq. (16) is clearly not always sharp, as shown by the case in which f is a continuous function, where the bound thus obtained is trivial. A natural class of functions for which one might expect optimality is supplied by *compensated pure jump functions*, that is, the functions f whose derivative f' in a distributional sense is of the form

$$\sum_{n=1}^{\infty}(a_n\delta_{x_n}+c_n). \tag{35}$$

In typical examples, the jump locations x_n form a dense subset of the ambient space. However, even in the case where f is a compensated pure jump function, the bound Eq. (16) need not be optimal, as shown by the following example of one-dimensional Davenport series:

$$f_\beta(x)=-\zeta(\beta)\{x\}+\sum_{n=1}^{\infty}\frac{\{nx\}}{n^\beta}, \tag{36}$$

where $\beta>3/2$. Indeed, the function f_β is continuous at zero and jumps at every nonvanishing rational p/q written in its irreducible form and the corresponding jump has magnitude $\Delta_{f_\beta}(p/q)$ equal to $\zeta(\beta)/q^\beta$. Thus, the bound Eq. (16) on its Hölder exponent at zero is realized by rational numbers of the form $1/q$, specifically,

$$h_{f_\beta}(0)\leq\liminf_{q\to\infty}\frac{\log\Delta_{f_\beta}(1/q)}{\log(1/q)}=\beta.$$

We shall prove the following result which shows that this bound is not optimal.

Proposition 8. *Let β be a real number larger than $3/2$ that is not an integer greater than or equal to 3. Then, the value of the Hölder exponent of f_β at zero is given by*

$$h_{f_\beta}(0)=\beta-1.$$

Proof. Given that the function f_β is odd, it is sufficient to study the increment $f_\beta(x)-f_\beta(0)$ for positive values of x only. If $x\in(0,1)$ and $n<1/x$, we have $\{nx\}=nx-1/2$. Letting $\lceil\cdot\rceil$ denote the ceiling function, we deduce that

$$\begin{aligned}f_\beta(x)-f_\beta(0)&=-\zeta(\beta)\left(x-\frac{1}{2}\right)+\sum_{n=1}^{\lceil 1/x\rceil-1}\frac{nx-1/2}{n^\beta}+\sum_{n=\lceil 1/x\rceil}^{\infty}\frac{\{nx\}}{n^\beta}\\&=-\zeta(\beta)x+x\sum_{n=1}^{\lceil 1/x\rceil-1}\frac{1}{n^{\beta-1}}+\sum_{n=\lceil 1/x\rceil}^{\infty}\frac{\{nx\}+1/2}{n^\beta}.\end{aligned}$$

Let us first assume that $\beta < 2$. While the first term is merely linear, it is easy to see that the second term is equivalent to $x^{\beta-1}/(2-\beta)$ as x goes to zero. Concerning the third term, its absolute value may be bounded by the sum of $1/n^{\beta}$ over $n \geq \lceil 1/x \rceil$, which is equivalent to $x^{\beta-1}/(\beta-1)$. Given that $\beta > 3/2$, the difference $|f_{\beta}(x) - f_{\beta}(0)|$ is thus of the order of $x^{\beta-1}$, and the result follows.

In the case where $\beta = 2$, the second term in the above decomposition is equivalent to $x \log(1/x)$, and the upper bound on the third term is equivalent to x, so the result follows as well.

Finally, let us consider the case in which $\beta > 2$. The upper bound on the third term above is again equivalent to $x^{\beta-1}/(\beta-1)$. Moreover, the second term may be rewritten as

$$x \sum_{n=1}^{\lceil 1/x \rceil - 1} \frac{1}{n^{\beta-1}} = \zeta(\beta-1)x - x \sum_{n=\lceil 1/x \rceil}^{\infty} \frac{1}{n^{\beta-1}},$$

and the second term of the latter expression is equivalent to $x^{\beta-1}/(\beta-2)$. The result now follows because β is not an integer. □

This example opens the possibility of studying the pointwise regularity of Davenport series for which the bound supplied by Lemma 2 is not optimal. Beyond the study of particular functions at particular points, natural general open questions are the following. Under simple assumptions on the Davenport coefficients, can one show that the bound is optimal at a given point? Everywhere? Outside a set of dimension zero? Or almost everywhere? Similar questions can also be raised in the more general setting of compensated pure jump functions.

9.2 *Hecke's Functions*

In dimension one, special attention has been paid to the study of very specific Davenport series, namely, Hecke's functions $\mathfrak{H}_{\beta}$, which depend on a parameter $\beta \in \mathbf{C}$ and are defined by Eq. (3). Note that they can actually turn out to be distributions when the real part $\Re\beta$ is sufficiently small.

These functions have a rich history. They were first considered as functions of the complex variable β; the real number x being merely a parameter. Hecke studied their analytic continuation, and his study was later extended by Hardy; these results showed that the range of analytic continuation depends on the Diophantine approximation properties of the parameter x. As a function of the real variable x, the spectrum of singularities was completely determined only in the case where $\Re\beta \geq 2$, which leaves open the case where $1 < \Re\beta < 2$, see [35] and Eq. (37) below. Note that the counterexample supplied in Sect. 9.1 is closely related with Hecke's functions, and we refer to Sect. 9.3 below for further connections. One could also consider multivariate extensions of these functions, specifically, the functions

$$x \mapsto \sum_{n \in \mathbf{Z}_*^d} \varepsilon_n \frac{\{n \cdot x\}}{|n|^{\beta}},$$

where $(\varepsilon_n)_{n \in \mathbf{Z}_*^d}$ is an odd sequence taking the values ± 1.

9.3 Spectrum of Singularities of Compensated Pure Jumps Functions

We now go back to the general setting supplied by the compensated pure jump functions. The examples of such functions whose multifractal properties are known may be separated into two large classes. The first class corresponds to the case where the jump locations x_n appearing in Eq. (35) are somehow homogeneously distributed; this is the case for many examples of Davenport series [35,38,39] or Lévy fields and processes [23,24,34]. The second class is composed of functions for which the jump locations x_n form a strongly inhomogeneous sequence; such examples have been investigated by Barral and Seuret and include Lévy subordinators in multifractal time [8] or heterogeneous sums of Dirac masses [5, 6, 9]. In the heterogeneous case, the obtained spectra strongly differ from those which have been exhibited in this chapter. Indeed, they are usually composed of two parts: a linear one (for sufficiently small values of the Hölder exponent h) followed by a strictly concave one. Note however that spectra of a different kind have been obtained in [10]; for some Markov processes which differ from Lévy processes, one meets spectra that are a superposition of linear functions with different slopes.

In the homogeneous case, the spectra that have been met up to now are linear. However, this is not a general rule, even in the particular case of one-variable Davenport series. Let us consider for instance the function f_β defined by Eq. (36) and suppose that β is a noninteger real number larger than two. The local spectrum of singularities of the corresponding Hecke function $\mathfrak{H}_\beta$ defined by Eq. (3) is then supported by the interval $[0,\beta/2]$ and satisfies

$$\forall h \in [0,\beta/2] \quad \forall W \neq \emptyset \text{ open} \qquad d_{\mathfrak{H}_\beta}(h,W) = \frac{2h}{\beta}; \tag{37}$$

this follows readily from the approach employed in [35] in order to compute the global spectrum of $\mathfrak{H}_\beta$, which corresponds to the case where W is equal to the whole real line. As shown by Proposition 8, subtracting the term $\zeta(\beta)\{x\}$ to Hecke's function $\mathfrak{H}_\beta(x)$ shifts the value of the Hölder exponent at the integers from zero to $\beta-1$. As a consequence, the local spectrum of singularities of the resulting function f_β is now supported in the set $[0,\beta/2]\cup\{\beta-1\}$. Moreover, the function f_β still satisfies Eq. (37) but, rather than being empty, its iso-Hölder set $E_{f_\beta}(\beta-1)$ is equal to $\mathbf{Z}$. Therefore, for any open subset W of $\mathbf{R}$ that contains an integer,

$$d_{f_\beta}(\beta-1,W) = 0.$$

In particular, the local spectrum of f_β depends on the particular region that is considered. Thus, unlike the corresponding Hecke function $\mathfrak{H}_\beta$, the function f_β is not a homogeneous multifractal function. Furthermore, unlike that of $\mathfrak{H}_\beta$, the global spectrum of singularities of f_β is not a linear function. More specifically, the graph of this spectrum is the union of a segment and an isolated point. The multifractal properties of f_β may be put in comparison with those of Riemann's function $\sum_n \sin(\pi n^2 x)/n^2$ whose global spectrum of singularities has exactly the same shape and which is a homogeneous multifractal function, see [32].

This example raises several questions. Is there a simple condition on the coefficients of a Davenport series which ensures that its spectrum is linear? What is the general form of the spectrum of a Davenport series? Similar questions can also be asked in the more general setting of compensated pure jumps functions. Note that some results on these problems have been obtained by J. Barral and S. Seuret in the slightly different setting supplied by the *large deviation spectrum*, see [7].

9.4 *p-Exponent*

Hölder pointwise regularity is defined only for locally bounded functions, which explains why we always made the assumption that the sequence of Davenport coefficients belongs to ℓ^1 when studying Hölder regularity. However, Proposition 7 shows that, even when the sequence of Davenport coefficients does not belong to ℓ^1, and therefore convergence in L^∞ is no more guaranteed, one can obtain convergence in L^2 (this corresponds to the Sobolev space H^s considered in Sect. 8 in the case where s is zero) and also for larger values of p; indeed when s is positive, the Sobolev embeddings imply that if f belongs to H^s, then it also belongs to L^p for all p smaller than the critical value p_0 defined by the condition

$$\frac{1}{p_0} = \frac{1}{2} - \frac{s}{d}.$$

Note also that the specific case of L^2 convergence of one-variable Davenport series has already been considered, see [15, 35].

In such situations, one can still perform a pointwise analysis of regularity, by using a definition of pointwise smoothness which is weaker than Hölder regularity and is compatible with functions that are not locally bounded: It is the notion of $T^p_\alpha(x_0)$ regularity, which was introduced by Calderón and Zygmund in 1961, see [18]. The next definition is an adaptation of Definition 2 to that setting.

Definition 9. Let f be a tempered distribution on $\mathbf{R}^d$, let $p \in [1,\infty)$, let $\alpha > -d/p$ and let $x_0 \in \mathbf{R}^d$. The distribution f belongs to $T^p_\alpha(x_0)$ if it coincides with an L^p function in the open ball $\mathrm{B}(x_0,R)$ for some real $R>0$, and if there exist a real $C>0$ and a polynomial P_{x_0} of degree less than α such that for all $r \in (0,R]$,

$$\left(\frac{1}{r^d}\int_{\mathrm{B}(x_0,r)} |f(x) - P_{x_0}(x)|^p \,\mathrm{d}x\right)^{1/p} \leq C r^\alpha.$$

The *p-exponent* of f at x_0 is then defined as

$$h^p_f(x_0) = \sup\{\alpha > -d/p \,|\, f \in T^p_\alpha(x_0)\}.$$

Note that the Hölder exponent corresponds to the case where $p = \infty$, and the condition on the degree of P_{x_0} implies its uniqueness. This definition is a natural substitute for pointwise Hölder regularity when functions in L^p_{loc} are considered. In particular, the p-exponent can take values down to $-d/p$, thereby allowing to take into account behaviors which are locally of the form $1/|x-x_0|^\alpha$ for $\alpha < d/p$.

Furthermore, similarly to Eqs. (26) and (27), we may define the analogs of the iso-Hölder sets and the local spectrum of singularities by

$$E^p_f(h) = \{x \in \mathbf{R}^d \,|\, h^p_f(x) = h\} \qquad \text{and} \qquad d^p_f(h,W) = \dim_{\mathrm{H}}(E^p_f(h) \cap W),$$

the latter quantity being referred to as the local p-spectrum of the distribution f.

Both in the univariate and the multivariate case, the subject of determining the p-exponents and the p-spectrum of a Davenport series with coefficients not belonging to ℓ^1 is completely open.

9.5 Directional Regularity

The notion of Hölder pointwise regularity given in Definition 2 does not take into account directional regularity but yields the worst possible regularity in all directions. Therefore, all the results obtained in this chapter do not take into account possible directional irregularity phenomena.

We now briefly discuss the notion of directional regularity. Let f be a locally bounded function defined on $\mathbf{R}^d$. In order to take into account directional behaviors, it is natural to define the Hölder regularity at x_0 in a direction $u \in \mathbf{R}^d \setminus \{0\}$ as the Hölder regularity at zero of the univariate function $t \mapsto f(x_0 + tu)$. This definition has several drawbacks which stem from the fact that the latter function is defined as the trace of f on a line, which is a set of measure zero, see [37] for a detailed discussion. Let us now give the definition of anisotropic smoothness which is currently used, see, e.g., [11, 37].

Definition 10. Let f be a real-valued function defined on $\mathbf{R}^d$ and bounded in a neighborhood of a point $x_0 \in \mathbf{R}^d$. Let $e = (e_1, \dots, e_d)$ be an orthonormal basis of $\mathbf{R}^d$ and let $\alpha = (\alpha_1, \dots, \alpha_d)$ be a d-tuple of nonnegative real numbers such that $\alpha_1 \geq \dots \geq \alpha_d$. The function f belongs to $C^\alpha(x_0, e)$ if there exist a real $C > 0$ and a polynomial P_{x_0} such that for all x in a neighborhood of x_0,

$$|f(x) - P_{x_0}(x)| \leq C \sum_{i=1}^{d} |(x - x_0) \cdot e_i|^{\alpha_i}.$$

Since, by construction, Davenport series display jumps along hyperplanes, thereby being extremely anisotropic by nature, a natural question is to determine their pointwise anisotropic regularity; the same remark is also relevant for Lévy fields, which present the same type of anisotropy, see [24]. These two examples would certainly be natural candidates to test possible definitions of *anisotropic spectra of singularities*.

Finally, note that an extension of pointwise smoothness combining anisotropy and the $T^p_\alpha(x_0)$ condition is proposed in [37]. This notion could be relevant in order to perform the study of the anisotropy of multivariate Davenport series with coefficients not belonging to ℓ^1.

10 Proof of Theorem 1

Let us begin by comparing the two lower limits appearing in the statement of the theorem. First, given that θ_a is bounded above by one, we have

$$\liminf_{\substack{q\to\infty\\ q\in\operatorname{supp}(A)}} \frac{\log|A_q|}{\log\delta_q^{\mathscr{P}}(x_0)} \leq \liminf_{\substack{q\to\infty\\ q\in\operatorname{supp}(\overline{a})}} \frac{\log|\overline{a}_q|}{\log\delta_q^{\mathscr{P}}(x_0)},$$

where $A = J(a)$ and $\overline{a} = M(a)$. In addition, replacing $\delta_q^{\mathscr{P}}(x_0)$ by $\delta_q(x_0)$ in the right-hand side does not change the value of the lower limit. Indeed, for any point q in the support of the sequence $\overline{a}$, the distance $\delta_q(x_0)$ is reached on a hyperplane of the form $H_{p,q}$ with $p \in \mathbf{Z}$, so that

$$\delta_q(x_0) = \frac{|q\cdot x_0 - p|}{|q|} = \frac{|q'\cdot x_0 - p'|}{|q'|} \geq \delta_{q'}^{\mathscr{P}}(x_0).$$

Here, $p' = p/r$ and $q' = q/r$, where r denotes the greatest common divisor of the integer p and the components of the vector q, so that $p' \in \mathscr{P}_{q'}$. Since the multiples of q are also multiples of q', the supremum over all integers $l \geq 1$ of $|a_{lq}|$ is at most that of $|a_{lq'}|$. As a consequence, for all $q \in \operatorname{supp}(\overline{a})$ large enough, there exists a point $q'|q$ such that

$$\frac{\log\overline{a}_{q'}}{\log\delta_{q'}^{\mathscr{P}}(x_0)} \leq \frac{\log\overline{a}_q}{\log\delta_q(x_0)} \leq \frac{\log\overline{a}_q}{\log\delta_q^{\mathscr{P}}(x_0)},$$

so the lower limit featuring $\delta_q^{\mathscr{P}}(x_0)$ in the denominator coincides with that featuring $\delta_q(x_0)$. Furthermore, $|a_q|$ is obviously bounded above by $\overline{a}_q$, so that

$$\frac{\log\overline{a}_q}{\log\delta_q(x_0)} \leq \frac{\log|a_q|}{\log\delta_q(x_0)}.$$

The above discussion, combined with Corollary 1, finally leads to Eq. (25). In particular, since $\delta_n(x_0)$ is bounded above by $1/|n|$ regardless of the value of x_0, we deduce the next uniform bound on the Hölder exponent:

$$h_f(x_0) \leq \gamma_a,$$

where γ_a is defined by Eq. (19). The remainder of the theorem then follows when the sequence of Davenport coefficients has slow decay, that is, when γ_a vanishes.

In order to finish the proof of the theorem, we thus may assume that γ_a is positive. It remains for us to establish that

$$h_f(x_0) \geq \liminf_{\substack{n\to\infty \\ n\in\operatorname{supp}(a)}} \frac{\log|a_n|}{\log\delta_n(x_0)} \tag{38}$$

for any point x_0 that does not belong to the set $D_{M(a)}$. Let us consider an integer $j_0 \geq 0$ and a point $x \in \mathbf{R}^d$ such that $2^{-(j_0+1)} \leq |x - x_0| < 2^{-j_0}$. Since the sequence a is in ℓ^1, we may define

$$\Sigma_{x_0,x}(Z) = \sum_{n\in Z} a_n (\{n\cdot x\} - \{n\cdot x_0\})$$

for any subset Z of $\mathbf{Z}^d$. Since the Davenport series converges normally, it is clear that its increment between x_0 and x may be written in the form

$$f(x) - f(x_0) = \Sigma_{x_0,x}(\mathbf{Z}^d).$$

Therefore, it suffices to handle the series $\Sigma_{x_0,x}(Z)$ for Z ranging over a collection of sets that form a partition of $\mathbf{Z}^d$.

To be specific, let us begin by giving an upper bound on $|\Sigma_{x_0,x}(\mathscr{C}_j \cap \mathbf{Z}^d)|$, where $\mathscr{C}_j$ is the domain defined by

$$\mathscr{C}_j = \left\{x \in \mathbf{R}^d \mid 2^j \leq |x| < 2^{j+1}\right\},$$

for any integer $j \geq 0$. For every fixed $\gamma \in (0, \gamma_a)$, there exists a constant $C_\gamma > 0$ such that $|a_n| \leq C_\gamma |n|^{-\gamma}$ for all $n \in \mathbf{Z}^d$. For the sake of simplicity, we merely write $|a_n| \ll |n|^{-\gamma}$ in such a situation, thereby making use of the Vinogradov symbol. Accordingly,

$$\left|\Sigma_{x_0,x}(\mathscr{C}_j \cap \mathbf{Z}^d)\right| \leq \sum_{n\in\mathscr{C}_j\cap\mathbf{Z}^d} |a_n| \ll \sum_{n\in\mathscr{C}_j\cap\mathbf{Z}^d} |n|^{-\gamma} \mathbb{1}_{\{n\in\operatorname{supp}(a)\}}.$$

Furthermore, for every fixed $\varepsilon > 0$, we deduce from the sparsity assumption bearing on the sequence a that $\#(\operatorname{supp}(a) \cap \mathscr{C}_j) \ll 2^{\varepsilon j}$. As a consequence,

$$\sum_{n\in\mathscr{C}_j\cap\mathbf{Z}^d} |n|^{-\gamma} \mathbb{1}_{\{n\in\operatorname{supp}(a)\}} \leq 2^{-\gamma j} \#(\operatorname{supp}(a) \cap \mathscr{C}_j) \ll 2^{(-\gamma+\varepsilon)j}.$$

The union over all integers $j \geq j_0$ is equal to the complement in $\mathbf{R}^d$ of the open ball centered at the origin with radius 2^{j_0}. Thus, summing over all these values of j, we obtain

$$\left|\Sigma_{x_0,x}(\mathbf{Z}^d \setminus \mathrm{B}(0, 2^{j_0}))\right| \ll 2^{(-\gamma+\varepsilon)j_0} \ll |x - x_0|^{\gamma-\varepsilon}.$$

Now, let $\mathscr{N}_{x_0,x}$ denote the set of all points $n \in \mathbf{Z}^d$ for which there exists an integer $k \in \mathbf{Z}$ satisfying either $n\cdot x_0 \leq k < n\cdot x$ or $n\cdot x \leq k < n\cdot x_0$, meaning that the hyperplane $H_{k,n}$ separates the points x_0 and x, and let $\mathscr{N}^{\mathrm{c}}_{x_0,x}$ denote its complement in $\mathbf{Z}^d$. For each $n \in \mathscr{N}^{\mathrm{c}}_{x_0,x}$, it is clear that $n \cdot x_0$ and $n \cdot x$ have the same integer part so that

$$\Sigma_{x_0,x}(\mathscr{N}^{\mathrm{c}}_{x_0,x} \cap \mathrm{B}(0,2^{j_0})) = (x - x_0) \cdot \sum_{\substack{n \in \mathscr{N}^{\mathrm{c}}_{x_0,x} \\ |n| < 2^{j_0}}} a_n n. \tag{39}$$

If γ_a is smaller than or equal to one, the modulus of the sum in Eq. (39) may be bounded above by the sum over $j \in \{0, \ldots, j_0 - 1\}$ of

$$\sum_{n \in \mathscr{C}_j \cap \mathbf{Z}^d} |a_n n| \ll \sum_{n \in \mathscr{C}_j \cap \mathbf{Z}^d} |n|^{1-\gamma} \mathbb{1}_{\{n \in \mathrm{supp}(a)\}} \ll 2^{(1-\gamma)j} \#(\mathrm{supp}\,(a) \cap \mathscr{C}_j) \ll 2^{(1-\gamma+\varepsilon)j}, \tag{40}$$

which entails that

$$\left|\Sigma_{x_0,x}(\mathscr{N}^{\mathrm{c}}_{x_0,x} \cap \mathrm{B}(0,2^{j_0}))\right| \ll |x - x_0|\, 2^{(1-\gamma+\varepsilon)j_0} \le |x - x_0|^{\gamma-\varepsilon}.$$

In the opposite case where γ_a is larger than one, we assume that $1 < \gamma < \gamma_a$ and rewrite Eq. (39) as a difference of two terms in the following form:

$$\Sigma_{x_0,x}(\mathscr{N}^{\mathrm{c}}_{x_0,x} \cap \mathrm{B}(0,2^{j_0})) = (x - x_0) \cdot \sum_{n \in \mathscr{N}^{\mathrm{c}}_{x_0,x}} a_n n - (x - x_0) \cdot \sum_{\substack{n \in \mathscr{N}^{\mathrm{c}}_{x_0,x} \\ |n| \ge 2^{j_0}}} a_n n.$$

The first series is normally convergent because, in view of Eq. (40), we have

$$\sum_{n \in \mathscr{N}^{\mathrm{c}}_{x_0,x}} |a_n n| \le \sum_{j=0}^{\infty} \sum_{n \in \mathscr{C}_j \cap \mathbf{Z}^d} |a_n n| \ll \sum_{j=0}^{\infty} 2^{(1-\gamma+\varepsilon)j} < \infty.$$

In order to handle the second term, we make use of Eq. (40) again; this implies that

$$\sum_{\substack{n \in \mathscr{N}^{\mathrm{c}}_{x_0,x} \\ |n| \ge 2^{j_0}}} |a_n n| \le \sum_{j=j_0}^{\infty} \sum_{n \in \mathscr{C}_j \cap \mathbf{Z}^d} |a_n n| \ll \sum_{j=j_0}^{\infty} 2^{(1-\gamma+\varepsilon)j} \ll |x - x_0|^{-1+\gamma-\varepsilon}. \tag{41}$$

We finally deduce that

$$\left|\Sigma_{x_0,x}(\mathscr{N}^{\mathrm{c}}_{x_0,x} \cap \mathrm{B}(0,2^{j_0})) - (x - x_0) \cdot \sum_{n \in \mathscr{N}^{\mathrm{c}}_{x_0,x}} a_n n\right| \ll |x - x_0|^{\gamma-\varepsilon}.$$

It remains for us to consider the behavior of $\Sigma_{x_0,x}$ on the set $\mathscr{N}_{x_0,x} \cap \mathrm{B}(0,2^{j_0})$. Let α denote the lower limit in the right-hand side of Eq. (38), which we may assume to be positive, and let $\alpha' \in (0,\alpha)$. By definition of α, we have $\delta_n(x_0) \ge |a_n|^{1/\alpha'}$ for $n \in \mathbf{Z}^d$ sufficiently far from the origin. In addition, a_n necessarily vanishes when the distance $\delta_n(x_0)$ is zero. As a matter of fact, in that situation, x_0 belongs to a hyperplane $H_{k,n}$ with $k \in \mathbf{Z}$. This hyperplane is represented by a unique pair $(p,q) \in \mathscr{H}_d$, and then p and q divide k and n, respectively. However, x_0 does not belong to

$D_{M(a)}$, so the index q cannot belong to the support of the sequence $M(a) = \overline{a}$, which entails that $|a_n| \leq \overline{a}_q = 0$. The upshot is that there exists a real $C_{\alpha'} > 0$ such that $\delta_n(x_0) \geq C_{\alpha'} |a_n|^{1/\alpha'}$ for all $n \in \mathbf{Z}^d$, which we write $|a_n| \ll \delta_n(x_0)^{\alpha'}$ still using the Vinogradov symbol. As a consequence,

$$\left|\Sigma_{x_0,x}(\mathscr{N}_{x_0,x} \cap \mathrm{B}(0,2^{j_0}))\right| \leq \sum_{\substack{n \in \mathscr{N}_{x_0,x} \\ |n|<2^{j_0}}} |a_n| \ll \sum_{\substack{n \in \mathscr{N}_{x_0,x} \\ |n|<2^{j_0}}} \delta_n(x_0)^{\alpha'} \mathbb{1}_{\{n \in \mathrm{supp}(a)\}}.$$

Moreover, when n belongs to $\mathscr{N}_{x_0,x}$, we have $|n \cdot x_0 - k| \leq |n \cdot (x - x_0)|$ for some integer $k \in \mathbf{Z}$, so that $\delta_n(x_0)$ is bounded above by $|x - x_0|$. Hence,

$$\sum_{\substack{n \in \mathscr{N}_{x_0,x} \\ |n|<2^{j_0}}} \delta_n(x_0)^{\alpha'} \mathbb{1}_{\{n \in \mathrm{supp}(a)\}} \leq |x - x_0|^{\alpha'} \#(\mathrm{supp}\,(a) \cap \mathrm{B}(0,2^{j_0})) \ll |x - x_0|^{\alpha'} 2^{\varepsilon j_0},$$

where the last bound follows from the sparsity assumption bearing on the sequence a. We deduce that

$$\left|\Sigma_{x_0,x}(\mathscr{N}_{x_0,x} \cap \mathrm{B}(0,2^{j_0}))\right| \ll |x - x_0|^{\alpha' - \varepsilon}.$$

The above approach also enables us to write that

$$\sum_{\substack{n \in \mathscr{N}_{x_0,x} \\ |n|<2^{j_0}}} |a_n n| \ll |x - x_0|^{\alpha'} \sum_{\substack{n \in \mathscr{N}_{x_0,x} \\ |n|<2^{j_0}}} |n| \mathbb{1}_{\{n \in \mathrm{supp}(a)\}} \ll |x - x_0|^{\alpha'} \sum_{j=0}^{j_0 - 1} 2^j \#(\mathrm{supp}\,(a) \cap \mathscr{C}_j),$$

where the last sum is bounded by $2^{(1+\varepsilon)j_0}$ up to a constant, in view of the sparsity of the sequence a. In addition, when $1 < \gamma < \gamma_a$, the bound given by Eq. (41) still holds when $\mathscr{N}_{x_0,x}^{\mathrm{c}}$ is replaced by $\mathscr{N}_{x_0,x}$. It follows that

$$\left| \sum_{n \in \mathscr{N}_{x_0,x}} a_n n \right| \ll |x - x_0|^{-1+\gamma-\varepsilon} + |x - x_0|^{-1+\alpha'-\varepsilon},$$

which readily implies that

$$\left| \Sigma_{x_0,x}(\mathscr{N}_{x_0,x} \cap \mathrm{B}(0,2^{j_0})) - (x - x_0) \cdot \sum_{n \in \mathscr{N}_{x_0,x}} a_n n \right| \ll |x - x_0|^{\gamma - \varepsilon} + |x - x_0|^{\alpha' - \varepsilon}.$$

Combining all the previously obtained bounds, we finally get

$$|f(x) - f(x_0)| \ll |x - x_0|^{\gamma - \varepsilon} + |x - x_0|^{\alpha' - \varepsilon}$$

when γ_a is smaller than or equal to one, and

$$\left| f(x) - f(x_0) - (x - x_0) \cdot \sum_{n \in \mathbf{Z}^d} a_n n \right| \ll |x - x_0|^{\gamma - \varepsilon} + |x - x_0|^{\alpha' - \varepsilon}$$

when γ_a is larger than one. In both cases, it appears that the Hölder exponent at x_0 of the Davenport series f is at least the minimum between $\gamma - \varepsilon$ and $\alpha' - \varepsilon$. The bound Eq. (38) finally follows from letting ε go to zero, γ to γ_a and α' to α.

11 Proof of Theorems 2 and 3

Throughout the section, f denotes a Davenport series with coefficients given by a sequence $a = (a_n)_{n \in \mathbf{Z}^d_*}$ in ℓ^1. We assume that the series is sparse and not asymptotically jump canceling and that γ_a is both positive and finite.

11.1 Locations of the Singularities

The first step to the proof of Theorems 2 and 3 consists in observing that the iso-Hölder sets and the singularity sets of the Davenport series f may be expressed in terms of the sets $L_a(\alpha)$ of all points that are at a distance less than $|a_n|^{1/\alpha}$ from a hyperplane $H_{k,n}$ defined as in Eq. (6) for infinitely many points n in the support of the sequence a. Put another way, a point x belongs to $L_a(\alpha)$ if and only if the distance $\delta_n(x)$ defined by Eq. (23) is less than $|a_n|^{1/\alpha}$ infinitely often. To be more specific, for any real $\alpha > 0$, the set $L_a(\alpha)$ is defined by

$$L_a(\alpha) = \left\{ x \in \mathbf{R}^d \mid |n \cdot x - k| < |n|\, |a_n|^{1/\alpha} \text{ for i.m. } (k,n) \in \mathbf{Z} \times \mathbf{Z}^d \right\}, \qquad (42)$$

where i.m. stands for "infinitely many". It is easy and useful to remark that the mapping $\alpha \mapsto L_a(\alpha)$ is nondecreasing.

The connection between the iso-Hölder and singularity sets and the sets $L_a(\alpha)$ is now given by the next result. It is a direct consequence of Theorem 1, along with the discussion made in Sect. 3 above according to which the Davenport series f is discontinuous on $D_{J(a)}$, thus having Hölder exponent zero thereon. In its statement, $M(a)$ denotes the image of the sequence a under the action of the maximal operator.

Lemma 4. *Let $h \in [0, \gamma_a]$. Then,*

$$E_f(h) \setminus D_{J(a)} \subseteq E'_f(h) \subseteq \left(D_{M(a)} \setminus D_{J(a)} \right) \cup \bigcap_{\alpha > h} L_a(\alpha). \qquad (43)$$

Moreover, $E_f(0) \supseteq D_{J(a)}$ and for the positive values of h,

$$E'_f(h) \supseteq L_a(h) \setminus D_{J(a)} \qquad \textit{and} \qquad E_f(h) \supseteq E'_f(h) \setminus \bigcup_{\alpha<h} L_a(\alpha). \tag{44}$$

Lemma 4 suggests that the proof of Theorems 2 and 3 will follow from a detailed understanding of the size and large intersection properties of the sets $L_a(\alpha)$. This is the purpose of the next subsection, but let us just point out here that

$$\bigcap_{\alpha>\gamma_a} L_a(\alpha) = \mathbf{R}^d. \tag{45}$$

Indeed, it is plain that $\delta_n(x) \leq 1/|n|$ for every $n \in \mathbf{Z}^d_*$ and every point $x \in \mathbf{R}^d$, and that $|a_n|^{1/\alpha}|n| \geq 1$ infinitely often, when $\alpha > \gamma_a$. We may thus restrict our attention to the case where $\alpha \leq \gamma_a$ in the study of the size and large intersection properties of $L_a(\alpha)$.

11.2 Size and Large Intersection Properties of the Sets $L_a(\alpha)$, Connection with the Duffin–Schaeffer and Catlin Conjectures

We shall investigate the size properties of the sets $L_a(\alpha)$ by estimating their Hausdorff measures for specific gauge functions. We call a gauge function any continuous nondecreasing function g which is defined on $[0,\varepsilon]$ for some $\varepsilon > 0$ and vanishes at zero. The Hausdorff measure associated with such a gauge function is then defined by

$$\mathscr{H}^g(E) = \lim_{\delta\downarrow 0} \uparrow \mathscr{H}^g_\delta(E) \qquad \text{with} \qquad \mathscr{H}^g_\delta(E) = \inf_{\substack{E\subseteq\bigcup_i U_i \\ |U_i|<\delta}} \sum_{i=1}^{\infty} g(|U_i|),$$

for any subset E of $\mathbf{R}^d$. Here, the infimum is taken over all sequences $(U_i)_{i\geq 1}$ of subsets of $\mathbf{R}^d$ satisfying $E \subseteq \bigcup_i U_i$ and $|U_i| < \delta$ for all i, where $|\cdot|$ denotes diameter. It is well-known that $\mathscr{H}^g$ is a Borel measure on $\mathbf{R}^d$, see, e.g., [44]. Moreover, the Hausdorff measure associated with the gauge function $r \mapsto r^s$ is called the s-dimensional Hausdorff measure and is denoted by $\mathscr{H}^s$; recall that such measures enable one to define the Hausdorff dimension of a nonempty set $E \subseteq \mathbf{R}^d$ by

$$\dim_{\mathrm{H}} E = \sup\{s \in (0,d) \mid \mathscr{H}^s(E) = \infty\} = \inf\{s \in (0,d) \mid \mathscr{H}^s(E) = 0\},$$

see Falconer's book [27] for instance.

Theorems 2 and 3 state that the iso-Hölder and the singularity sets of the Davenport series f all have Hausdorff dimension between $d-1$ and d. Therefore, on our way to the proof of these results, we may restrict our attention to the gauge functions of the form $r \mapsto r^{d-1+s}$, with $0 \leq s \leq 1$, as well as slight corrections thereof. These corrections are obtained by replacing r^s in the previous expression by more general functions, specifically, the continuous nondecreasing functions φ defined on $[0,\varepsilon]$ for some $\varepsilon > 0$ which vanish at zero, for which $r \mapsto \varphi(r)/r$ is nonincreasing and positive, and for which the limit

$$s_\varphi = \lim_{r\to 0} \frac{\log \varphi(r)}{\log r}$$

exists (this limit is then between zero and one). The collection of all such functions is denoted by Φ, and clearly contains the functions $r \mapsto r^s$, for $0 \leq s \leq 1$. For any $\varphi \in \Phi$, it is now plain that the function $r \mapsto r^{d-1}\varphi(r)$ is a gauge; the corresponding Hausdorff measure is denoted by $\mathscr{H}^{d-1,\varphi}$, and the value that it assigns to the set $L_a(\alpha)$ is discussed in the next statement.

Lemma 5. *For any real $\alpha > 0$ and any function $\varphi \in \Phi$,*

$$\sum_{n\in\mathbf{Z}^d} |n|\varphi(|a_n|^{1/\alpha}) < \infty \qquad \Longrightarrow \qquad \mathscr{H}^{d-1,\varphi}(L_a(\alpha)) = 0.$$

Proof. Let $\rho_n = |a_n|^{1/\alpha}$ for any $n \in \mathbf{Z}^d$, and let us consider two real numbers $A > 1$ and $\delta \in (0,1]$. Let us assume that the series appearing in the statement of the lemma converges. So, there necessarily exists an integer $\eta_0 \geq 1$ such that $4\rho_n < \delta$ for all $n \in \mathbf{Z}^d$ with $|n| \geq \eta_0$. Then, for any $\eta_1 \geq \eta_0$,

$$L_a(\alpha) \cap \mathrm{B}(0,A-1) \subseteq \bigcup_{\substack{n\in\mathrm{supp}(a)\\|n|\geq\eta_1}} \bigcup_{\substack{k\in\mathbf{Z}\\|k|<A|n|}} \left\{x \in \mathrm{B}(0,A) \,\middle|\, \mathrm{dist}(x,H_{k,n}) < \rho_n\right\}.$$

Moreover, each set in the union above may be covered by $(2\lfloor 2A\sqrt{d}/\rho_n\rfloor)^{d-1}$ open balls with radius $2\rho_n$. Therefore,

$$\begin{aligned}\mathscr{H}_\delta^{r\mapsto r^{d-1}\varphi(r)}(L_a(\alpha)\cap \mathrm{B}(0,A-1)) &\leq \sum_{\substack{n\in\mathrm{supp}(a)\\|n|\geq\eta_1}} 2A|n| \left(\frac{4A\sqrt{d}}{\rho_n}\right)^{d-1} (4\rho_n)^{d-1}\varphi(4\rho_n)\\ &\leq 8A(16A\sqrt{d})^{d-1} \sum_{\substack{n\in\mathbf{Z}^d\\|n|\geq\eta_1}} |n|\varphi(|a_n|^{1/\alpha}).\end{aligned}$$

Letting $\eta_1 \to \infty$ and $\delta \to 0$, we deduce that $\mathscr{H}^{d-1,\varphi}(L_a(\alpha)\cap \mathrm{B}(0,A-1)) = 0$. This holds for all integers $A \geq 1$, so the result follows. $\square$

In particular, letting φ be the identity function in the statement of Lemma 5 and letting $\mathscr{L}^d$ be the Lebesgue measure in $\mathbf{R}^d$, we deduce that

$$\sum_{n\in\mathbf{Z}^d} |n|\,|a_n|^{1/\alpha} < \infty \qquad \Longrightarrow \qquad \mathscr{L}^d(L_a(\alpha)) = 0. \tag{46}$$

Owing to the alternate expression (20) of γ_a, it is easy to see that the above series converges once α is less than γ_a. Owing to Lemma 4 and the fact that $D_{M(a)}$ is a countable union of hyperplanes, we deduce that the iso-Hölder sets $E_f(h)$ and the singularity sets $E'_f(h)$ have Lebesgue measure zero when $h < \gamma_a$. Together with Corollary 3, this implies that the sets $E_f(\gamma_a)$ and $E'_f(\gamma_a)$ both have full Lebesgue measure. Therefore, $h_f(x_0) = \gamma_a$ for $\mathscr{L}^d$-almost every $x_0 \in \mathbf{R}^d$.

To proceed with the proof of Theorems 2 and 3, we shall need a kind of converse to Lemma 5, which ensures that $L_a(\alpha)$ has a positive Hausdorff measure for specific gauge functions. The study of the size properties of various classical sets arising in the metric theory of Diophantine approximation suggests that such a converse should look like the following:

$$\sum_{n\in\mathbf{Z}^d} |n|\varphi(|a_n|^{1/\alpha}) = \infty \Longrightarrow \forall W \text{ open} \quad \mathscr{H}^{d-1,\varphi}(L_a(\alpha)\cap W) = \mathscr{H}^{d-1,\varphi}(W), \tag{47}$$

see for instance [14, 21] and references therein. Given that $L_a(\alpha)$ is of the form

$$K_d(\psi) = \left\{x \in \mathbf{R}^d \;\middle|\; |n\cdot x - k| < \psi(n) \text{ for i.m. } (k,n) \in \mathbf{Z}\times\mathbf{Z}^d\right\},$$

where $\psi : \mathbf{Z}^d \to [0,\infty)$ is a multivariate approximating function, and thanks to the mass transference principle of [12] and the slicing technique of [13], this would follow from the next general statement: The set $K_d(\psi)$ has full Lebesgue measure in $\mathbf{R}^d$ if the series $\sum_n \psi(n)$ diverges. However, such a statement is known to be false and we refer to Sect. 5 in [14] for a counterexample. The expected result is actually given by a generalization of the Catlin conjecture to the case of dual approximation, which has been formulated by Beresnevich, Bernik, Dodson and Velani [14], and consists in replacing the previous series by

$$\sum_{q\in\mathbf{Z}^d_*} \phi_d(q) \max_{t\geq 1} \frac{\psi(tq)}{t|q|_\infty},$$

where $|\cdot|_\infty$ denotes the supremum norm and $\phi_d(q)$ is the number of positive integers less than or equal to $|q|_\infty$ which are coprime with the components of q. We refer to [14] for a motivation of this conjecture and for its relationship with the dual form of the famous Duffin–Schaeffer conjecture. The upshot is that it seems rather difficult to provide a converse to Lemma 5 in the form (47).

As shown by the statement of Theorems 2 and 3, we ultimately describe the size of the iso-Hölder and singularity sets in terms of Hausdorff dimension, rather than using general Hausdorff measures. Thus, we do not need to call upon such precise results as those mentioned just above. In fact, we only need to prove that appropriate modifications of the set

$$L_a(\gamma_a) = \left\{x \in \mathbf{R}^d \;\middle|\; |n\cdot x - k| < |n|\,|a_n|^{1/\gamma_a} \text{ for i.m. } (k,n) \in \mathbf{Z}\times\mathbf{Z}^d\right\},$$

obtained by letting $\alpha = \gamma_a$ in Eq. (42), have full Lebesgue measure in $\mathbf{R}^d$. Note that the set $L_a(\alpha)$ has full Lebesgue measure in $\mathbf{R}^d$ when $\alpha > \gamma_a$ due to Eq. (45), and has Lebesgue measure zero when $\alpha < \gamma_a$ by virtue of Lemma 5. The study of its Lebesgue measure for the critical value $\alpha = \gamma_a$ is more delicate. Indeed, if true, Eq. (47) would imply that $L_a(\gamma_a)$ has full Lebesgue measure when the series $\sum_n |n|\,|a_n|^{1/\gamma_a}$ diverges. However, the coefficients of the Davenport f series may be chosen in such a way that the series converges, e.g., when the nonvanishing coefficients are given by $|a_{\lambda_m}| = (m^2|\lambda_m|)^{-\gamma}$ for some positive real γ and some sparse injective sequence $(\lambda_m)_{m\geq 1}$, in which case $L_a(\gamma_a)$ has Lebesgue measure zero, by Lemma 5. This means that we shall have to slightly reshape this set in order to ensure that we work with a set with full Lebesgue measure. Actually, as shown by the next lemma, a slight modification of $L_a(\gamma_a)$ enables one to recover the whole space. This modification is written in the form

$$L_a^{(\varphi,i)}(\gamma_a) = \left\{x \in \mathbf{R}^d \;\middle|\; |n\cdot x - k| < |n|\,\varphi(|a_n|^{1/\gamma_a}) \text{ for i.m. } (k,n) \in \mathbf{Z}\times\mathscr{N}_i\right\},$$

where φ and i are appropriately chosen in Φ and $\{1,\ldots,d\}$, respectively. Here, $\mathscr{N}_i$ denotes the set of all points $n = (n_1,\ldots,n_d)$ in $\mathbf{Z}^d$ such that $|n|_\infty = |n_i|$; note that the sets $\mathscr{N}_i$ obviously form a covering of $\mathbf{Z}^d$. Whereas the function φ is crucial in order to enlarge the set $L_a(\gamma_a)$ and then to recover the whole space, the index i, which is used to retain only some specific frequencies among the support of a, is introduced merely for technical reasons appearing in the proof of Lemma 7 below. In the following, $\Phi_\star$ denotes the collection of all functions $\varphi \in \Phi$ with $s_\varphi = 1$ for which at least one of the sets $L_a^{(\varphi,1)}(\gamma_a),\ldots,L_a^{(\varphi,d)}(\gamma_a)$ has full Lebesgue measure in $\mathbf{R}^d$. The next lemma shows that $\Phi_\star$ is nonempty.

Lemma 6. *There exist an index $i_\star$ and a function $\varphi_\star$ with $s_{\varphi_\star} = 1$ such that*

$$|n|\,\varphi_\star(|a_n|^{1/\gamma_a}) \geq 1 \qquad \text{for i.m.} \quad n \in \mathscr{N}_{i_\star}.$$

In particular, the set $L_a^{(\varphi_\star,i_\star)}(\gamma_a)$ is equal to the whole space $\mathbf{R}^d$, and $\varphi_\star$ is in $\Phi_\star$.

Proof. Let $(\lambda_m)_{m\geq 1}$ denote an enumeration of the support of the sequence a. Given that the Davenport series is sparse, the sequence $(\lambda_m)_{m\geq 1}$ is sparse and injective, and, up to rearranging its terms, we may assume that the sequence $(|\lambda_m|)_{m\geq 1}$ is nondecreasing. Then, let $\rho_m = |a_{\lambda_m}|^{1/\gamma_a}$ and $u_m = 1/|\lambda_m|$ for any integer $m \geq 1$. The case in which the sequence $(u_m/\rho_m)_{m\geq 1}$ does not diverge to infinity is elementary. Indeed, in that situation, there exists a constant $C > 0$ such that $u_m \leq C\rho_m$ for all m belonging to some infinite subset $\mathscr{M}$ of $\mathbf{N}$. Then, the function defined by $\varphi_\star(r) = Cr$ clearly satisfies the required properties, and it suffices to choose $i_\star$ in such a way that $\mathscr{N}_{i_\star}$ contains infinitely many points λ_m with $m \in \mathscr{M}$.

We may therefore suppose from now on that $(u_m/\rho_m)_{m\geq 1}$ diverges to infinity. The definition (19) of γ_a, combined with the observation that $\mathbf{Z}^d$ is covered by the sets $\mathscr{N}_i$, ensures the existence of an index $i_\star$ such that

$$\limsup_{\substack{m\to\infty \\ \lambda_m\in\mathcal{N}_{i_\star}}} \frac{\log u_m}{\log \rho_m} = 1.$$

In addition, both u_m and ρ_m tend to zero as $m\to\infty$. Thus, we may find a sequence of indices $(m_k)_{k\geq 1}$ in $\mathbf{N}$ along which all the following properties hold: The sequence $(u_{m_k}/\rho_{m_k})_{k\geq 1}$ diverges to infinity monotonically; $\log u_{m_k}/\log \rho_{m_k}$ tends to one as k goes to infinity; for all $k\geq 1$,

$$\begin{cases} \lambda_m \in \mathcal{N}_{i_\star}, \\ \rho_{m_{k+1}} \leq (\rho_{m_k})^k, \\ u_{m_{k+1}} \leq (u_{m_k})^k. \end{cases}$$

It is now straightforward to check that any logarithmic interpolation of the points (ρ_{m_k}, u_{m_k}) yields a suitable function $\varphi_\star$. To be specific, any function $\varphi_\star$ defined on $[0,\infty)$ for which

$$\log\varphi_\star(r) = \log u_{m_k} - \frac{\log\rho_{m_k} - \log r}{\log\rho_{m_k} - \log\rho_{m_{k+1}}}(\log u_{m_k} - \log u_{m_{k+1}}),$$

for all $r\in(\rho_{m_{k+1}}, \rho_{m_k}]$ and $k\geq 1$, clearly belongs to the set Φ and meets all our requirements. □

For any $\alpha\in(0,\gamma_a]$, we now define a mapping T_α on the set $\Phi_\star$ by letting

$$T_\alpha\varphi : r\mapsto \varphi(r^{\alpha/\gamma_a})$$

for any function $\varphi\in\Phi_\star$. All the functions $T_\alpha\varphi$ belong to Φ and satisfy $s_{T_\alpha\varphi} = \alpha/\gamma_a$, so they roughly behave like r^{α/γ_a} near the origin. Furthermore, T_{γ_a} is the identity mapping. Now, recall that Lemma 6 yields a function $\varphi_\star$ for which $|n|\,\varphi_\star(|a_n|^{1/\gamma_a})\geq 1$ infinitely often; the series $\sum_n |n| T_\alpha\varphi_\star(|a_n|^{1/\alpha})$ thus diverges. We are in a situation where the assumption of Lemma 5 fails and, in fact, the set $L_a(\alpha)$ does not necessarily have a vanishing $\mathcal{H}^{d-1,T_\alpha\varphi_\star}$-mass. On the contrary, it should be regarded as large and omnipresent in $\mathbf{R}^d$ in terms of $(d-1+\alpha/\gamma_a)$-dimensional Hausdorff measure, in the sense that it belongs to Falconer's class $\mathcal{G}^{d-1+\alpha/\gamma_a}$ of sets with large intersection defined above.

This follows from our next result, namely, Lemma 7, which describes the large intersection properties of the sets $L_a(\alpha)$. The properties are expressed by means of the classes $\mathrm{G}^g(W)$ that were introduced in [21] in order to extend Falconer's classes to general gauge functions g and open sets $W\subseteq\mathbf{R}^d$, with a view to establishing a suitable framework to describe precisely the large intersection properties of various sets arising in the metric theory of Diophantine approximation. In what follows, we shall restrict our attention to the classes $\mathrm{G}^{d-1,\varphi}(\mathbf{R}^d)$ of sets with large intersection in the whole space $\mathbf{R}^d$ with respect to gauge functions of the form $r\mapsto r^{d-1}\varphi(r)$ with $\varphi\in\Phi$. We refer to [21] for a precise definition of those classes and a description of their main properties, and we content ourselves here with recalling that the class $\mathrm{G}^{d-1,\varphi}(\mathbf{R}^d)$ is closed under countable intersections and bi-Lipschitz mappings, and is formed of G_δ-subsets E of $\mathbf{R}^d$ satisfying

$$\forall W \neq \emptyset \text{ open} \qquad \mathscr{H}^{d-1,\psi}(E \cap W) = \infty \tag{48}$$

for any function $\psi \in \Phi$ growing faster than φ at zero, in the sense that ψ/φ tends to infinity monotonically, in which case we write $\psi \prec \varphi$. A straightforward consequence of these properties is the fact that, when $d-1+s_\varphi$ is positive, the class $\mathrm{G}^{d-1,\varphi}(\mathbf{R}^d)$ is included in Falconer's class $\mathscr{G}^{d-1+s_\varphi}$. Let us now describe the large intersection properties of the sets $L_a(\alpha)$; in the next statement, Φ_α denotes the collection of functions $\varphi \in \Phi$ satisfying $\varphi \prec T_\alpha \varphi_\star$ for some $\varphi_\star \in \Phi_\star$.

Lemma 7. *For any real* $\alpha \in (0, \gamma_a]$,

$$L_a(\alpha) \in \bigcap_{\varphi \in \Phi_\alpha} \mathrm{G}^{d-1,\varphi}(\mathbf{R}^d) \subseteq \mathscr{G}^{d-1+\alpha/\gamma_a}.$$

Proof. Let us consider a function $\varphi \in \Phi_\alpha$ and a function $\varphi_\star \in \Phi_\star$ for which $\varphi \prec T_\alpha \varphi_\star$. We shall make use of a ubiquity result, which enables one to deduce the large intersection properties of the set $L_a(\alpha)$ from the sole fact that a corresponding enlarged set, namely, one of the sets $L_a^{(\varphi_\star,1)}(\gamma_a), \ldots, L_a^{(\varphi_\star,d)}(\gamma_a)$, has full Lebesgue measure in $\mathbf{R}^d$. Indeed, there exists an index i such that Lebesgue-almost every point $x \in \mathbf{R}^d$ belongs to $L_a^{(\varphi_\star,i)}(\gamma_a)$, that is, satisfies

$$\operatorname{dist}(x, H_{k,n}) < T_\alpha \varphi_\star(|a_n|^{1/\alpha}) \qquad \text{for i.m.} \quad (k,n) \in \mathbf{Z} \times \mathscr{N}_i.$$

Moreover, letting U_i denote the line spanned by the ith vector of the canonical basis of $\mathbf{R}^d$, we see that the hyperplanes $H_{k,n}$ are such that

$$\sup_{\substack{n \in \mathscr{N}_i \\ k \in \mathbf{Z}}} |\{x \in U_i \,|\, \operatorname{dist}(x, H_{k,n}) < 1\}| < \infty.$$

We may therefore apply Theorem 3.6 in [22] and deduce that $L_a(\alpha) \in \mathrm{G}^{d-1,\varphi}(\mathbf{R}^d)$. To finish the proof, it suffices to consider a function $\varphi \in \Phi_\alpha$ with $s_\varphi = \alpha/\gamma_a$ and to recall that $\mathrm{G}^{d-1,\varphi}(\mathbf{R}^d) \subseteq \mathscr{G}^{d-1+s_\varphi}$. Such functions exist; as a matter of fact, one may take $\varphi(r) = T_\alpha \varphi_\star(r) \log(1/T_\alpha \varphi_\star(r))$ where $\varphi_\star$ is given by Lemma 6. □

The above results being established, we are now in position to prove Theorems 2 and 3. This is the purpose of the last part of this section.

11.3 End of the Proof

Recall that we have to establish the following properties: The iso-Hölder sets $E_f(h)$ and the singularity sets $E'_f(h)$ of the Davenport series f have Hausdorff dimension equal to $d-1+h/\gamma_a$ in every nonempty open set. We also need to show that the latter sets belong to the classes $\overline{\mathscr{G}}^{d-1+h/\gamma_a}$ when h is positive. By virtue of Theorem D in [26], this implies that the singularity sets have packing dimension equal to d in every nonempty open set, a feature that is also mentioned in the statement of Theorem 3. In view of various remarks written above, it only remains to consider the case where $h<\gamma_a$ and to establish the following three propositions.

Proposition 9. *For any real number $h\in[0,\gamma_a)$,*

$$\max\{\dim_{\mathrm{H}} E_f(h), \dim_{\mathrm{H}} E'_f(h)\} \le d-1+\frac{h}{\gamma_a}.$$

Proof. We begin by making use of Lemma 4. The two inclusions (43), combined with the fact that the sets $D_{J(a)}$ and $D_{M(a)}$ are countable unions of hyperplanes, imply that

$$\begin{cases} \dim_{\mathrm{H}} E_f(h) \le \max\{d-1, \dim_{\mathrm{H}} E'_f(h)\}, \\ \dim_{\mathrm{H}} E'_f(h) \le \max\left\{d-1, \inf\limits_{\alpha>h} \dim_{\mathrm{H}} L_a(\alpha)\right\} \end{cases}.$$

To conclude, it suffices to apply Lemma 5 which, along with the alternate expression (20) of γ_a, ensures that the dimension of $L_a(\alpha)$ is bounded above by $d-1+\alpha/\gamma_a$. □

In the next statement, $\overline{\mathrm{G}}^{d-1,\varphi}(\mathbf{R}^d)$ denotes the extended class of sets with large intersection that is defined in terms of the initial class $\mathrm{G}^{d-1,\varphi}(\mathbf{R}^d)$ by the following condition: For all $E\subseteq\mathbf{R}^d$,

$$E\in\overline{\mathrm{G}}^{d-1,\varphi}(\mathbf{R}^d) \qquad \Longleftrightarrow \qquad \exists E'\in\mathrm{G}^{d-1,\varphi}(\mathbf{R}^d) \quad E'\subseteq E.$$

The purpose of this extension is to avoid checking that the singularity sets are G_δ-sets, which is inessential here. It is easy to see that the extended class $\overline{\mathrm{G}}^{d-1,\varphi}(\mathbf{R}^d)$ contains the initial class $\mathrm{G}^{d-1,\varphi}(\mathbf{R}^d)$ and coincides with the latter on the G_δ-sets, in view of [21, Proposition 1(e)]. Moreover, the extended class enjoys the same remarkable properties as the initial class: $\overline{\mathrm{G}}^{d-1,\varphi}(\mathbf{R}^d)$ is closed under countable intersections and bi-Lipschitz mappings, and its members satisfy Eq. (48). Moreover, when $d-1+s_\varphi$ is positive, the class $\overline{\mathrm{G}}^{d-1,\varphi}(\mathbf{R}^d)$ is included in the corresponding extended version $\overline{\mathscr{G}}^{d-1+s_\varphi}$ of Falconer's class.

Proposition 10. *Let us consider a real number $h\in(0,\gamma_a)$. Then,*

$$E'_f(h) \in \bigcap_{\varphi \in \Phi_h} \overline{\mathrm{G}}^{d-1,\varphi}(\mathbf{R}^d) \subseteq \overline{\mathscr{G}}^{d-1+h/\gamma_a}.$$

Proof. Lemma 7 ensures that the set $L_a(h)$ belongs to the classes $\mathrm{G}^{d-1,\varphi}(\mathbf{R}^d)$ associated with the functions $\varphi \in \Phi_h$. The same property holds for the set $\mathbf{R}^d \setminus D_{J(a)}$; indeed, being the complement of a countable union of hyperplanes, this set is a G_δ-set with full Lebesgue measure in $\mathbf{R}^d$, and such a set belongs to all the classes $\mathrm{G}^g(\mathbf{R}^d)$, see [21, Proposition 11]. Using the stability under intersection of the classes of sets with large intersection, we deduce that

$$L_a(h) \setminus D_{J(a)} \in \bigcap_{\varphi \in \Phi_h} \mathrm{G}^{d-1,\varphi}(\mathbf{R}^d) \subseteq \mathscr{G}^{d-1+h/\gamma_a},$$

where the last inclusion also follows from Lemma 7. To conclude, it suffices to make use of Lemma 4, which ensures that $L_a(h) \setminus D_{J(a)}$ is a subset of $E'_f(h)$, see the first inclusion in Eq. (44). □

Our last statement gives a lower bound on the Hausdorff dimension of the sets $E_f(h)$ and $E'_f(h)$ in every nonempty open subset of $\mathbf{R}^d$. Recall that $E_f(0)$ contains the set $D_{J(a)}$, by virtue of Lemma 4. As γ_a is finite, the latter set is a dense countable union of hyperplanes, thereby having Hausdorff dimension at least $d-1$ in every nonempty open set. We may therefore restrict our attention to the positive values of h.

Proposition 11. *Let us consider a real number $h \in (0, \gamma_a)$ and a nonempty open subset W of $\mathbf{R}^d$. Then,*

$$\min\{\dim_{\mathrm{H}}(E_f(h) \cap W), \dim_{\mathrm{H}}(E'_f(h) \cap W)\} \geq d - 1 + \frac{h}{\gamma_a}.$$

Proof. It follows from Proposition 10 that the singularity set $E'_f(h)$ satisfies Eq. (48) for any function $\psi \in \Phi$ such that $\psi \prec \varphi$ for some $\varphi \in \Phi_h$. Moreover, choosing a function ψ for which $s_\psi = h/\gamma_a$, we also deduce from Lemma 5 that all the sets $L_a(\alpha)$, for $\alpha < h$, have a vanishing $\mathscr{H}^{d-1,\psi}$-mass. Such a function ψ exists: It suffices to take $\psi(r) = T_h\varphi_\star(r)\,(\log(1/T_h\varphi_\star(r)))^2$, and also $\varphi(r) = T_h\varphi_\star(r)\log(1/T_h\varphi_\star(r))$, where $\varphi_\star$ is given by Lemma 6. We finally get

$$\mathscr{H}^{d-1,\psi}\left(E_f(h) \cap W\right) \geq \mathscr{H}^{d-1,\psi}\left(E'_f(h) \cap W\right) = \infty,$$

where the first inequality is due to the second inclusion in Eq. (44), which appears in the statement of Lemma 4. The result follows. □

Acknowledgements The authors are grateful to Julien Brémont for pointing out a mistake in a first version of this chapter and to the anonymous referee for the careful reading and many valuable remarks.

References

1. Abry, P., Jaffard, S., Wendt, H.: Irregularities and scaling in signal and image processing: multifractal analysis. In: Frame, M. (ed.) Benoit Mandelbrot: A Life in Many Dimensions. World Scientific (2013)
2. Abry, P., Wendt, H., Jaffard, S.: When Van Gogh meets Mandelbrot: multifractal classification of painting's texture. Signal Process. **93**(3), 554–572 (2013) doi:10.1016/j.sigpro.2012.01.016
3. Apostol, T.M.: Introduction to analytic number theory. Undergraduate Texts in Mathematics. Springer, New York (1976)
4. Aubry, J.-M., Jaffard, S.: Random wavelet series. Comm. Math. Phys. **227**(3), 483–514 (2002)
5. Barral, J., Seuret, S.: Function series with multifractal variations. Math. Nachr. **274/275**, 3–18 (2004)
6. Barral, J., Seuret, S.: Combining multifractal additive and multiplicative chaos. Comm. Math. Phys. **257**(2), 473–497 (2005)
7. Barral, J., Seuret, S.: Information parameters and large deviation spectrum of discontinuous measures. Real Anal. Exchange **32**(2), 429–454 (2007)
8. Barral, J., Seuret, S.: The singularity spectrum of Lévy processes in multifractal time. Adv. Math. **214**(1), 437–468 (2007)
9. Barral, J., Seuret, S.: The multifractal nature of heterogeneous sums of Dirac masses. Math. Proc. Cambridge Philos. Soc. **144**(3), 707–727 (2008)
10. Barral, J., Fournier, N., Jaffard, S., Seuret, S.: A pure jump Markov process with a random singularity spectrum. Ann. Probab. **38**(5), 1924–1946 (2010)
11. Ben Slimane, M.: Multifractal formalism and anisotropic self-similar functions. Math. Proc. Cambridge Philos. Soc. **124**(2), 329–363 (1998)
12. Beresnevich, V., Velani, S.: A mass transference principle and the Duffin-Schaeffer conjecture for Hausdorff measures. Ann. Math. (2) **164**(3), 971–992 (2006)
13. Beresnevich, V., Velani, S.: Schmidt's theorem, Hausdorff measures and slicing. Int. Math. Res. Not. Article ID 48794, 1–24 (2006)
14. Beresnevich, V., Bernik, V., Dodson, M., Velani, S.: Classical metric Diophantine approximation revisited. In: Analytic Number Theory, pp. 38–61. Cambridge University Press, Cambridge (2009)
15. Brémont, J.: Davenport series and almost-sure convergence. Q. J. Math. **62**(4), 825–843 (2011)
16. de la Bretèche, R.: Estimation de sommes multiples de fonctions arithmétiques. Compos. Math. **128**(3), 261–298 (2001)
17. de la Bretèche, R., Tenenbaum, G.: Séries trigonométriques à coefficients arithmétiques. J. Anal. Math. **92**, 1–79 (2004)
18. Calderón, A.-P., Zygmund, A.: Local properties of solutions of elliptic partial differential equations. Studia Math. **20**, 171–225 (1961)
19. Davenport, H.: On some infinite series involving arithmetical functions. Quart. J. Math. (Oxford Ser.) **8**, 8–13 (1937)
20. Davenport, H.: On some infinite series involving arithmetical functions. II. Quart. J. Math. (Oxford Ser.) **8**, 313–320 (1937)
21. Durand, A.: Sets with large intersection and ubiquity. Math. Proc. Cambridge Philos. Soc. **144**(1), 119–144 (2008)
22. Durand, A.: Large intersection properties in Diophantine approximation and dynamical systems. J. London Math. Soc. (2) **79**(2), 377–398 (2009)
23. Durand, A.: Singularity sets of Lévy processes. Probab. Theory Relat. Fields **143**(3–4), 517–544 (2009)
24. Durand, A., Jaffard, S.: Multifractal analysis of Lévy fields. Probab. Theory Relat. Fields **153**(1-2), 45–96 (2012) doi:10.1007/s00440-011-0340-0.
25. Essouabri, D., Matsumoto, K., Tsumura, H.: Multiple zeta-functions associated with linear recurrence sequences and the vectorial sum formula. Canad. J. Math. **63**(2), 241–276 (2011)

26. Falconer, K.: Sets with large intersection properties. J. London Math. Soc. (2) **49**(2), 267–280 (1994)
27. Falconer, K.: Fractal Geometry: Mathematical Foundations and Applications, 2nd edn. Wiley, Chichester (2003)
28. Hecke, E.: Über analytische Funktionen und die Verteilung von Zahlen mod. eins. Hamb. Abh. **1**, 54–76 (1921)
29. Hooley, C.: On the difference of consecutive numbers prime to n. Acta Arith. **8**, 343–347 (1962/1963)
30. Hooley, C.: On the difference between consecutive numbers prime to n. II. Publ. Math. Debrecen **12**, 39–49 (1965)
31. Hooley, C.: On the difference between consecutive numbers prime to n. III. Math. Z. **90**, 355–364 (1965)
32. Jaffard, S.: The spectrum of singularities of Riemann's function. Rev. Mat. Iberoamericana **12**(2), 441–460 (1996)
33. Jaffard, S.: Old friends revisited: the multifractal nature of some classical functions. J. Fourier Anal. Appl. **3**(1), 1–22 (1997)
34. Jaffard, S.: The multifractal nature of Lévy processes. Probab. Theory Relat. Fields **114**(2), 207–227 (1999)
35. Jaffard, S.: On Davenport expansions. In: Fractal geometry and applications: a jubilee of Benoît Mandelbrot, Part 1. In: Proceedings of Symposia in Pure Mathematics, vol. 72, pp. 273–303. American Mathematical Society, Providence (2004)
36. Jaffard, S.: Wavelet techniques in multifractal analysis. In: Fractal Geometry and Applications: A Jubilee of Benoît Mandelbrot, Part 2. In: Proceedings of Symposia in Pure Mathematics, vol. 72, pp. 91–151. American Mathematical Society, Providence (2004)
37. Jaffard, S.: Pointwise and directional regularity of nonharmonic Fourier series. Appl. Comput. Harmon. Anal. **28**(3), 251–266 (2010)
38. Jaffard, S., Nicolay, S.: Pointwise smoothness of space-filling functions. Appl. Comput. Harmon. Anal. **26**(2), 181–199 (2009)
39. Jaffard, S., Nicolay, S.: Space-filling functions and Davenport series. In: Barral, J., Seuret, S. (eds.) Recent Developments in Fractals and Related Fields. Applied and Numerical Harmonic Analysis, pp. 19–34. Birkhäuser, Boston (2010)
40. Kahane, J.-P., Lemarié-Rieusset, P.-G.: Fourier Series and Wavelets. Gordon & Breach, Reading (1996)
41. Oppenheim, H.: Ondelettes et Multifractals: Application à une fonction de Riemann en dimension 2. Ph.D. thesis, Université Paris IX Dauphine (1997)
42. Pólya, G., Szego, G.: Problems and Theorems in Analysis. Vol. I Series, Integral Calculus, Theory of Functions. Springer, New York (1972). Translated from the German by D. Aeppli, Die Grundlehren der mathematischen Wissenschaften, Band 193.
43. Riemann, B.: Über die Darstellbarkeit einer Funktion durch eine trigonometrische Reihe (Habilitationsschrift, 1854). Abhandlungen der Königlichen Gesellschaft der Wissenschaften zu Göttingen, 13 (1868)
44. Rogers, C.: Hausdorff Measures. Cambridge University Press, Cambridge (1970)
45. Sándor, J., Mitrinović, D., Crstici, B.: Handbook of number theory. Mathematics and its Applications, vol. 351. Kluwer, Dordrecht (1996)
46. Young, R.M.: An introduction to nonharmonic Fourier series. Pure and Applied Mathematics, vol. 93. Academic [Harcourt Brace Jovanovich Publishers], New York (1980)
47. Zhou, D.: Certaines études sur la minimalité et la propriété chaotique de dynamiques p-adiques et la régularité locale des séries de Davenport avec translation de phase. Ph.D. thesis, Université Paris-Est (2009)

Dimensions of Self-affine Sets: A Survey

Kenneth Falconer

Abstract Self-affine sets may be expressed as unions of reduced scale affine copies of themselves. We survey general and specific constructions of self-affine sets and in particular the problem of finding or estimating their Hausdorff or box-counting dimensions. The structure and dimensional properties of self-affine sets are somewhat subtle, for example, their dimensions need not vary continuously in the defining transformations.

1 Introduction

Many familiar fractals are made up of smaller copies of themselves in some sense. Probably best known are self-similar sets such as the Sierpiński triangle or von Koch curve. However, there are many other possibilities. For example, self-conformal sets are made up of conformal images of themselves, and statistically self-similar sets comprise scaled down components with the same statistical distribution as the whole. Here we will consider self-affine sets, which are composed of scaled down affine copies of themselves. We will review both general and specific constructions of self-affine sets and in particular the problem of finding the Hausdorff or box-counting dimensions of such sets.

K. Falconer (✉)
Mathematical Institute, University of St Andrews, North Haugh, St Andrews, Fife, KY16 9SS, Scotland
e-mail: kjf@st-and.ac.uk

J. Barral and S. Seuret (eds.), *Further Developments in Fractals and Related Fields*, Trends in Mathematics, DOI 10.1007/978-0-8176-8400-6_6,

1.1 Basic Definitions

The "iterated function system" framework that has been universally adopted for representing sets that are unions of smaller copies was introduced by John Hutchinson [35] and promoted by Michael Barnsley [3] and others.

A family $\{S_1,\ldots,S_m\}$ of contractions on $\mathbb{R}^N$, that is, mappings $\mathbb{R}^N \to \mathbb{R}^N$ satisfying

$$|S_i(x) - S_i(y)| \leq c_i|x-y|, \quad x,y \in \mathbb{R}^N, \quad 0 < c_i < 1$$

is called an *iterated function system* (IFS). The fundamental property of an IFS is that there exists a unique, non-empty compact $E \subseteq \mathbb{R}^N$ such that

$$E = \bigcup_{i=1}^{m} S_i(E), \tag{1}$$

called the *attractor* of the IFS. This may be proved elegantly by applying Banach's fixed-point theorem to the mapping $A \mapsto \bigcup_{i=1}^m S_i(A)$ on the complete metric space of non-empty compact subsets of $\mathbb{R}^N$ endowed with the Hausdorff metric (see [4, 24, 35]). If the S_i are similarity transformations, E is called *self-similar*. If the $S_i = T_i + \omega_i$ are affine contractions, where the T_i are non-singular contracting linear mappings on $\mathbb{R}^N$ and $\omega_i \in \mathbb{R}^N$ are translation vectors, E is called *self-affine*.

Taking a (large enough) compact domain B such that $S_i(B) \subseteq B$ for all i, we get an iterated construction of E:

$$E = \bigcap_{k=0}^{\infty} \bigcup_{i_1,\ldots,i_k} S_{i_1} \circ \cdots \circ S_{i_k}(B);$$

see Fig. 1. We also get a coding of points of E: if $\mathbf{i} = (i_1, i_2, \ldots) \in \{1,2,\ldots,m\}^{\mathbb{N}}$, let

$$X(\mathbf{i}) \equiv X(i_1,i_2,\ldots) := \lim_{k\to\infty} S_{i_1} \circ S_{i_2} \circ \cdots \circ S_{i_k}(0) \tag{2}$$

(the limit exists since the S_i are contractions). Then

$$E = \bigcup_{i_1,i_2,\ldots} X(i_1,i_2,\ldots),$$

though the points of E do not necessarily have a unique coding.

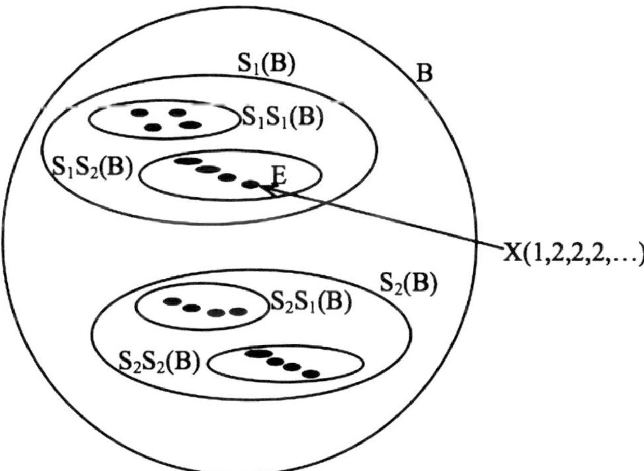

Fig. 1 The hierarchy of regions in the construction of the attractor

We will be particularly interested in the dimensions of attractors. We write $\dim_{\mathrm{H}}$ for Hausdorff dimension, $\underline{\dim}_{\mathrm{B}}$ and $\overline{\dim}_{\mathrm{B}}$ for lower and upper box-counting dimension and, if these are equal, $\dim_{\mathrm{B}}$ for the box dimension or Minkowski dimension.

For a self-similar set E, where each S_i is a similarity transformation of ratio r_i,

$$\dim_{\mathrm{H}} E = \underline{\dim}_{\mathrm{B}} E = \overline{\dim}_{\mathrm{B}} E = s \quad \text{where} \quad 1 = \sum_{i=1}^{m} r_i^s, \tag{3}$$

provided the *open set condition* holds, that is, there exists a non-empty open set O such that $\bigcup_{i=1}^{m} S_i(O) \subseteq O$ with this union disjoint; this ensures that the union in Eq. (1) is "almost disjoint". The *similarity dimension* of the self-similar set E is the value of s that satisfies Eq. (3) which depends only on the scaling ratios of the similarities in the IFS.

For self-affine sets, things are more awkward. One of the difficulties is that the dimensions are not continuous in the contractions in the IFS even when the $S_i(E)$ are well separated. For example, the attractor of the two affine maps in Fig. 2 has dimension 1 if the displacement $\lambda > 0$, since its projection onto the horizontal axis has positive length, whilst if $\lambda = 0$ the attractor is a middle-third Cantor set in the vertical axis, which has dimension $\log 2/\log 3 < 1$. In other examples, such as the carpets discussed in Sect. 3, the different types of dimension may take different values.

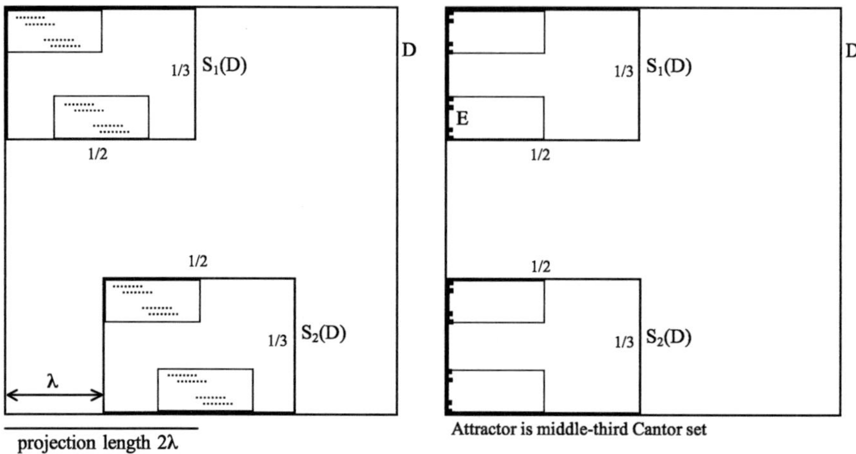

Fig. 2 The attractor on the left where $\lambda > 0$ has dimension 1, whilst that on the right where $\lambda = 0$ has dimension $\log 2/\log 3 < 1$

There are two approaches to finding dimensions of self-affine sets:

(a) Finding *generic formulae* for dimensions that hold almost always (in some sense)
(b) Seeking the dimension of *specific sets*

We will consider both approaches in the succeeding sections.

Note that several papers and books contain overviews of various aspects of self-affine sets as well as many further references; see, for example, [10, 11, 16, 49, 50].

2 The Affinity Dimension

It is generally easier to obtain upper bounds for dimensions than lower estimates. For self-affine sets the "affinity dimension", which is defined in terms of "singular value functions", always gives an upper bound and, in many generic situations, gives the actual value of the dimensions.

2.1 *Cutting up Ellipses*

To find the Hausdorff dimension of a set $E \subseteq \mathbb{R}^N$ we need to consider sums $\sum_i |U_i|^s$ where $\{U_i\}$ is a δ-cover of E, that is, a cover by small sets of diameter at most δ, and $|\cdot|$ denotes diameter. For the case where $N = 2$, suppose some covering set U_i is

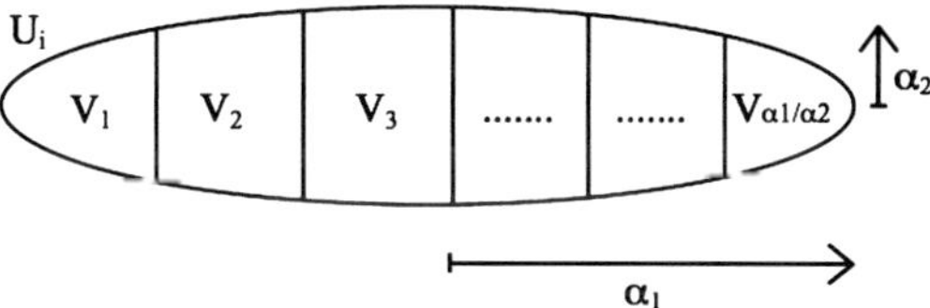

Fig. 3 The ellipse U_i may be cut into roughly α_1/α_2 pieces V_i of diameters about α_2 and these may contribute to a more efficient cover of the attractor than the single set U_i

"long and thin", for example, an ellipse with semi-axes $\alpha_1 \geq \alpha_2$. The contribution to $\sum |U_i|^s$ from U_i is $\approx \alpha_1^s$. For an alternative covering we can partition or "cut up" U_i into about α_1/α_2 pieces $\{V_j\}_{j=1}^{\alpha_1/\alpha_2}$ that are roughly square with side α_2, so that the single term $|U_i|^s \approx \alpha_1^s$ in the sum may be replaced by

$$\sum_{j=1}^{\alpha_1/\alpha_2} |V_j|^s \approx \frac{\alpha_1}{\alpha_2}\alpha_2^s = \alpha_1\alpha_2^{s-1},$$

and this may be much less than α_1^s for small α_1 if $s > 1$ see (Fig. 3).

Thus we define the *singular values* $\alpha_1 \geq \alpha_2 \geq 0$ of a linear mapping $T : \mathbb{R}^2 \to \mathbb{R}^2$ to be the positive square roots of the eigenvalues of TT^*. Equivalently the α_i are the semi-axis lengths of the ellipse $T(B)$ where B is the unit ball. We then define the *singular value function* of T by

$$\phi^s(T) = \begin{cases} \alpha_1^s & (0 \leq s \leq 1) \\ \alpha_1\alpha_2^{s-1} & (1 \leq s \leq 2) \end{cases}.$$

More generally for $T : \mathbb{R}^N \to \mathbb{R}^N$,

$$\phi^s(T) = \alpha_1 \dots \alpha_{p-1}\alpha_p^{s-p+1}, \tag{4}$$

where α_i is the ith singular value (arranged in decreasing order) and p is the integer such that $p-1 \leq s \leq p$.

There are two important properties of ϕ^s. Firstly it is *submultiplicative*, that is,

$$\phi^s(T_1T_2) \leq \phi^s(T_1)\phi^s(T_2), \tag{5}$$

and secondly, if T is a contracting linear map, then $\phi^s(T)$ is continuous and *decreasing* in s.

We may now get an upper bound for the dimensions of self-affine sets. Let

$$S_i(x) = T_i(x) + \omega_i \quad i = 1, 2, \dots, m \tag{6}$$

be an affine IFS, where the T_i are linear contractions on $\mathbb{R}^N$ and the ω_i are translation vectors, and let E be its self-affine attractor. Let B be a large disc with $S_i(B) \subseteq (B)$ for all i. Then for each k we get a covering of E by ellipses:

$$E \subseteq \bigcup_{i_1 \dots i_k} S_{i_1} \circ \cdots \circ S_{i_k}(B),$$

where each $S_{i_1} \circ \cdots \circ S_{i_k}(B)$ is an ellipse with semi-axes given by the singular values of $T_{i_1} \circ \cdots \circ T_{i_k}$ (see Fig. 1)

Thus, given $\delta > 0$, if k is large enough, we may use these ellipses (if $0 < s \leq 1$) or partitioned ellipses (if $1 < s \leq 2$) to get a δ-cover of E that gives an upper bound for the s-dimensional Hausdorff premeasure $\mathscr{H}^s_\delta(E)$ of E:

$$\begin{aligned}
\mathscr{H}^s_\delta(E) &= \inf\{\textstyle\sum |U_i|^s : \{U_i\} \text{ is a } \delta\text{-cover of } E\} \\
&\leq \begin{cases} c\sum_{i_1 \dots i_k} \alpha_1(T_{i_1} \circ \cdots \circ T_{i_k})^s & (0 \leq s \leq 1) \\ c\sum_{i_1 \dots i_k} \alpha_1(T_{i_1} \circ \cdots \circ T_{i_k})\alpha_2(T_{i_1} \circ \cdots \circ T_{i_k})^{s-1} & (1 \leq s \leq 2) \end{cases} \\
&= c \sum_{i_1 \dots i_k} \phi^s(T_{i_1} \circ \cdots \circ T_{i_k})
\end{aligned}$$

where c does not depend on k or δ. Hence, writing

$$\Phi^s_k := \sum_{i_1 \dots i_k} \phi^s(T_{i_1} \circ \cdots \circ T_{i_k}),$$

we have $\mathscr{H}^s_\delta(E) \leq c\Phi^s_k$ if k is sufficiently large. It follows from Eq. (5) that Φ^s_k itself is also submultiplicative, that is, $\Phi^s_{k+l} \leq \Phi^s_k \Phi^s_l$, so, by the standard property of submultiplicative sequences, the limit

$$\Phi^s := \lim_{k \to \infty} (\Phi^s_k)^{1/k}$$

exists and is decreasing in s. Thus if $\Phi^s < 1$ then the Hausdorff measure $\mathscr{H}^s(E) = \lim_{\delta \to 0} \mathscr{H}^s_\delta(E) \leq \lim_{k \to \infty} \Phi^s_k = 0$, so

$$\dim_{\mathrm{H}} E \leq s \text{ where } s \text{ satisfies } \Phi^s = 1.$$

A slight refinement of this argument choosing a covering by pieces all about the same size also gives an upper bound for the box dimensions:

$$\dim_{\mathrm{H}} E \leq \underline{\dim}_{\mathrm{B}} E \leq \overline{\dim}_{\mathrm{B}} E \leq s \text{ where } s \text{ satisfies } \Phi^s = 1. \tag{7}$$

The argument easily extends to $\mathbb{R}^N$ with Eq. (7) holding where the singular value function is defined by Eq. (4).

2.2 The Affinity Dimension

The *affinity dimension* $\mathrm{d}_{\mathrm{aff}} \equiv \mathrm{d}_{\mathrm{aff}}(T_1, \cdots, T_m)$ of the self-affine set E of the IFS of affine maps $S_i = T_i + \omega_i$ is the value of s satisfying

$$\Phi^s(T_1, \ldots, T_m) \equiv \Phi^s = \lim_{k\to\infty} \Big(\sum_{i_1 \ldots i_k} \phi^s(T_{i_1} \circ \cdots \circ T_{i_k}) \Big)^{1/k} = 1; \tag{8}$$

notice that the affinity dimension depends only on the linear parts of the IFS functions. Thus we showed above:

Proposition 1. *Let E be the self-affine attractor of the IFS consisting of affine mappings $S_i(x) = T_i(x) + \omega_i$. Then*

$$\dim_{\mathrm{H}} E \leq \underline{\dim}_{\mathrm{B}} E \leq \overline{\dim}_{\mathrm{B}} E \leq \mathrm{d}_{\mathrm{aff}}(T_1, \cdots, T_m). \tag{9}$$

We shall see that we “often” get equality in Eq. (9).

We first note some general features of $\Phi^s \equiv \Phi^s(T_1, \ldots, T_m)$ for affine IFSs on $\mathbb{R}^N$ which have consequences for the affinity dimension:

(a) Φ^s is continuous and strictly decreasing in s.
(b) Φ^s is piecewise convex in s, being convex between integral values of s but not in general differentiable when s is an integer.
(c) For IFSs on $\mathbb{R}^2$, $\Phi^s(T_1, \ldots, T_m)$ is continuous in $T_1, \ldots, T_m$ except (perhaps) when $T_1, \ldots, T_m$ have a common real eigenvector. More generally for IFSs on $\mathbb{R}^N$, $\Phi^s(T_1, \ldots, T_m)$ is continuous except (perhaps) on a set V of T_i which may be expressed as a finite union of algebraic hypersurfaces [29].
(d) Apart from the exceptional $\{T_i\}$ in (c), convergence in Eq. (8) is controlled by

$$\Phi^s \leq (\Phi_k^s)^{1/k} \leq c a^{1/k} \Phi^s$$

for some constant c.

In the special case where we can choose a coordinate basis of $\mathbb{R}^N$ such that $T_1, \ldots, T_m$ are all upper triangular, it may be shown that

$$\Phi^s(T_1, \ldots, T_m) = \Phi^s(T_1^D, \ldots, T_m^D), \tag{10}$$

where T_i^D is the diagonal matrix with the same diagonal elements as T_i. Since the T_i^D commute, $\Phi^s(T_1^D, \ldots, T_m^D)$ is easy to calculate and can be expressed as the maximum of a finite number of explicit expressions, allowing the affinity dimension to be found (see [2, 27]).

2.3 Generic Results

We now seek lower estimates of dimensions of self-affine sets. In this section we obtain some generic or almost sure estimates and in particular show that self-affine sets often have Hausdorff and box dimensions equal to the affinity dimension. The dimensions of sets constructed from affine maps with the same linear parts but with different translations were addressed by Falconer [19] in the case where $\|T_i\| < \frac{1}{3}$ and extended to $\|T_i\| < \frac{1}{2}$ by Solomyak [53]. As before we consider affine IFS maps $S_i(x) = T_i(x) + \omega_i$ on $\mathbb{R}^N$. We write $\omega := (\omega_1, \ldots, \omega_m)$, and denote the attractor by E_ω when we need to emphasize its dependence on the translations.

Theorem 1. *If $\|T_i\| < \frac{1}{2}$ for all i then*

$$\dim_{\mathrm{H}} E_\omega = \dim_{\mathrm{B}} E_\omega = \min\{d_{\mathrm{aff}}(T_1, \ldots, T_m), N\}$$

for almost all $\omega = (\omega_1, \ldots, \omega_m) \in \mathbb{R}^{Nm}$ w.r.t. Nm-dimensional Lebesgue measure.

Sketch of Proof. The upper bound follows from Proposition 1. For the lower bound, code points of the attractor E_ω as $X_\omega(\mathbf{i}) = \lim_{k\to\infty} S_{i_1} \circ \cdots \circ S_{i_k}(0)$. Let A be a ball in $\mathbb{R}^N$, and think of $\omega = (\omega_1, \ldots, \omega_m)$ as a random point of A^m with respect to a probability measure given by normalized Lebesgue measure. Let $\mathbb{E}$ denote expectation. A geometrical calculation shows that, if $0 < s < N$,

$$\mathbb{E}\,|X_\omega(\mathbf{i}) - X_\omega(\mathbf{j})|^{-s} \leq \frac{c}{\phi^s(T_{\mathbf{i}\wedge\mathbf{j}})}$$

for a constant c independent of $\mathbf{i}, \mathbf{j}$, where $\mathbf{i} \wedge \mathbf{j}$ is the common initial word of $\mathbf{i} = i_1, i_2, \ldots$ and $\mathbf{j} = j_1, j_2, \ldots$.

If $\Phi^s > 1$ one may construct a measure μ on $\{1, 2, \ldots, m\}^{\mathbb{N}}$ such that $\mu(C_{\mathbf{i}}) \leq c_1 \phi^s(T_{\mathbf{i}})$ for all $\mathbf{i}$ (where $C_{\mathbf{i}} = \{\mathbf{ij} : \mathbf{j} \in \{1, 2, \ldots, m\}^{\mathbb{N}}\}$ are the cylinder sets). This ensures that

$$\mathbb{E} \int\!\!\int \frac{\mathrm{d}\mu(\mathbf{i})\mathrm{d}\mu(\mathbf{j})}{|X_\omega(\mathbf{i}) - X_\omega(\mathbf{j})|^s} \leq c \int\!\!\int \frac{\mathrm{d}\mu(\mathbf{i})\mathrm{d}\mu(\mathbf{j})}{\phi^s(T_{\mathbf{i}\wedge\mathbf{j}})} < \infty. \tag{11}$$

For each ω, the attractor E_ω supports a measure μ_ω given by the projection of μ under $\mathbf{i} \mapsto X_\omega(\mathbf{i})$, that is with $\int f(x)\mathrm{d}\mu_\omega = \int f(X_\omega(\mathbf{i}))\mathrm{d}\mu(\mathbf{i})$ for continuous functions $f : \mathbb{R}^N \to \mathbb{R}$. It follows from Eq. (11) that for almost all $\omega \in A^N$

$$\int_{E_\omega}\int_{E_\omega} \frac{\mathrm{d}\mu_\omega(x)\mathrm{d}\mu_\omega(y)}{|x-y|^s} = \int\!\!\int \frac{\mathrm{d}\mu(\mathbf{i})\mathrm{d}\mu(\mathbf{j})}{|X_\omega(\mathbf{i}) - X_\omega(\mathbf{j})|^s} < \infty$$

which implies that $\dim_{\mathrm{H}} E_\omega \geq s$ by the energy criterion for Hausdorff dimension. □

In a recent variant, Käenmäki and Vilppolainen [41] obtained a similar result to Theorem 1 for *sub-self-affine sets*, that is, compact sets E satisfying $E \subseteq \bigcup_{i=1}^k S_i(E)$

where the S_i are self-affine mappings. The Hausdorff and box dimensions both equal the zero of a pressure function almost surely.

Unfortunately, Theorem 1 gives little information about which translations $(\omega_1,\ldots,\omega_m)$ give an attractor with dimensions equal to the affinity dimension. Nevertheless, the set of exceptional dimensions cannot be too big. With s as the affinity dimension of $\{T_1,\ldots,T_m\}$ and $t \leq s$, write

$$E(t)=\{(\omega_1,\ldots,\omega_m)\subseteq \mathbb{R}^{Nm} : \dim_{\mathrm{H}} E_\omega < t\},$$

for the set of translates where the dimension of the attractor is exceptionally small. Recall that the *Fourier dimension* of $A \subseteq \mathbb{R}^{Nm}$ is defined by

$$\dim_{\mathrm{F}} A = \sup\{t: \text{ there exists a measure } \mu \text{ on } A \text{ s.t. } \hat{\mu}(\mathbf{x}) = O(|\mathbf{x}|^{-t/2}) \text{ as } |\mathbf{x}|\to\infty\}.$$

Theorem 2 ([28]).

(a) $\dim_{\mathrm{H}} E(t) \leq Nm - c(s-t)$ *for a constant* $c>0$.
(b) $\dim_{\mathrm{F}} E(t) \leq 2t$.

A further drawback of Theorem 1 is that it only holds when $\|T_i\| < \frac{1}{2}$ for all i; indeed examples show that the conclusion can fail if $\|T_i\| = \frac{1}{2}$ for all i (see [18,53]). One approach to getting around this restriction is to introduce more randomness by allowing perturbative translations at each stage of the construction. The sets obtained are no longer strictly self-affine, but "almost self-affine".

Recall that for self-affine sets $E_\omega = \bigcup_{\mathbf{i}} X_\omega(\mathbf{i})$ where

$$\begin{aligned} X_\omega(\mathbf{i}) &= \lim_{k\to\infty} S_{i_1}\circ S_{i_2}\circ\cdots\circ S_{i_k}(0) \\ &= \lim_{k\to\infty} (T_{i_1}+\omega_{i_1})(T_{i_2}+\omega_{i_2})(T_{i_3}+\omega_{i_3})\cdots(T_{i_k}+\omega_{i_k})(0) \\ &= \lim_{k\to\infty} \omega_{i_1} + T_{i_1}\omega_{i_2} + T_{i_1}T_{i_2}\omega_{i_3} + T_{i_1}T_{i_2}T_{i_3}\omega_{i_4}+\cdots \end{aligned}$$

Now introduce a *random perturbation at each stage* of the construction:

$$\begin{aligned} X_\omega(\mathbf{i}) &= \lim_{k\to\infty} (T_{i_1}+\omega_{i_1})(T_{i_2}+\omega_{i_1,i_2})(T_{i_3}+\omega_{i_1,i_2,i_3})\cdots(T_{i_k}+\omega_{i_1,i_2,\ldots i_k})(0) \\ &= \lim_{k\to\infty} \omega_{i_1} + T_{i_1}\omega_{i_1,i_2} + T_{i_1}T_{i_2}\omega_{i_1,i_2,i_3} + T_{i_1}T_{i_2}T_{i_3}\omega_{i_1,i_2,i_3,i_4}\cdots, \end{aligned}$$

where $\omega_{i_1,i_2,\ldots,i_k}$ are independent and identically distributed "perturbations". We then call the set

$$E_\omega = \bigcup_{\mathbf{i}} X_\omega(\mathbf{i})$$

almost self-affine. These sets were investigated by Jordan, Pollicott and Simon.

Theorem 3 ([36]). *For almost self-affine sets E_ω,*

$$\dim_{\rm H} E_\omega = \dim_{\rm B} E_\omega = \min\{d_{\rm aff}(T_1,\ldots,T_m),N\}$$

for almost all ω. (Here we require only that $\|T_i\| < 1$ for all i.)

2.4 Sets with Dimension Attaining the Affinity Dimension

There are a number of situations where we can be sure that the dimension of a self-affine set equals its affinity dimension.

For example, plane self-affine sets whose projections onto lines have uniformly positive length have box dimension equal to the affinity dimension.

Theorem 4 ([20]). *Let $S_i(x) = T_i(x) + \omega_i$ be an IFS of affine contractions on $\mathbb{R}^2$ with attractor E. Suppose that:*

(a) The open set condition holds.
(b) There is a $c > 0$ such that the Lebesgue measure of the projection of E onto every line is at least c. (Note that this is the case if E contains a connected component other than a line segment.)

Then

$$\dim_{\rm B} E = d_{\rm aff}(T_1,\ldots,T_m).$$

Similar results hold for affine IFSs on $\mathbb{R}^N$.

Sets satisfying Theorem 4 include generalized Sierpínski triangles [26]. Here the IFS consists of three affine transformations mapping a right-angled isosceles triangle with unit perpendicular sides to subtriangles of base lengths $a, 1-a, 1-b$ and heights $b, b, 1-b$, respectively, where $0 < a,b < 1$ (see Fig. 4). We may apply Theorem 4 and, noting that the transformations are upper triangular, obtain from Eq. (10) that

$$\begin{aligned}\dim_{\rm B} E &= d_{\rm aff}(T_1,T_2,T_3)\\ &= \max\left\{s,t : b^s + (1-b)^s = 1, b\big((a^{t-1} + (1-a)^{t-1}\big) + (1-b)^t = 1\right\}.\end{aligned}$$

There are many further examples of sets satisfying Theorem 4 with the linear parts of the affine transformations representable by upper triangular matrices for which the box dimension may be found using Eq. (10) such as the example in Fig. 5.

Käenmäki and Shmerkin [40] introduced a class of self-affine sets of “Kakeya type”. Here the linear part of each IFS transformation and its adjoint (1) maps a certain cone into itself, with these cones being essentially disjoint for different IFS transformations, and (2) satisfies a projection condition. Then the box dimension of

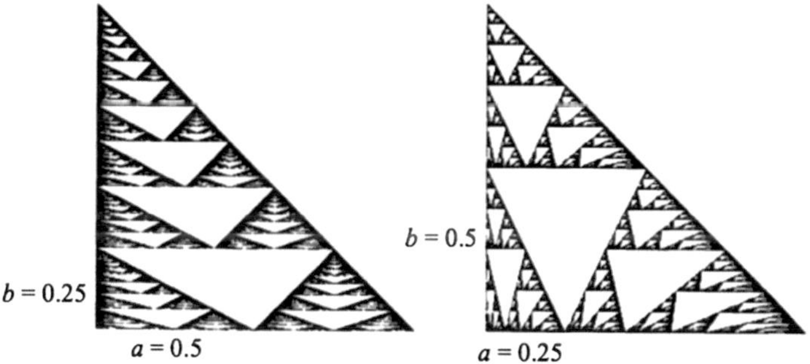

Fig. 4 Affine Sierpiński triangles for different a and b

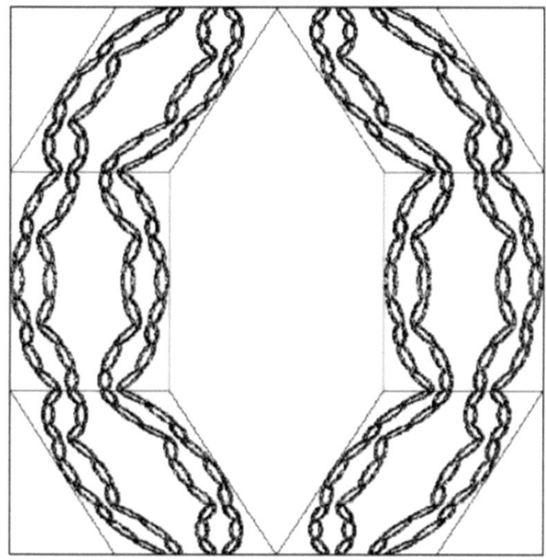

Fig. 5 A self-affine set with affine transformations given by upper triangular matrices

Kakeya-type affine sets equals the affinity dimension, even if overlapping occurs in the construction.

Situations which guarantee that the Hausdorff dimension is given by the affinity dimension are harder to identify. However, Hueter and Lalley [34] have shown that this is the case for certain affine IFSs where the inverses of the linear parts map a quadrant of the plane into itself.

Theorem 5. *Suppose that* $S_i = T_i + \omega_i : \mathbb{R}^2 \to \mathbb{R}^2$ $(i = 1, \ldots, m)$ *satisfy:*

(a) $\alpha_1(T_i)^2 < \alpha_2(T_i) \leq \alpha_1(T_i)$ *for all i.*
(b) $T_i^{-1}(Q \setminus \{0\}) \subseteq \mathrm{int}Q$ *for all i where Q is the second quadrant of the plane.*
(c) The S_i *satisfy the strong separation condition, that is, the images* $S_i(E)$ *are disjoint.*

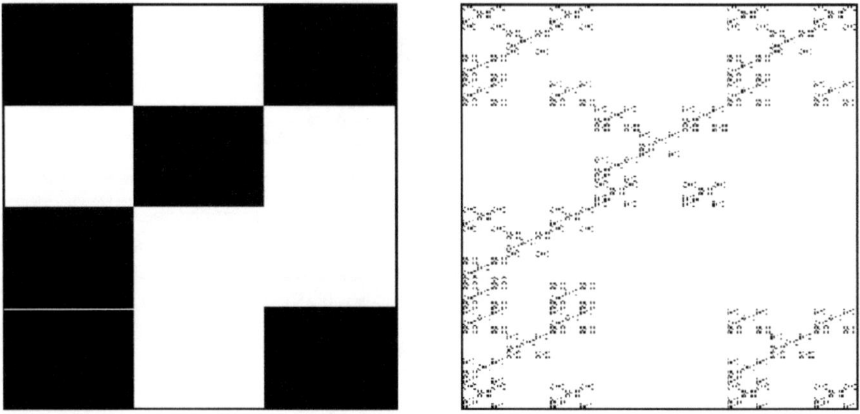

Fig. 6 A Bedford–McMullen carpet with its template

Then

$$\dim_{\mathrm{H}} E = \dim_{\mathrm{B}} E = \mathrm{d}_{\mathrm{aff}}(T_1, \ldots, T_m).$$

3 Self-affine Carpets

We will call a plane self-affine set a *carpet* if it is the attractor of an affine IFS of maps $S_i = T_i + \omega_i$ such that there is an orthonormal coordinate basis of $\mathbb{R}^2$ such that each T_i maps the positive coordinate axes onto themselves, in other words each T_i has a positive diagonal matrix with respect to this basis. Thus there are coordinate rectangles $R_1, \ldots, R_m$ such that $S_i : [0,1]^2 \to R_i$ are *direct* affine maps, that is, mappings that do not involve rotation or reflection. We term such $R_1, \ldots, R_m$ a "template" for the self-affine set, which can be constructed by repeated substitution of the template in itself. Carpets often have dimensions strictly less than the affinity dimension, often with different Hausdorff and box dimensions.

3.1 *Bedford–McMullen Carpets*

The first carpets were analysed independently by Bedford [12] and McMullen [46]. Here the unit square is divided into $p \times q$ equal rectangles of sides $\frac{1}{p} \times \frac{1}{q}$, where $2 \le p < q$. A subcollection $R_1, \ldots, R_m$ of these rectangles is selected for the template, with the IFS consisting of direct affine mappings $S_i : [0,1]^2 \to R_i$ (see Fig. 6).

Suppose that there are N_j-selected rectangles in the jth column for $j = 1, 2, \ldots, p$ and p_1 of the columns contain at least one selected rectangle. Then

$$\dim_{\mathrm{H}} E = \frac{1}{\log p} \log\Big(\sum_{j=1}^{p} N_j^{\log p/\log q}\Big),$$

$$\dim_{\mathrm{B}} E = \frac{\log p_1}{\log p} + \log\Big(\frac{1}{p_1}\sum_{j=1}^{p} N_j\Big)\frac{1}{\log q}.$$

Note that the dimensions depend on the positions of the rectangles selected, and $\dim_{\mathrm{H}} E$ and $\dim_{\mathrm{B}} E$ differ from each other and from the affinity dimension unless the same number of rectangles are selected in every column. These formulae were extended to higher dimensions, where the sets are called "self-affine sponges", by Kenyon and Peres [38].

3.2 *Other Carpets*

A detailed analysis of carpets base on two congruent rectangles was given by Pollicott and Weiss [51]. Gatzouras and Lalley [32] found the Hausdorff and box dimensions of a generalization of Bedford–McMullen carpets. Here the template is based on nonoverlapping columns with each column containing several rectangles each with width at least equal to the height (see Fig. 7).

A further generalization was given by Barański [1] who found the Hausdorff and box dimensions of "aligned rectangle" constructions. In this case the unit square is divided into rectangles by (finite) sets of parallel horizontal and vertical lines, not necessarily equally spaced. The template comprises a set of rectangles $R_1,\dots,R_m$ selected from the resulting grid.

Another variant was provided by Feng and Wang [30] where the IFS is given by direct affine maps of the unit square onto arbitrary coordinate rectangles $R_1,\dots,R_m$ with disjoint interiors (or more generally satisfy a "rectangular open set condition"). In this case the box dimension of the attractor E is given in terms of the dimensions of projections of E onto the two coordinate axes, though in general these are not always easy to calculate explicitly.

In general the formulae for the dimensions of these carpets assume that the rectangles R_i do not have overlapping interiors. However, Shmerkin [54] considers carpets where overlapping is permitted, obtaining an expression that gives the Hausdorff and box dimensions of almost all self-affine sets in certain parameterized families.

3.3 *Box-Like Sets*

"Box-like sets", though not strictly carpets, were considered by Fraser [31]. As before, $R_1,\dots,R_k$ are coordinate rectangles with disjoint interiors in the unit square, and $S_i : [0,1]^2 \to R_i$ are affine mappings. Here, however, the S_i may incorporate any of the symmetries of the unit square before scaling, that is, rotations of $90°, 180°$ and $270°$ and reflections about horizontal, vertical and diagonal axes (Fig. 8).

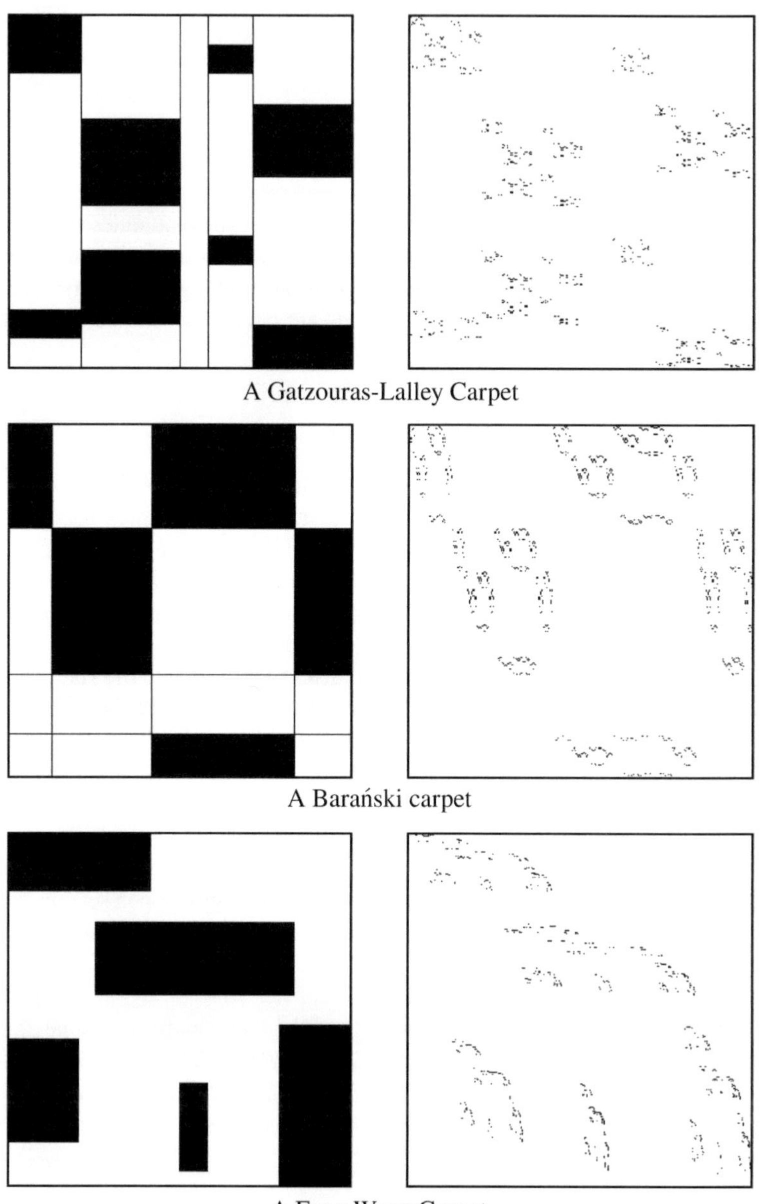

A Gatzouras-Lalley Carpet

A Barański carpet

A Feng-Wang Carpet

Fig. 7 Varieties of carpet with their templates

Assume that at least one $S_i = T_i + \omega_i$ involves a 90° or 270° rotation or a diagonal reflection prior to coordinate scalings. Let $p = \dim_B \mathrm{proj}E$ where proj denotes projection onto one of the principal axes (the rotation or reflection ensures that the

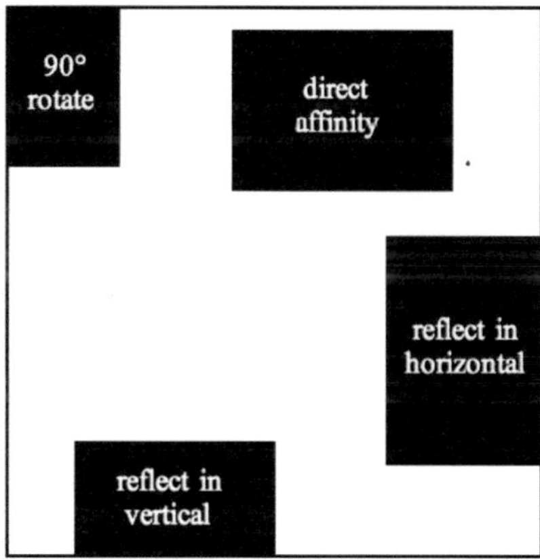

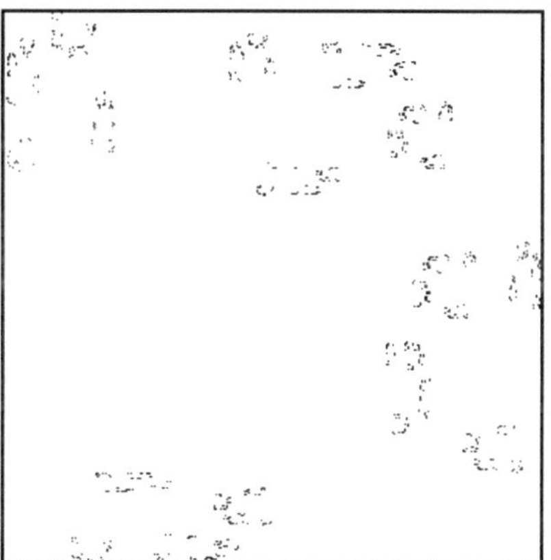

Fig. 8 A box-like set with its template

dimension of the projection is the same for both axes). Define a *modified singular value function* by setting

$$\phi_0^s(T) = \alpha_1^p \alpha_2^{s-p},$$

where $\alpha_1 \geq \alpha_2$ are the singular values of T, and similarly to before let

$$\Phi_0^s = \lim_{k\to\infty} \Big(\sum_{i_1 \ldots i_k} \phi_0^s(T_{i_1} \circ \cdots \circ T_{i_k}) \Big)^{1/k}.$$

Then $\dim_B \mathrm{proj} E = s$ where s is the unique positive solution of $\Phi_0^s = 1$. To apply this formula one needs to know $\dim_B \mathrm{proj} E$, though if the projections onto the axes have positive length or dimension 1 then s is just the affinity dimension. The projections may be represented as a pair of graph-directed self-similar sets, which under certain special conditions satisfy an open set condition in which case their dimension is given by the spectral radius of a certain matrix.

4 Self-affine Functions

Dimensions of graphs of (usually continuous) self-affine functions have also been well studied. One basic model is an IFS on a strip in the coordinate plane $D := \{(x,y) : 0 \leq x \leq 1\}$ with the IFS consisting of affine maps $S_1, \ldots, S_m$ of the form

$$S_i(x,y) = (x/m + (i-1)/m, a_i x + c_i y + b_i) \quad i = 1, \ldots, m,$$

where $m \geq 2$ is an integer and $a_i, c_i, b_i \in \mathbb{R}$. Thus the S_i preserve vertical lines. We assume that $1/m < c_i < 1$ so that the S_i contract more in the x-direction than in the y-direction. Writing $p_1 = (0, b_1/(1-c_1))$ and $p_m = (1, (a_m + b_m)/(1-c_m))$

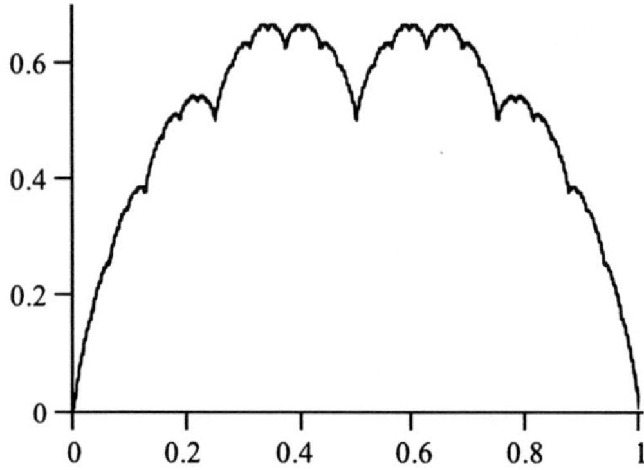

Fig. 9 The Takagi or "Blancmange" function

for the fixed points of S_1 and S_m, we also assume that $S_i(p_m) = S_{i+1}(p_1)$ for $i = 1,\dots,m-1$, which ensures that the IFS attractor is the graph of a continuous function. Then, except for the degenerate case when $S_1(p_1),\dots,S_m(p_1),p_m$ are collinear, the graph has box dimension $1+\log(c_1+\cdots+c_m)$ which is just the affinity dimension in the case of triangular matrices. There are many variants on this model, for example with varying increment lengths in the x-direction. Earlier papers on dimensions of self-affine functions [3, 13, 42] concentrated on the box dimensions of the graphs, but the Hausdorff dimension has since been addressed in various ways, with conditions given for the Hausdorff and box dimensions to be equal [14, 56, 57]. Ledrappier [43] found the Hausdorff dimension of the graph of the Takagi "Blancmange" function, given by $\sum_{n=0}^{\infty} 2^{-n}h(2^n x)$, where $h(y)$ is the distance of y from the nearest integer (see Fig. 9).

Barnsley (see [3–5]) proposed using self-affine functions for "fractal interpolation", that is, finding a function that takes prescribed values at certain points, with the function required to have a suitable degree of irregularity, which can be interpreted as the graph having a specified dimension. There is now a very considerable literature on fractal interpolation, its variants and applications.

5 Related Topics

We have restricted attention above to self-affine sets, but there are many natural generalizations and extensions, some of which we mention briefly here.

5.1 Multifractal Analysis of Measures on Self-affine Sets

Dimension questions for fractal sets frequently have multifractal analogues. Thus given a probability measure μ with support on one of the self-affine sets constructed above, it is natural to consider the generalized q-dimensions and multifractal spectra of μ (see [24, 50]). Cases of particular interest are where μ is the projection of a Bernoulli or Gibbs measure on $\{1,2,\ldots,m\}^{\mathbb{N}}$ under the map (2).

For $q > 0$ there is a natural generalization of Eq. (8) suited to multifractal analysis given by

$$\Phi_q^s = \lim_{k\to\infty}\Big(\sum_{i_1\ldots i_k}\phi^s(T_{i_1}\circ T_{i_2}\circ\cdots\circ T_{i_k})^{1-q}\mu(C_{i_1,i_2,\ldots,i_k})^q\Big)^{1/k},$$

where $C_{\mathbf{i}} = \{\mathbf{ij} : \mathbf{j}\in\{1,2,\ldots,m\}^{\mathbb{N}}\}$ are the cylinder sets. The generalized q-dimensions of μ are bounded above by s_q, the solution of $\Phi_q^{s_q} = 1$.

Under certain conditions the generalized q-dimension equals s_q. This is the case for Bernoulli measures on almost all self-affine sets in the sense of Theorem 1 if $\|T_i\| < \frac{1}{2}$ and $1 < q \leq 2$, see [23], and for Bernoulli measures on almost all almost self-affine sets if $q > 1$, see [25]. Recently results on fine multifractal analysis of such measures on self-affine sets have been obtained in certain cases [6].

There is a substantial literature on multifractal analysis of Bernoulli measures on self-affine carpets. King [39] gave a detailed multifractal analysis for measures on Bedford–McMullen carpets, and this was extended to Gibbs measures by Barral and Mensi [7] and to higher dimensional self-affine sponges by Olsen [47, 48]. However, all these analyses required a very strong separation condition between the rectangles of the carpets (or blocks of the sponges), but this requirement was dispensed with recently, at least in the plane case, by Jordan and Rams [37].

The work on the dimension of Feng–Wang carpets [30] also includes an analysis of measures on these carpets.

5.2 Nonlinear Analogues

Another major extension is to attractors (1) of IFSs where the S_i are nonlinear mappings. These occur in the context of dynamical systems where the S_i may be branches of the inverse of some mapping f of a domain, with the IFS attractor being a hyperbolic dynamical repeller of f. Assuming that the S_i are differentiable, such mappings may be regarded as "locally affine".

Self-conformal sets, that is, where the derivatives of the contractions S_i are similarity mappings, are fairly well understood. The thermodynamic formalism enables the self-similar theory to be extended to self-conformal sets, with Bowen's pressure formula giving the Hausdorff and box dimension of the attractors (see, e.g., [15, 22, 50]). The associated dynamical repellers include many Julia sets of complex dynamics.

For nonconformal IFSs and repellers of nonconformal dynamical systems, the theory is far from complete. It is often possible to get upper bounds (in the plane case) by iterating a cover of discs to get a cover by ellipses which can be cut up into roughly round pieces, a technique that goes back to Douady and Oesterlé [17] (see also [22, 55]), but finding exact formulae is difficult.

A subadditive version of the thermodynamic formalism (see the book by Barreira [10]) enables one to define a pressure-type expression Φ^s involving singular value functions of the derivatives at fixed points of the iterated mappings such that the solution of $\Phi^s = 1$ might be a good candidate for the dimension. This was used in [21] to obtain a nonlinear analogue of Theorem 4 for the box dimension in the nonconformal case under a "1-bunched" or "bounded distortion" condition. Examples involving "triangular maps" [45] showed this condition to be necessary. For the case of a product of expanding maps, see [33]. Luzia [44] gave a nonlinear version of Theorem 5 which applies to both box-counting and Hausdorff dimensions. Other estimates for dimensions of nonlinear attractors or repellers are given in [8–10] and there are two recent surveys [11, 16] containing many further references.

There is a substantial body of literature on the dimension of functional attractors or repellers of differential equations; such estimates are important in applications in estimating the extent of chaotic regimes and for applying embedding results (see the books [52, 55]).

References

1. Barański, K.: Hausdorff dimension of the limit sets of some planar geometric constructions. Adv. Math. **210**, 215–245 (2007)
2. Barański, K.: Hausdorff dimension of self-affine limit sets with an invariant direction. Discrete Contin. Dyn. Syst. **21**, 1015–1023 (2008)
3. Barnsley, M.F.: Fractal functions and interpolation. Constr. Approx. **2**, 303–329 (1986)
4. Barnsley, M.F.: Fractals Everywhere, 2nd edn. Academic Press Professional, Boston (1993)
5. Barnsley, M.F., Harrington, A.N.: The calculus of fractal interpolation functions. J. Approx. Theory **57**, 14–34 (1989)
6. Barral, J., Feng, D.-J.: Multifractal formalism for almost all self-affine measures, to appear, Comm. Math. Phys.
7. Barral, J., Mensi, M.: Gibbs measures on self-affine Sierpiński carpets and their singularity spectrum. Ergodic Theory Dynam. Syst. **27**, 1419–1443 (2007)
8. Barreira, L: A non-additive thermodynamic formalism and applications to dimension theory of hyperbolic dynamical systems. Ergodic Theory Dynam. Syst. **16**, 871–927 (1996)
9. Barreira, L: Dimension estimates in nonconformal hyperbolic dynamics. Nonlinearity **16**, 1657–1672 (2003)
10. Barreira, L.: Thermodynamic Formalism and Applications to Dimension Theory. Birkhäuser, Basel (2011)
11. Barreira, L., Gelfert, K.: Dimension estimates in smooth dynamics: a survey of recent results. Ergodic Theory Dynam. Syst. **31**, 641–671 (2011)
12. Bedford, T.: Crinkly curves, Markov partitions and box dimensions in self-similar sets. PhD thesis, University of Warwick (1984)
13. Bedford, T.: The box dimension of self-affine graphs and repellers. Nonlinearity **2**, 53–71 (1989)
14. Bedford, T., Urbański, M.: The box and Hausdorff dimension of self-affine sets. Ergodic Theory Dynam. Syst. **10**, 627–644 (1990)

15. Bowen, R.: Hausdorff dimension of quasi-circles. Publ. Math. IHES **50**, 11–26 (1979)
16. Chen, J., Pesin, Y.: Dimension of non-conformal repellers: a survey. Nonlinearity **23**, R93–R114 (2010)
17. Douady, A., Oesterlé, J.: Dimension de Hausdorff des attracteurs. C. R. Acad. Sci. Paris Sr. A **290**, 1135–1138 (1980)
18. Edgar, G.A.: Fractal dimension of self-affine sets: some examples. Rend. Circ. Mat. Palermo (2) Suppl. **28**, 341–358 (1988)
19. Falconer, K.J.: The Hausdorff dimension of self-affine fractals. Math. Proc. Cambridge Philos. Soc. **103**, 339–350 (1988)
20. Falconer, K.J.: The dimension of self-affine fractals II. Math. Proc. Cambridge Philos. Soc. **111**, 169–179 (1992)
21. Falconer, K.J.: Bounded distortion and dimension for non-conformal repellers. Math. Proc. Cambridge Philos. Soc. **115**, 315–334 (1994)
22. Falconer, K.J.: Techniques in Fractal Geometry. Wiley, Chichester (1997)
23. Falconer, K.J.: Generalized dimensions of measures on self-affine sets. Nonlinearity **12**, 877–891 (1999)
24. Falconer, K.J.: Fractal Geometry—Mathematical Foundations and Applications, 2nd edn. Wiley, Chichester (2003)
25. Falconer, K.J.: Generalised dimensions of measures on almost self-affine sets. Nonlinearity **23**, 1047–1069 (2010)
26. Falconer, K.J., Lammering, B.: Fractal properties of general Sirepiński triangles. Fractals **6**, 31–41 (1998)
27. Falconer, K.J., Miao, J.: Dimensions of self-affine fractals and multifractals generated by upper-triangular matrices. Fractals **15**, 289–299 (2007)
28. Falconer, K.J., Miao, J.: Exceptional sets for self-affine fractals. Math. Proc. Cambridge Philos. Soc. **145**, 669–684 (2008)
29. Falconer, K.J., Sloan, A.: Continuity of subadditive pressure for self-affine sets. R. Anal. Exchange **34**, 413–428 (2009)
30. Feng, D.-J., Wang, Y.: A class of self-affine sets and self-affine measures. J. Fourier Anal. Appl. **11**, 107–124 (2005)
31. Fraser, J.M.: On the packing dimension of box-like self-affine sets in the plane, Nonlinearity **25**, 2075–2092 (2012)
32. Gatzouras, D., Lalley, S.P.: Hausdorff and box dimensions of certain self-affine fractals. Indiana Univ. Math. J. **41**, 533–568 (1992)
33. Gatzouras, D., Peres, Y.: Invariant measures of full dimension for some expanding maps. Ergodic Theory Dynam. Syst. **17**, 147–67 (1997)
34. Hueter, I., Lalley, S.P.: Falconer's formula for the Hausdorff dimension of a self-affine set in $\mathbb{R}^2$. Ergodic Theory Dynam. Syst. **15**, 77–97 (1995)
35. Hutchinson, J.E.: Fractals and self-similarity. Indiana Univ. Math. J. **30**, 713–747 (1981)
36. Jordan, T., Pollicott, M., Simon, K.: Hausdorff dimension for randomly perturbed self affine attractors. Commun. Math. Phys. **270**, 519–544 (2007)
37. Jordan, T., Rams, M.: Multifractal analysis for Bedford-McMullen carpets. Math. Proc. Cambridge Philos. Soc. **150**, 147–156 (2011)
38. Kenyon, R., Peres, Y.: Hausdorff dimensions of sofic affine-invariant sets. Israel J. Math. **94**, 157–178 (1996)
39. King, J.F.: The singularity spectrum for general Sierpiński carpets. Adv. Math. **116**, 1–11 (1995)
40. Käenmäki, A., Shmerkin, P.: Overlapping self-affine sets of Kakeya type. Ergodic Theory Dynam. Syst. **29**, 941–965 (2009)
41. Käenmäki, A., Vilppolainen, M.: Dimension and measures on sub-self-affine sets. Monatsh. Math. **161**, 271–293 (2010)
42. Kono, N.: On self-affine functions. Japan J. Appl. Math. **3**, 259–269 (1986)
43. Ledrappier, F.: On the dimension of some graphs. In: Symbolic Dynamics and Its Applications. Contemp. Math. vol. 135, pp. 285–293. Amer. Math. Soc., Providence (1992)

44. Luzia, N.: Hausdorff dimension for an open class of repellers in $\mathbb{R}^2$. Nonlinearity **19**, 2895–2908 (2006)
45. Manning, A., Simon, K.: Subadditive pressure for triangular maps. Nonlinearity **20**, 133–149 (2007)
46. McMullen, C.: The Hausdorff dimension of general Sierpiński carpets. Nagoya Math. J. **96**, 1–9 (1984)
47. Olsen, L.: Self-affine multifractal Sierpiński sponges in $\mathbb{R}^d$. Pacific J. Math. **183**, 143–199 (1998)
48. Olsen, L.: Symbolic and geometric local dimensions of self-affine multifractal Sierpiński sponges in $\mathbb{R}^d$. Stoch. Dyn. **7**, 37–51 (2007)
49. Peres, Y., Solomyak, B.: Problems on self-similar sets and self-affine sets: an update. In: Bandt, C., Graf, S., Zähle, M. (eds.) Fractal Geometry and Stochastics II. Progress in Probability, vol. 46, pp. 95–106. Birkhäuser, Basel (2000)
50. Pesin, Y.B.: Dimension Theory in Dynamical Systems. University of Chicago Press, Chicago (1997)
51. Pollicott, M., Weiss, H.: The dimensions of some self-affine limit sets in the plane. J. Stat. Phys. **77** (1994), 841–866.
52. Robinson, J.C.: Dimensions, Embeddings, and Attractors. Cambridge Tracts in Mathematics, vol. 186. Cambridge University Press, Cambridge (2011)
53. Solomyak, B.: Measure and dimensions for some fractal families. Math. Proc. Cambridge Philos. Soc. **124**, 531–546 (1998)
54. Shmerkin, P.: Overlapping self-affine sets. Indiana Univ. Math. J. **55**, 1291–1331 (2006)
55. Temam, R.: Infinite-dimensional dynamical systems in mechanics and physics. Applied Mathematical Sciences, vol. 68, 2nd edn. Springer, New York (1997)
56. Urbański, M.: The probability distribution and Hausdorff dimension of self-affine functions. Probab. Theory Related Fields **84**, 377–391 (1990)
57. Urbański, M.: The Hausdorff dimension of the graphs of continuous self-affine functions. Proc. Amer. Math. Soc. **108**, 921–930 (1990)

The Multifractal Spectra of V-Statistics

Ai-hua Fan, Jörg Schmeling, and Meng Wu

Abstract Let (X,T) be a topological dynamical system and let $\Phi : X^r \to \mathbb{R}$ be a continuous function on the product space $X^r = X \times \cdots \times X$ $(r \geq 1)$. We are interested in the limit of V-statistics taking Φ as kernel:

$$\lim_{n\to\infty} n^{-r} \sum_{1\leq i_1,\cdots,i_r\leq n} \Phi(T^{i_1}x,\cdots,T^{i_r}x).$$

The multifractal spectrum of topological entropy of the above limit is expressed by a variational principle when the system satisfies the specification property. Unlike the classical case $(r = 1)$ where the spectrum is an analytic function when Φ is Hölder continuous, the spectrum of the limit of higher-order V-statistics $(r \geq 2)$ may be discontinuous even for very nice kernel Φ.

A.-h. Fan (✉)
LAMFA, UMR 7352 CNRS, University of Picardie Jules Verne 33, rue Saint Leu, 80039 Amiens Cedex, France
e-mail: ai-hua.fan@u-picardie.fr

J. Schmeling
Mathematics Centre for Mathematical Sciences, Lund Institute of Technology, Lund University Box 118 SE-221 00 Lund, Sweden
e-mail: joerg@maths.lth.se

M. Wu
Laboratoire Amiénois. de Mathématique Fondamentale et Appliquée, UMR 7352 CNRS, University of Picardie, 33 rue Saint Leu, 80039 Amiens Cedex, France
e-mail: meng.wu@u-picardie.fr

J. Barral and S. Seuret (eds.), *Further Developments in Fractals and Related Fields*, Trends in Mathematics, DOI 10.1007/978-0-8176-8400-6_7,

1 Introduction

Consider a topological dynamical system (X,T), where $T:X\to X$ is a continuous transformation on a compact metric space X with metric d. For $r\geq 1$, let $X^r = X\times\cdots\times X$ (product of r copies of X) and let $C(X^r)$ be the space of continuous functions $\Phi : X^r\to\mathbb{R}$.

For $\Phi\in C(X^r)$ and $n\geq 1$, let

$$V_\Phi(n,x) = n^{-r}\sum_{1\leq i_1,\cdots,i_r\leq n}\Phi(T^{i_1}x,\cdots,T^{i_r}x)$$

and $V_\Phi(x) = \lim_{n\to\infty} V_\Phi(n,x)$ if the limit exists. For $\alpha\in\mathbb{R}$, define

$$E_\Phi(\alpha) = \left\{x\in X : \lim_{n\to\infty} V_\Phi(n,x) = \alpha\right\}.$$

The problem treated in the present chapter is to measure the sizes of the sets $E_\Phi(\alpha)$. To measure the sizes of the sets $E_\Phi(\alpha)$, we adopt the notion of topological entropy introduced by Bowen ([8]), denoted by h_{top}. We denote by $\mathcal{M}_{\text{inv}}$ the set of all T-invariant probability Borel measures on X and by $\mathcal{M}_{\text{erg}}$ its subset of all ergodic measures. The measure-theoretic entropy of μ in $\mathcal{M}_{\text{inv}}$ is denoted by h_μ.

For $\mu\in\mathcal{M}_{\text{inv}}$, the set G_μ of μ-*generic points* is defined by

$$G_\mu := \left\{x\in X : \frac{1}{n}\sum_{j=0}^{n-1}\delta_{T^jx}\xrightarrow{w^*}\mu\right\},$$

where $\xrightarrow{w^*}$ stands for the weak star convergence of measures. Bowen ([8]) proved that on any dynamical system, we have $h_{\text{top}}(G_\mu)\leq h_\mu$ for any $\mu\in\mathcal{M}_{\text{inv}}$. For ergodic measure μ, we get equality. But in general, the equality does not hold. A dynamical system (X,T) is said to be *saturated* if for any $\mu\in\mathcal{M}_{\text{inv}}$, we have $h_{\text{top}}(G_\mu) = h_\mu$. It is proved in [13] that systems of specification are saturated.

In this chapter, we shall prove a variational principle which relates the topological entropy $h_{\text{top}}(E_\Phi(\alpha))$ to the measure-theoretic entropies of invariant measures in the following set, called (Φ,α)-fiber,

$$\mathcal{M}_\Phi(\alpha) = \left\{\mu\in\mathcal{M}_{\text{inv}} : \int_{X^r}\Phi \mathrm{d}\mu^{\otimes r} = \alpha\right\},$$

where $\mu^{\otimes r} = \mu\times\cdots\times\mu$ is the product of r copies of μ.

Theorem 1. *Suppose that the dynamical system (X,T) is saturated. Let $\Phi\in C(X^r)$ $(r\geq 1)$. If $\mathcal{M}_\Phi(\alpha)=\emptyset$, we have $E_\Phi(\alpha)=\emptyset$. If $\mathcal{M}_\Phi(\alpha)\neq\emptyset$, we have*

$$h_{\text{top}}(E_\Phi(\alpha)) = \sup_{\mu\in\mathcal{M}_\Phi(\alpha)} h_\mu. \tag{1}$$

Theorem 1 is well known when $r=1$ (see, e.g., [3, 4, 11, 13]). In particular, it is known that for regular potential Φ, $\alpha \mapsto h_{\text{top}}(E_\Phi(\alpha))$ is an analytic function (see, e.g., [10, 19]). But as we shall see, when $r \geq 2$, this function can admit discontinuity even for "very regular" potentials.

The above consideration was motivated by the following problem. Recently the multiple ergodic limit

$$M_\Phi(x) := \lim_{n\to\infty} \frac{1}{n} \sum_{i=0}^{n-1} \Phi(\sigma^i x, \sigma^{2i} x, \cdots, T^{ri} x) \tag{2}$$

has been studied by Furstenberg ([15]), Bergelson ([5]), Bourgain ([7]), Assani ([2]), Host and Kra ([16]), and others. Fan, Liao, and Ma proposed in [12] to give a multifractal analysis of the multiple ergodic average M_Φ, in other words, to determine the Hausdorff dimensions of the level sets

$$L_\Phi(\alpha) = \{x \in X : M_\Phi(x) = \alpha\}.$$

This problem in its generality remains open.

However, there are two results for the shift dynamics on symbolic space and for some special potentials Φ. The first one concerns the case where $X = \{-1,1\}^{\mathbb{N}}$, T is the shift, and $\Phi(x_1,\cdots,x_r) = x_1^{(1)}\cdots x_r^{(1)}$ ($x_i^{(1)}$ being the first coordinate of x_i). By using Riesz products, the authors in [12] proved that for $\alpha \in [-1,1]$ we have

$$\dim L_\Phi(\alpha) = 1 - \frac{1}{r} - \frac{1}{r}\left(\frac{1-\alpha}{2}\log_2\frac{1-\alpha}{2} + \frac{1+\alpha}{2}\log_2\frac{1+\alpha}{2}\right).$$

The second one concerns the case where $X = \{0,1\}^{\mathbb{N}}$, T is the shift, and $\Phi(x_1,x_2) = F(x_1^{(1)},x_2^{(1)})$ is a function depending only on the first coordinates $x_1^{(1)}$ and $x_2^{(1)}$ of x_1 and x_2. The multifractal analysis of these double ergodic average was determined in [14]. A related work was done in [17] answering a question in [12] about the Hausdorff dimension of a subset of $L_\Phi(\alpha)$ for extremal values of α.

As shown in [14], the dimension of the "mixing part" of $L_\Phi(\alpha)$ which is defined by

$$\sup\{\dim\mu : \mu(L_\Phi(\alpha)) = 1,\ \mu \text{ is mixing}\}$$

is equal to

$$\sup\left\{\dim\mu : \int \Phi \mathrm{d}\mu^{\otimes r} = \alpha,\ \mu \text{ is mixing}\right\}.$$

This equality is very similar to the variational principle stated in Theorem 1.

In Sect. 2, we recall some facts about V-statistics. In Sect. 3, we recall some notions like topological entropy, generic points, and specification property. The main theorem, Theorem 1, is proved in Sect. 4. In Sect. 5, we examine the special case of full shift together with some examples. We will see that, even for very regular function Φ, the function $\alpha \to h_{\text{top}}(L_\Phi(\alpha))$ may admit discontinuity.

To finish this introduction, we emphasize that the problem of multifractal analysis of multiple ergodic limits remains largely open.

2 V-Statistics

V-statistics are tightly related to U-statistics which are well known in statistics. Let μ be a probability law on $\mathbb{R}$. A U-parameter of μ is defined through a function called kernel $h : \mathbb{R}^d \to \mathbb{R}$ by

$$\theta(\mu) = \theta_h(\mu) = \int_{\mathbb{R}^d} h d\mu^{\otimes d},$$

where $\mu^{\otimes d}$ is the product measure $\mu \times \cdots \times \mu$ (d times) on $\mathbb{R}^d$. This U-statistics is well defined for all μ such that the integral exists.

In statistics, U-parameters are also called estimable parameters and they constitute the set of all parameters that can be estimated in an unbiased fashion. A fundamental problem in statistics is the estimation of a parameter $\theta(\mu)$ for an unknown probability law μ. To estimate a U-parameter θ_h, people employ the U-statistics for θ_h:

$$U_h(X_1, \cdots, X_n) = \frac{(n-d)!}{n!} \sum h(X_{i_1}, \cdots, X_{i_d})$$

where the sum is taken over all $(i_1, \cdots, i_d)$ with i_j's distinct and $1 \le i_j \le n$, where $X_1, \cdots, X_d$ is a sequence of observations of μ. Closely related to U-statistics is the V-statistics (von Mises statistics):

$$V_h(X_1, \cdots, X_n) = n^{-d} \sum_{1 \le i_1, \cdots, i_d \le n} h(X_{i_1}, \cdots, X_{i_d}).$$

People expect that $U_h(X_1, \cdots, X_n)$ converges almost surely to θ_h. This fact, if it holds, allows one to estimate θ_h using observations. If it is the case, we say the U-parameter strong law of large numbers (SLLN) holds. The U-statistics SLLN had been well studied for independent observations. In [1], the authors have studied the U-statistics SLLN for ergodic stationary process (X_n), i.e., $X_n = f \circ T^n$ where T is ergodic measure-preserving transformation on a probability space $(\Omega, \mathscr{A}, \mathbb{P})$, $f : \Omega \to \mathbb{R}$ is a measurable function, and X_1 admits μ as probability law.

If h is a kernel bounded by a integrable function and if (X_n) is ergodic, it can be proved (see [1]) that almost surely

$$\lim_{n \to \infty} |U_h(X_1, \cdots, X_n) - V_h(X_1, \cdots, X_n)| = 0.$$

It is also proved in [1] that the U-statistics SLLN holds if the kernel h is continuous. In the following, we consider only V-statistics.

3 Topological Entropy

For any integer $n \geq 1$, the Bowen metric d_n on X is defined by

$$d_n(x,y) = \max_{0 \leq j < n} d(T^j x, T^j y).$$

For any $\varepsilon > 0$, we will denote by $B_n(x,\varepsilon)$ the open d_n-ball centered at x of radius ε.

Let $Z \subset X$ be a subset of X. Let $\varepsilon > 0$. A cover is a collection of Bowen balls (at most countable) $R = \{B_{n_i}(x_i,\varepsilon)\}$ such that $Z \subset \bigcup_i B_{n_i}(x_i,\varepsilon)$. For such a cover R, we put $n(R) = \min_i n_i$. Let $s \geq 0$. Define

$$H_n^s(Z,\varepsilon) = \inf_R \sum_i \exp(-sn_i),$$

where the infimum is taken over all covers R of Z with $n(R) \geq n$. The quantity $H_n^s(Z,\varepsilon)$ being a nondecreasing function of n, the following limit exists:

$$H^s(Z,\varepsilon) = \lim_{n \to \infty} H_n^s(Z,\varepsilon).$$

Considering the quantity $H^s(Z,\varepsilon)$ as a function of s, there exists a critical value, which we denote by $h_{\text{top}}(Z,\varepsilon)$, such that

$$H^s(Z,\varepsilon) = \begin{cases} +\infty, & s < h_{\text{top}}(Z,\varepsilon) \\ 0, & s > h_{\text{top}}(Z,\varepsilon). \end{cases}$$

The following limit exists:

$$h_{\text{top}}(Z) = \lim_{\varepsilon \to 0} h_{\text{top}}(Z,\varepsilon).$$

The limit $h_{\text{top}}(Z)$ is called the *topological entropy* of Z ([8]).

For $x \in X$, we denote by $V(x)$ the set of all weak limits of the sequence of probability measures $n^{-1}\sum_{j=0}^{n-1} \delta_{T^j x}$. Recall that X is compact. It is clear then that for any x we have

$$\emptyset \neq V(x) \subset \mathscr{M}_{\text{inv}}.$$

The following lemma is due to Bowen ([8]).

Lemma 1. *For $t \geq 0$, we have $h_{\text{top}}(B^{(t)}) \leq t$ where*

$$B^{(t)} = \left\{x \in X : \exists\, \mu \in V(x) \text{ satisfying } h_\mu \leq t\right\}.$$

The set G_μ of μ-generic points is the set of all x such that $V(x) = \{\mu\}$. The Bowen lemma implies that

$$h_{\text{top}}(G_\mu) \leq h_\mu$$

for any invariant measure μ. It is simply because $x \in G_\mu$ implies $\mu \in V(x)$. Bowen also proved that the inequality becomes equality when μ is ergodic. However, in general, we do not have the equality and it is even possible that $G_\mu = \emptyset$. In fact, $\mu(G_\mu) = 1$ or 0 according to whether μ is ergodic or not (see [9]).

The equality $h_{\text{top}}(G_\mu) = h_\mu$ does hold for any invariant probability measure in any dynamical system with specification [13].

Lemma 2. *Any dynamical system with specification (X,T) is saturated. In other words, $h_{\text{top}}(G_\mu) = h_\mu$ for any $\mu \in \mathscr{M}_{\text{inv}}$.*

A dynamical system (X,T) is said to satisfy the *specification property* if for any $\varepsilon > 0$ there exists an integer $m(\varepsilon) \geq 1$ having the property that for any integer $k \geq 2$, for any k points $x_1, \ldots, x_k$ in X, and for any integers

$$a_1 \leq b_1 < a_2 \leq b_2 < \cdots < a_k \leq b_k$$

with $a_i - b_{i-1} \geq m(\varepsilon) \quad (\forall 2 \leq i \leq k)$, there exists a point $y \in X$ such that

$$d(T^{a_i+n}y, T^n x_i) < \varepsilon \qquad (\forall\, 0 \leq n \leq b_i - a_i, \quad \forall 1 \leq i \leq k).$$

The specification property implies the topological mixing. Blokh ([6]) proved that these two properties are equivalent for continuous interval transformations. Mixing subshifts of finite type satisfy the specification property. In general, a subshift satisfies the specification if for any admissible words u and v there exists a word w with $|w| \leq k$ (some constant k) such that uwv is admissible. For β-shifts defined by $T_\beta x = \beta x (\mod 1)$, there is only a countable number of β's such that the β-shifts admit Markov partition (i.e., subshifts of finite type), but an uncountable number of β's such that the β-shifts satisfy the specification property [20].

We finish this section by mentioning that continuous functions on X^r can be uniformly approximated by tensor functions. It is a consequence of the Stone-Weierstrass theorem.

Lemma 3. *Let $F \in C(X^r)$. For any $\varepsilon > 0$, there exists a function of the form*

$$\widetilde{F}(x_1, \cdots, x_r) = \sum_{j=1}^{n} f_j^{(1)}(x_1) f_j^{(2)}(x_2) \cdots f_j^{(r)}(x_r),$$

where $f_j^{(i)} \in C(X)$, such that $\|F - \widetilde{F}\|_\infty < \varepsilon$.

We will write

$$\widetilde{F} = \sum_{j=1}^{n} f_j^{(1)} \otimes f_j^{(2)} \otimes \cdots \otimes f_j^{(r)}.$$

4 Proof of Theorem 1

We can actually consider Banach-valued V-statistics. More than Theorem 1 can be proved.

Let $\mathbb{B}$ be a real Banach space and $\mathbb{B}^*$ its dual space. The duality will be denoted by $\langle y,x\rangle$ ($x\in\mathbb{B}, y\in\mathbb{B}^*$). We consider $\mathbb{B}^*$ as a locally convex topological space with the weak star topology $\sigma(\mathbb{B}^*,\mathbb{B})$. For any $\mathbb{B}^*$-valued continuous function $\Phi : X\to\mathbb{B}^*$, we consider its V-statistics $V_\Phi(n,x)$ as before, formally in the same way.

Fix a subset $W\subset\mathbb{B}$. For a sequence $\{\xi_n\}\subset\mathbb{B}^*$ and a point $\xi\in\mathbb{B}^*$, we denote by $\limsup_{n\to\infty}\xi_n \overset{W}{\leq} \xi$ the fact

$$\limsup_{n\to\infty}\langle \xi_n, w\rangle \leq \langle \xi, w\rangle \ \text{ for all } w\in W.$$

It is clear that $\limsup_{n\to\infty}\xi_n \overset{\mathbb{B}}{\leq} \xi$ means ξ_n converges to ξ in the weak star topology $\sigma(\mathbb{B}^*,\mathbb{B})$.

Given $\alpha\in\mathbb{B}^*$ and $W\subset\mathbb{B}$. We define

$$E_\Phi(\alpha,W) = \left\{x\in X : \limsup_{n\to\infty} V_\Phi(n,x) \overset{W}{\leq} \alpha\right\}$$

$$\mathscr{M}_\Phi(\alpha,W) = \left\{\mu\in\mathscr{M}_{\text{inv}} : \int \Phi d\mu \overset{W}{\leq} \alpha\right\},$$

where $\int\Phi \mathrm{d}\mu$ denotes the vector-valued integral in Pettis' sense (see [18]) and the inequality "$\overset{W}{\leq}$" means $\int\langle\Phi,w\rangle \mathrm{d}\mu \leq \langle\alpha,w\rangle$ for all $w\in W$.

Theorem 2. *Suppose that the dynamical system (X,T) is saturated. If $\mathscr{M}_\Phi(\alpha,W)=\emptyset$, we have $E_\Phi(\alpha,W)=\emptyset$. If $\mathscr{M}_\Phi(\alpha,W)\neq\emptyset$, we have*

$$h_{\text{top}}(E_\Phi(\alpha,W)) = \sup_{\mu\in\mathscr{M}_\Phi(\alpha,W)} h_\mu. \tag{3}$$

Proof. We prove the first assertion by showing that $E_\Phi(\alpha,W)\neq\emptyset$ implies $\mathscr{M}_\Phi(\alpha,W)\neq\emptyset$. Let $x\in E_\Phi(\alpha,W)$. There exists a measure $\mu\in V(x)\subset\mathscr{M}_{\text{inv}}$ and a sequence of integers (n_k) such that

$$\mu = w^* - \lim_{k\to\infty}\frac{1}{n_k}\sum_{j=1}^{n_k}\delta_{T^jx}. \tag{4}$$

We are going to show that $\mu\in\mathscr{M}_\Phi(\alpha,W)$.

Let $w\in W$. Then $\langle\Phi,w\rangle$ is a continuous function on X. For an arbitrarily small $\varepsilon>0$, by the Stone-Weierstrass theorem (see Lemma3), there exists a function $\widetilde{\Phi}$ of the form

$$\widetilde{\Phi} = \sum_j f_j^{(1)} \otimes f_j^{(2)} \otimes \cdots \otimes f_j^{(r)}$$

(finite sum of tensor products) such that

$$\|\langle \Phi, w\rangle - \widetilde{\Phi}\|_\infty \le \varepsilon.$$

Notice that

$$V_{\widetilde{\Phi}}(n,x) = \sum_j \prod_{i=1}^r \frac{S_n f_j^{(i)}(x)}{n},$$

where

$$S_n f(x) = \sum_{k=1}^n f(T^k x)$$

denotes the ergodic sum for a given function f. According to Eq. (4), we have

$$\lim_{k\to\infty} V_{\widetilde{\Phi}}(n_k,x) = \sum_j \prod_{i=1}^r \int_X f_j^{(i)} \mathrm{d}\mu = \int_{X^r} \widetilde{\Phi} \mathrm{d}\mu^{\otimes d}. \tag{5}$$

On the other hand, we write

$$\int \langle \Phi, w\rangle \mathrm{d}\mu^{\otimes d} - \langle \alpha, w\rangle = \sigma_1 + \sigma_2 + \sigma_3 + \sigma_4,$$

where

$$\begin{aligned}
\sigma_1 &= \int (\langle \Phi, w\rangle - \widetilde{\Phi}) \mathrm{d}\mu^{\otimes d}, \\
\sigma_2 &= \int \widetilde{\Phi} \mathrm{d}\mu^{\otimes d} - V_{\widetilde{\Phi}}(n_k,x), \\
\sigma_3 &= V_{\widetilde{\Phi}}(n_k,x) - V_{\langle \Phi, w\rangle}(n_k,x), \\
\sigma_4 &= V_{\langle \Phi, w\rangle}(n_k,x) - \langle \alpha, w\rangle.
\end{aligned}$$

We have

$$|\sigma_1| \le \varepsilon, \quad |\sigma_3| \le \varepsilon, \quad \lim \sigma_2 = 0, \quad \limsup \sigma_4 \le 0.$$

So, we get

$$\int \langle \Phi, w\rangle \mathrm{d}\mu^{\otimes d} \le \langle \alpha, w\rangle + 2\varepsilon.$$

Since ε is arbitrary, we have thus proved that $\mu \in \mathscr{M}_{\text{inv}}(\alpha, W)$. The first assertion is then proved.

Prove now the second assertion. What we have just proved also implies

$$E_\Phi(\alpha, W) \subset B^{(t)} = \{x \in X : \exists \mu \in V(x) \text{ such that } h_\mu \le t\},$$

where $t = \sup_{\mu \in \mathscr{M}_\Phi(\alpha,W)} h_\mu$. By the Bowen lemma (Lemma 1), we get

$$h_{\text{top}}(E_\Phi(\alpha)) \leq \sup_{\mu \in \mathscr{M}_\Phi(\alpha)} h_\mu. \tag{6}$$

To finish the proof of the second assertion, it suffices to prove the reverse inequality of Eq. (6). Let $\mu \in \mathscr{M}_\Phi(\alpha)$. Let $x \in G_\mu$. For any $\varepsilon > 0$ and any $w \in W$, consider $\widetilde{\Phi}$ as above. We have

$$\lim_{n\to\infty} V_{\widetilde{\Phi}}(n,x) = \int_{X^r} \widetilde{\Phi} \mathrm{d}\mu^{\otimes r}.$$

It follows that

$$\begin{aligned}
\limsup_{n\to\infty} V_{\langle \Phi, w\rangle}(n,x) &\leq \lim_{n\to\infty} V_{\widetilde{\Phi}}(n,x) + \varepsilon \\
&= \int_{X^r} \widetilde{\Phi} \mathrm{d}\mu^{\otimes r} + \varepsilon \\
&\leq \int_{X^r} \langle \Phi, w\rangle \mathrm{d}\mu^{\otimes r} + 2\varepsilon \leq \langle \alpha, w\rangle + 2\varepsilon.
\end{aligned}$$

Letting $\varepsilon \to 0$ we get

$$\limsup_{n\to\infty} \langle V_\Phi(n,x), w\rangle \leq \langle \alpha, w\rangle.$$

In other words, we have proved $G_\mu \subset E_\Phi(\alpha, W)$ for all $\mu \in \mathscr{M}_{\text{inv}}(\alpha, W)$. So,

$$h_{\text{top}}(E_\Phi(\alpha)) \geq h_{\text{top}}(G_\mu).$$

By Lemma 2, $h_{\text{top}}(G_\mu) = h_\mu$. Taking the supremum over $\mu \in \mathscr{M}_{\text{inv}}(\alpha, W)$ leads to the reverse inequality of Eq. (6).

5 Example: Shift Dynamics

Let $(X,T) = (\Sigma_m, \sigma)$ with $m \geq 2$, where $\sigma\colon \Sigma_m \to \Sigma_m$ is the shift on the space $\Sigma_m = \{0, 1, \cdots, m-1\}^{\mathbb{N}}$.

Let

$$L_{(\Phi,W)} = \{\alpha \in \mathbb{B}^* : E_\Phi(\alpha, W) \neq \emptyset\}.$$

If $W = \mathbb{B}$, we write $L_\Phi = L_{(\Phi,W)}$. Define $f_{(\Phi,W)} : L_{(\Phi,W)} \to \mathbb{R}$ by

$$f_{(\Phi,W)}(\alpha) = h_{\text{top}}(E_\Phi(\alpha, W)).$$

Theorem 3. *$f_{(\Phi,W)} : L_{(\Phi,W)} \to \mathbb{R}$ is upper semi-continuous.*

Proof. Let $\alpha_n, \alpha \in L_{(\Phi,W)}$. Suppose $\alpha_n \to \alpha$ in the weak star topology. We have to show that

$$\limsup_n f_{(\Phi,W)}(\alpha_n) \leq f_{(\Phi,W)}(\alpha).$$

Since each fiber like $\mathscr{M}_{\mathrm{inv}}(\alpha, W)$ is compact, there are maximizing measures $\mu_{\alpha_n} \in \mathscr{M}_{\mathrm{inv}}(\alpha_n, W)$ and $\mu_\alpha \in \mathscr{M}_{\mathrm{inv}}(\alpha, W)$ such that

$$f_{(\Phi,W)}(\alpha_n) = h_{\alpha_n}, \qquad f_{(\Phi,W)}(\alpha) = h_\alpha. \tag{7}$$

Without loss of generality, we can assume that μ_{α_n} converges weakly, say to μ^*. Since

$$\forall w \in W, \quad \int \langle \Phi, w \rangle \mathrm{d}\mu_n \leq \langle \alpha_n, w \rangle,$$

taking limit shows that $\mu^* \in \mathscr{M}_{\mathrm{inv}}(\alpha, W)$. It follows that

$$h_{\mu^*} \leq h_{\mu_\alpha}. \tag{8}$$

On the other hand, recall that for the shift dynamics, the entropy function $\mu \mapsto h_\mu$ is upper semi-continuous. So,

$$\limsup_n h_{\alpha_n} \leq h_{\mu^*}. \tag{9}$$

We combine Eqs. (7), (8), and (9) to finish the proof.

Theorem 4. *Assume that Φ is a function defined on Σ_m^r $(r \geq 1)$ which depends only on the first k coordinates of each of its variables $(k \geq 1)$. Then the supremum in the variational principle (3) is attained by a $(k-1)$-Markov measure.*

Proof. This is just because the integral $\int \Phi \mathrm{d}\mu^{\otimes r}$ depends only on the values $\mu([a_1, \cdots, a_k])$ of the measure μ on cylinders $[a_1, \cdots, a_k]$ and there exists a $(k-1)$-Markov measure ν such that

$$\mu([a_1, \cdots, a_k]) = \nu([a_1, \cdots, a_k])$$

for all cylinders $[a_1, \cdots, a_k]$ and such that $h_\nu \geq h_\mu$.

In particular, if $k = 1$, maximizing measures are Bernoulli measures. For the Bernoulli measure μ_p determined by a probability vector $p = (p_0, \cdots, p_{m-1})$, we have $h_{\mu_p} = H_1(p)$ where

$$H_1(p) = -\sum_{j=0}^{m-1} p_j \log p_j.$$

Suppose that the function Φ is a product of r functions and each of its factor depends only on the first coordinate, i.e.,

$$\Phi(x^{(1)}, \cdots, x^{(r)}) = \phi_1(x_1^{(1)}) \cdots \phi_r(x_1^{(r)}).$$

Let

$$A(p)=\int_{\Sigma_m^r}\Phi(x^{(1)},\cdots,x^{(r)})\mathrm{d}\mu_p(x^{(1)})\cdots\mathrm{d}\mu_p(x^{(r)}).$$

Notice that $E_\Phi(\alpha)\neq\emptyset$ iff $\alpha=A(p)$ for some probability vector $p=(p_0,\cdots,p_{m-1})$. The following result is a direct consequence of the last theorem.

Theorem 5. *Let $\Phi(x^{(1)},\cdots,x^{(r)})=\phi_1(x_1^{(1)})\cdots\phi_r(x_1^{(r)})$. We have*

$$A(p)=\prod_{k=1}^{r}\sum_{j=0}^{m-1}\phi_k(j)p_j.$$

For any α satisfying $E_\Phi(\alpha)\neq\emptyset$, we have

$$h_{\mathrm{top}}(E_\Phi(\alpha))=\max_{A(p)=\alpha}H_1(p), \tag{10}$$

where the maximum is taken over all probability vectors p satisfying $A(p)=\alpha$.

If $k=2$, maximizing measures are Markov measures. A Markov measure $\mu_{p,P}$ is determined by a probability vector p and a transition matrix P. Its entropy is equal to

$$H_2(p,P)=-\sum_{i=0}^{m-1}p_i\sum_{j=0}^{m-1}p_{i,j}\log p_{i,j}.$$

Suppose $\Phi(x^{(1)},\cdots,x^{(r)})$ is of the form $\phi_1(x_1^{(1)},x_2^{(1)})\cdots\phi_r(x_1^{(r)},x_2^{(r)})$. Let

$$A(p,P)=\int_{\Sigma_m^r}\Phi(x^{(1)},\cdots,x^{(r)})\mathrm{d}\mu_{p,P}(x^{(1)})\cdots\mathrm{d}\mu_{p,P}(x^{(r)}).$$

Theorem 6. *Let $\Phi(x^{(1)},\cdots,x^{(r)})=\phi_1(x_1^{(1)},x_2^{(1)})\cdots\phi_r(x_1^{(r)},x_2^{(r)})$. We have*

$$A(p,P)=\prod_{k=1}^{r}\sum_{i,j=0}^{m-1}\phi_k(i,j)p_ip_{i,j}.$$

For any α satisfying $E_\Phi(\alpha)\neq\emptyset$, we have

$$h_{\mathrm{top}}(E_\Phi(\alpha))=\max_{A(p,P)=\alpha}H_2(p,P), \tag{11}$$

where the maximum is taken over all couples (p,P) satisfying $A(p,P)=\alpha$.

Let us consider two examples. We will use the following trivial property of the entropy function $H(x)=-x\log x-(1-x)\log(1-x)$.

Lemma 4. *Given two numbers $p_1,p_2\in[0,1]$. We have*

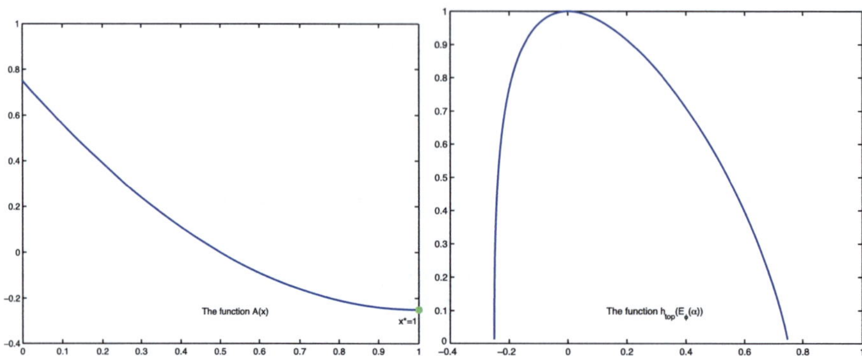

Fig. 1 Case $x^* = 1$ (with $a = 0.5, b = 1.5$)

$$H(p_1) < H(p_2) \text{ iff } |p_1 - 1/2| > |p_2 - 1/2|.$$

We have $H(p_1) = H(p_2)$ *iff* $|p_1 - 1/2| = |p_2 - 1/2|$.

Example 1. Consider the case $m = 2$, $k = 1$, and $r = 2$. Let $x = p_1$. Then $p_0 = 1 - x$ and we have

$$A(p) = [\phi_1(0)(1-x) + \phi_1(1)x][\phi_2(0)(1-x) + \phi_2(1)x].$$

For simplicity, we write $A(x)$ for $A(p)$. Suppose that $\phi_1(0) \neq \phi_1(1)$ and $\phi_2(0) \neq \phi_2(1)$. Otherwise, the question is trivial. By multiplying ϕ by a constant we can suppose that $A(x)$ is of the form

$$A(x) = (x-a)(x-b).$$

Let $x = x^*$ be the critical point of the quadratic function A (i.e., $x^* = \frac{a+b}{2}$).

Using the last lemma, it is easy to find the unique point x_α such that

$$A(x_\alpha) = \alpha, \ \ h_{\text{top}}(E_\Phi(\alpha)) = H(x_\alpha).$$

The point x_α is the closest to $1/2$ among those x such that $A(x) = \alpha$.

We distinguish three cases.

Case I. $x^* \leq 0$ or $x^* \geq 1$ (see Fig. 1).

1. $A(x)$ is strictly monotonic in the interval $[0,1]$.
2. L_Φ is the interval with end points ab and $(1-a)(1-b)$.
3. For any $\alpha \in L_\Phi$, $A(x_\alpha) = \alpha$ admits a unique solution x_α in $[0,1]$.

Case II. $0 < x^* \leq 1/2$ (see Fig. 2).

1. $A(x)$ is strictly monotonic in the intervals $[x^*,1]$.

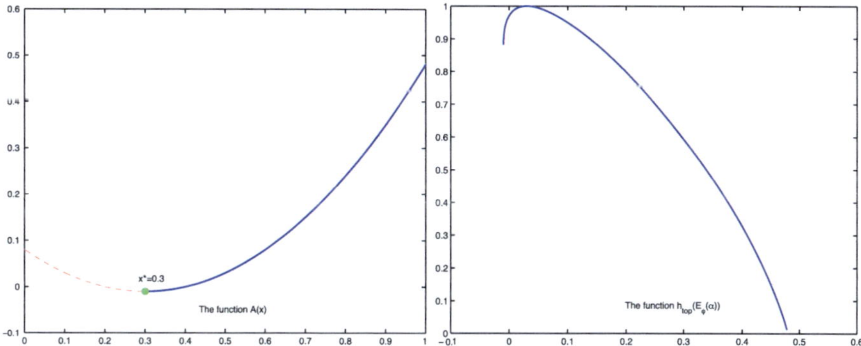

Fig. 2 Case $0 < x^* < 1/2$ (with $a = 0.2, b = 0.4$)

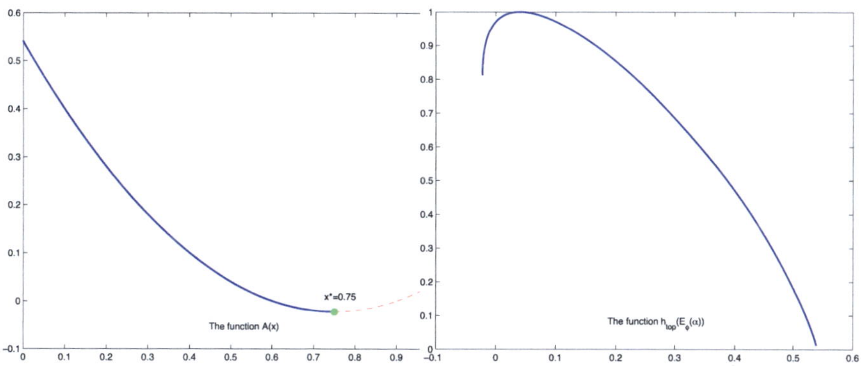

Fig. 3 Case $1/2 < x^* < 1$ (with $a = 0.6, b = 0.9$)

2. L_Φ is the interval with end points $A(x^*)$ and $(1-a)(1-b)$.
3. For any $\alpha \in L_\Phi$, $A(x_\alpha) = \alpha$ admits a unique solution x_α in $[x^*, 1]$.

Case III. $1/2 \le x^* < 1$ (see Fig. 3).

1. $A(x)$ is strictly increasing in the interval $[0, x^*]$.
2. L_Φ is the interval with end points ab and $A(x^*)$.
3. For any $\alpha \in L_\Phi$, $A(x_\alpha) = \alpha$ admits a unique solution x_α in $[0, x^*]$.

Remark 1. We can see in the case $m = 2$, $k = 1$, and $r = 2$ the spectrums are always continuous (in fact, they are differentiable in the interior of L_Φ). In the following examples we will see that this is no longer the case when $m = 2$, $k = 1$, and $r = 3$.

Example 2. Consider the case $m = 2$, $k = 1$ and $r = 3$. We have

$$A(x) = [\phi_1(0)(1-x) + \phi_1(1)x][\phi_2(0)(1-x) + \phi_2(1)x][\phi_3(0)(1-x) + \phi_3(1)x].$$

By multiplying ϕ by a constant, we can always suppose that A is of the form

$$A(x) = (x-a)(x-b)(x-c).$$

This cubic polynomial function is either increasing or admit a local maximal point $x_{\max}$ and a local minimal point $x_{\min}$ and then we must have $x_{\max} < x_{\min}$. As we will see, the continuity of the spectrum depends on the location of $x_{\max}$ and $x_{\min}$.

When A is increasing or when $x_{\max}, x_{\min} \notin (0,1)$, L_Φ is the interval with $-abc$ and $(1-a)(1-b)(1-c)$ as end points. For any α in the interval, $A(x_\alpha) = \alpha$ admits a unique solution x_α in $[0,1]$ and $h_{\text{top}}(E_\Phi(\alpha)) = H(x_\alpha)$. In this case the spectrum is continuous (and differentiable).

Suppose now that $A(x)$ admits a local maximal point $x_{\max}$ and a local minimal point $x_{\min}$ (with $x_{\max} < x_{\min}$). Then there exist a unique $x' > x_{\min}$ and a unique $x'' < x_{\max}$ such that

$$A(x') = A(x_{\max}), \quad A(x'') = A(x_{\min}).$$

We point out that there are three possible situations: the spectrum is continuous, admits one discontinuous point, or admits two discontinuous points. Before presenting in detail these three situations we prove the following lemma which will be useful for our discussion.

Lemma 5. *Let P be a polynomial of degree 3 with positive leading coefficient. Suppose that P admits a local maximal point $x_{\max}$ and a local minimal point $x_{\min}$. Then $x_{\max} < x_{\min}$ and*

$$x_1 < x_{\max} < x_2, |x_1 - x_{\max}| = |x_2 - x_{\max}| \Rightarrow P(x_1) < P(x_2)$$

$$y_1 < x_{\min} < y_2, |y_1 - x_{\min}| = |y_2 - x_{\min}| \Rightarrow P(y_1) < P(y_2).$$

Proof. The fact $x_{\max} < x_{\min}$ follows from $P(-\infty) = -\infty$ and $P(+\infty) = +\infty$. By the existence of the extremal points, we can write

$$P'(x) = \lambda(x - x_{\max})(x - x_{\min})$$

with $\lambda > 0$. It follows that

$$u < x_{\max} < v, x_{\max} - u = v - x_{\max} \Rightarrow \frac{|P'(u)|}{|P'(v)|} = \frac{|u - x_{\min}|}{|v - x_{\min}|} > 1.$$

This means that for two equidistant points from $x_{\max}$, the left point climbs quicker than the right point descents. By integration, we get

$$P(x_1) = P(x_{\max}) + \int_{x_{\max}}^{x_1} P'(u)\mathrm{d}u, \; P(x_2) = P(x_{\max}) + \int_{x_{\max}}^{x_2} P'(u)\mathrm{d}u.$$

Making the change of variable $v - x_{\max} = x_{\max} - u$, we obtain

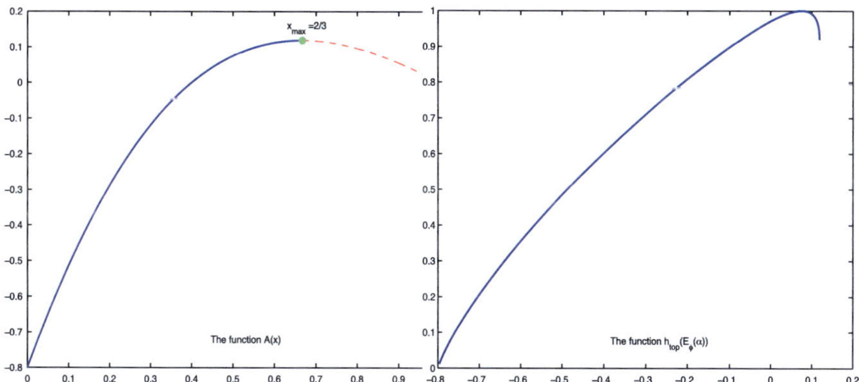

Fig. 4 Situation $1/2 < x_{\max} < 1 < x_{\min}$ ($a = 0.4$, $b = 1$, and $c = 2$)

$$\int_{x_{\max}}^{x_1} P'(u)\mathrm{d}u = -\int_{x_1}^{x_{\max}} |P'(u)|\mathrm{d}u < -\int_{x_{\max}}^{x_2} |P'(v)|\mathrm{d}v \le P(x_2) - P(x_{\max}).$$

The first equality holds since P' is positive in $(x_1, x_{\max})$. Hence $P(x_1) < P(x_2)$. We prove $P(y_1) < P(y_2)$ in the same way.

In the following we present three situations. We use the last two lemmas. In each situation, there is a unique point x_α such that

$$A(x_\alpha) = \alpha,\ h_{\text{top}}(E_\Phi(\alpha)) = H(x_\alpha).$$

We call x_α the maximizing point. For every $\alpha \in L_\Phi$, there could be one, two, or three points x such that $A(x) = \alpha$. The maximizing point x_α is the one which is the nearest to 1/2. In Figs. 4–6, those parts of graph of A corresponding to the maximizing points will be traced by solid lines, and other parts will be traced by dotted lines.

Situation I. $1/2 \le x_{\max} < 1 < x_{\min}$ (see Fig. 4).
Let $a = 0.4$, $b = 1$, and $c = 2$. Then $x_{\max} = 2/3$ and $x_{\min} = 1.6$. The spectrum is continuous. The following hold:

1. $L_\Phi = [A(0), A(x_{\max})]$.
2. The maximizing points lie in $[0, x_{\max}]$.
3. $A(x)$ is strictly monotonic in $[0, x_{\max}]$.

Situation II. $1/2 \le x_{\max} < x_{\min} < 1$ (see Fig. 5).
Let $a = 0.4$, $b = 0.7$, and $c = 0.8$. Then $x_{\max} = 0.5131$, $x_{\min} = 0.7375$, and $x' = 0.8737$. The spectrum admits one discontinuous point. The following hold:

1. $L_\Phi = [A(0), A(1)]$.
2. The maximizing points lie in $[0, x_{\max}] \cup (x', 1]$.
3. $A(x)$ is strictly monotonic in each of above two intervals.
4. The spectrum has one discontinuous point at $A(x_{\max})(= A(x'))$, the entropy jumps from $H(x_{\max})$ to $H(x')$.

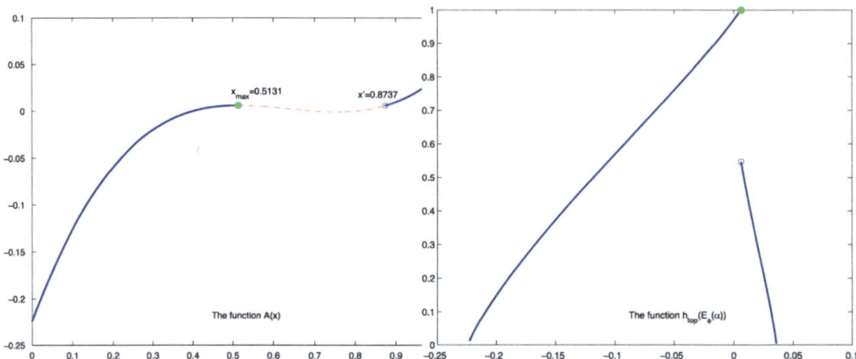

Fig. 5 Situation $1/2 < x_{\max} < x_{\min} < 1$ ($a = 0.4$, $b = 0.7$, and $c = 0.8$)

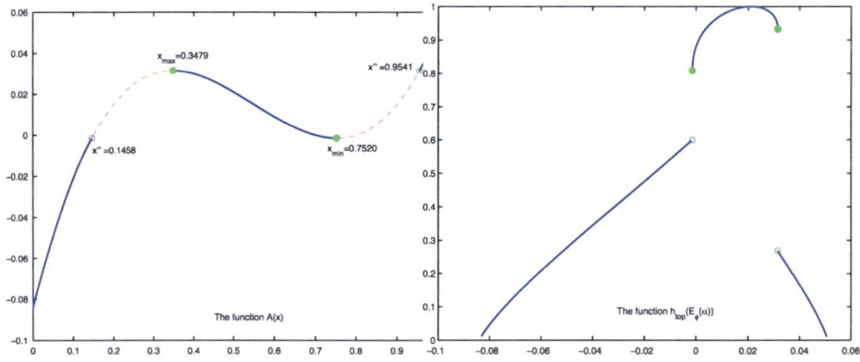

Fig. 6 Situation $0 < x_{\max} < 1/2 < x_{\min} < 1$ ($a = 0.15$, $b = 0.7$, and $c = 0.8$)

Situation III. $0 < x_{\max} < 1/2 < x_{\min} < 1$ (see Fig. 6).

Let $a = 0.15$, $b = 0.7$, and $c = 0.8$. Then $x_{\max} = 0.3479$, $x_{\min} = 0.7520$, $x' = 0.9541$, and $x'' = 0.1458$. The spectrum admits two discontinuous points. The following hold:

1. $L_\Phi = [A(0), A(1)]$.
2. The maximizing points lie in the intervals $[0, x'') \cup [x_{\max}, x_{\min}] \cup (x', 1]$.
3. $A(x)$ is strictly monotonic in each of above three intervals.
4. The spectrum has two discontinuity points. One is $A(x'')(= A(x_{\min}))$, where the entropy jumps from $H(x'')$ to $H(x_{\min})$; the other is $A(x_{\max})(= A(x'))$, where the entropy jumps from $H(x_{\max})$ to $H(x')$.

References

1. Aaronson, J., Burton, R., Dehling, H., Gilat, D., Hill, T., Weiss, B.: Strong laws for L- and U-statistics. Trans. Amer. Math. Soc. **348**, 2845–2866 (1996)
2. Assani, I.: Multiple recurrence and almost sure convergence for weakly mixing dynamical systems. Israel. J. Math. **1–3**, 111–124 (1987)
3. Barreira, L.: "Dimension and recurrence in hyperbolic dynamics." Progress in Mathematics. Soc. **272**. Birkhäuser, Basel (2008)
4. Barreira, L., Saussol, B., Schmeling, J.: Higher-dimensional multifractal analysis. J. Math. Pures Apple. **81**, 67–91 (2002)
5. Bergelson, V.: Weakly mixing PET. Ergod. Th. Dynam. Sys. **3**, 337–349 (1987)
6. Blokh, A.M.: Decomposition of dynamical systems on an interval. Usp. Mat. Nauk. **38**, 179–180 (1983)
7. Bourgain, J.: Double recurrence and almost sure convergence. J. Reine Angew. Math. **404**, 140–161 (1990)
8. Bowen, R.: Topological entropy for noncompact sets. Trans. Amer. Math. Soc. **184**, 125–136 (1973)
9. Denker, M., Grillenberger, C., Sigmund, K.: Ergodic Theory on Compact Spaces. Springer, Berlin (1976)
10. Fan, A.H.: Sur les dimensions de mesures. Studia Math. **111**, 1–17 (1994)
11. Fan, A.H., Feng, D.J., Wu, J.: Recurrence, entropy and dimension. J. London Math. Soc. **64**, 229–244 (2001)
12. Fan, A.H., Liao, L.M., Ma, J.H.: Level sets of multiple ergodic averages. Monatshefte für Mathematik (2011) online
13. Fan, A.H., Liao, L.M., Peyrière, J.: Generic points in systems of specification and Banach valued Birkhoff ergodic average. DCDS **21**, 1103–1128 (2008)
14. Fan, A.H., Schmeling, J., Wu, M.: Multifractal analysis of multiple ergodic averages. C. R. Mathématique. **349**(17–18), 961–964 (2011)
15. Furstenberg, H.: Ergodic behavior of diagonal measures and a theorem of Szemerédi on arithmetic progressions. J. d'Analyse Math. **31**, 204–256 (1977)
16. Host, B., Kra, B.: Nonconventional ergodic averages and nilmanifolds. Ann. Math. **161**, 397–488 (2005)
17. Kenyon, R., Peres, Y., Solomyak, B.: Hausdorff dimension of the multiplicative golden mean shift. C. R. Mathematique **349**(11–12), 625–628 (2011)
18. Rudin, W.: Functional Analysis. McGraw-Hill, New York (1973)
19. Ruelle, D.: Thermodynamic formalism. The mathematical structures of classical equilibrium statistical mechanics. Encyclopedia of Mathematics and Its Applications, vol. 5. Addison-Wesley, Boston (1978)
20. Schmeling, J.: Symbolic dynamics for β-shifts and self-normal numbers. Ergod. Th. Dynam. Sys. **17**, 675–694 (1997)

Projections of Measures Invariant Under the Geodesic Flow

Maarit Järvenpää

Abstract We discuss projection properties of measures which are invariant under the geodesic flow and describe their connection to quantum unique ergodicity. This overview is based on collaboration with R. Hovila, E. Järvenpää, F. Ledrappier, and M. Leikas.

1 Introduction

Dimensional properties of projections of sets and measures have been a subject of intensive investigation for decades. The study of the behaviour of Hausdorff dimension under projections dates back to the 1950s when Marstrand [16] proved a well-known preservation theorem according to which the Hausdorff dimension of a planar set is preserved under typical orthogonal projections. In [12] Kaufman verified the same preservation result using potential theoretical methods, and in [17] Mattila considered the higher-dimensional case. For measures the following analogous principle has been discovered in various contexts (see, e.g., Kaufman [12], Mattila [17], Hu and Taylor [8], and Falconer and Mattila [3]): Let m and n be integers such that $0 < m < n$ and let μ_V be the image of a compactly supported Radon measure μ on $\mathbf{R}^n$ under the orthogonal projection onto an m-plane V. Then for almost all m-planes V we have

$$\dim_H \mu_V = \dim_H \mu \text{ provided that } \dim_H \mu \leq m. \tag{1}$$

On the other hand, for almost all m-planes V,

$$\mu_V \ll \mathscr{L}^m \text{ provided that } \dim_H \mu > m. \tag{2}$$

M. Järvenpää (✉)
Department of Mathematical Sciences, University of Oulu, P.O. Box 3000, 90014 Oulu, Finland
e-mail: maarit.jarvenpaa@oulu.fi

J. Barral and S. Seuret (eds.), *Further Developments in Fractals and Related Fields*, Trends in Mathematics, DOI 10.1007/978-0-8176-8400-6_8,

Here $\dim_H$ is Hausdorff dimension, $\mathscr{L}^m$ is m-dimensional Lebesgue measure, and the symbol $\ll$ denotes the absolute continuity. Moreover, we use the orthogonally invariant Radon probability measure on the Grassmann manifold of all m-dimensional linear subspaces of $\mathbf{R}^n$. In the case that μ has finite m-energy, that is,

$$I_m(\mu) = \int\int |x-y|^{-m}\,\mathrm{d}\mu(x)\,\mathrm{d}\mu(y) < \infty,$$

a substantially stronger form of theorem 2 holds: we have for all typical m-planes that

$$\mu_V \ll \mathscr{L}^m \text{with Radon} - \text{Nikodym derivative in } L^2. \tag{3}$$

Analogies of the above results have been investigated for typical smooth mappings in the sense of prevalence and for infinite dimensional spaces in [9, 25], and [10]. In [21] Peres and Schlag verified an elegant extension of the projection formalism to parametrized families of transversal mappings and Sobolev dimensions of measures on compact metric spaces.

For the purpose of studying projection properties of measures which are invariant under the geodesic flow, we may restrict our consideration to the one-dimensional parameter space and define transversality in the following context: Let (Y,d) be a compact metric space, let $J \subset \mathbf{R}$ be an open interval, and let $P : J \times Y \to \mathbf{R}$ be a continuous function. Assume that for any $l = 0,1,\ldots$ there is a constant $\widetilde{C}_l \geq 1$ such that

$$|\partial_t^l P(t,y)| \leq \widetilde{C}_l \tag{4}$$

for all $t \in J$ and $y \in Y$. Here ∂_t^l is the lth partial derivative with respect to t. Moreover, for all $t \in J$ and x, $y \in Y$ with $x \neq y$, define

$$T_t(x,y) = \frac{P(t,x) - P(t,y)}{d(x,y)}.$$

Transversality is defined in the following manner:

Definition 1. The mapping P satisfying Eq. 4 is transversal if there is a constant C_T such that for all $t \in J$ and for all x, $y \in Y$ with $x \neq y$ the condition $|T_t(x,y)| \leq C_T$ implies that

$$|\partial_t T_t(x,y)| \geq C_T,$$

and furthermore, for all $l = 0,1,\ldots$, there exists a constant C_l such that

$$|\partial_t^l T_t(x,y)| \leq C_l$$

for all $t \in J$ and $x,y \in Y$ with $x \neq y$.

The methods of [21] play an important role in the results discussed in Sect. 2. For our purposes, a significant difference between the earlier results and those of [21] is that the absolute continuity result Eq. 3 is generalized in terms of fractional derivatives by showing that densities of typical projections onto m-dimensional spaces have fractional derivatives of order $\varepsilon/2$ in L^2 provided that the original measure has finite $(m+\varepsilon)$-energy. For detailed information about a variety of related contributions, see [19, 21].

This chapter is organized as follows: In Sect. 2 we introduce our setting and discuss projection properties of measures which are invariant under the geodesic flow. Special emphasis is given to the role of the machinery of [21]. Quite surprisingly, it turns out that the preservation result is valid for two-dimensional base manifolds only; in higher-dimensional case such resOult fails. Section 3 is dedicated to the relation between quantum unique ergodicity and projection properties of measures invariant under the geodesic flow.

2 Projections of Measures Invariant Under the Geodesic Flow

Assume that M is a smooth compact Riemann surface. Denoting by T^1M the unit tangent bundle and by $\varphi = \varphi_t$, $t \in \mathbf{R}$, the geodesic flow on T^1M, let μ be a Radon probability measure on T^1M which is invariant under the geodesic flow, that is, $\varphi_* \mu = \mu$. Here $\varphi_* \mu$ is the image measure of μ under φ. Finally, let $\Pi : T^1M \to M$ be the natural projection.

The projection theorems mentioned in Sect. 1, in particular, results Eqs. 1, 2, and 3, are genuinely "almost all" results, and therefore, they do not provide information about any specified projection. However, similar methods are applicable for the purpose of studying the Hausdorff dimension of the image $\Pi_* \mu$ of μ under Π. This interesting feature was discovered by Ledrappier and Lindenstrauss in [13]. The following analogues of Eqs. 1 and 2 hold (see [13]):

$$\dim_H \Pi_* \mu = \dim_H \mu \text{ provided that } \dim_H \mu \leq 2 \tag{5}$$

and

$$\Pi_* \mu \ll \mathscr{L}^2 \text{ provided that } \dim_H \mu > 2. \tag{6}$$

Analogously to Eq. 3, Ledrappier and Lindenstrauss proved that

$$\text{the Radon} - \text{Nikodym derivative is a } L^2\text{–function if } I_\alpha(\mu) < \infty \text{ for } \alpha > 2. \tag{7}$$

The methods of [13] are based on the Kaufman-type potential theoretic techniques [12].

In [11] the question of whether Eq. 7 could be further generalized in terms of fractional derivatives is investigated. In addition to giving a positive answer to this question by employing the techniques from [21], the validity of the results Eqs. 5 and 6 is studied for higher-dimensional base manifolds. Quite surprisingly, it appears that the Hausdorff dimension is not necessarily preserved in the higher-dimensional case. Indeed, for any $n \geq 3$, there exist an n-dimensional compact smooth Riemann manifold M and a measure μ on T^1M such that μ is locally invariant and the Hausdorff dimension of μ decreases under the projection $\Pi : T^1M \to M$. For the construction, see [11].

The failure of the preservation can be understood in terms of the formalism introduced in [21]—the reason behind the failure being that for n-dimensional base manifolds the local invariance produces parametrized family of projections onto $(n-1)$-dimensional planes in $2(n-1)$-dimensional space. The parameter is given by the time coordinate, and therefore the family is one-dimensional. Since the dimension of the space of $(n-1)$-planes in $2(n-1)$-dimensional space is greater than 1, if $n \geq 3$, the transversality condition cannot hold.

The methods in [11] are based on the techniques of [21]. In general, the measure μ is complicated to handle. However, the fact that μ is invariant under the geodesic flow implies that locally a suitable restriction of μ is roughly of the form $\nu \times \mathscr{L}^1$ where ν is a measure on a two-dimensional square. In [11] we verify that locally the projection of the restriction of μ is in a certain sense of the form $\nu_t \times \mathscr{L}^1$ where ν_t is a projection of ν onto one-dimensional space. In this we obtain a family of projections parametrized by the time coordinate t. If the base manifold is two-dimensional the family turns out to be transversal in the sense of definition 1. This enables us to reprove theorems 5 and 6 by employing the techniques in [21]. As mentioned above, the novelty of our proof is that it illustrates the reason behind the failure of the preservation in higher-dimensional case. For details, see [11].

In [11] theorem 7 is extended by showing that $\Pi_*\mu$ has fractional derivatives of order γ in L^2 for all $\gamma < (\alpha-2)/2$ provided that $I_\alpha(\mu) < \infty$ for $\alpha > 2$. The proof relies again on the machinery of [21].

The behaviour of projections of measures which are invariant under the geodesic flow has been studied using other concepts of dimension. In [14] Leikas computed the packing dimension of $\Pi_*\mu$ and gave an example illustrating that the packing dimension can decrease even in the two-dimensional case. By employing the techniques of [4], Hovila [5] considered the lower and upper dimension spectra and parametrized families of transversal mappings between smooth manifolds and computed for $1 < q \leq 2$ the lower and upper q-dimensions of $\Pi_*\mu$.

3 Quantum Unique Ergodicity

According to Eq. 6, the canonical projection of a φ-invariant measure of dimension greater than 2 is absolutely continuous with respect to the Lebesgue measure. In this section we address the question of whether the absolute continuity condition holds at the threshold 2 and give a negative answer to this question. Our discussion is motivated by quantum unique ergodicity.

Letting ψ_n be a sequence of orthonormal eigenfunctions of the Laplacian on M, the associated eigenvalues converge to infinity. The aim of quantum unique ergodicity is to describe the possible weak* limits of the probability measures with density $|\psi_n|^2$ as n tends to infinity.

The problem is solved by Lindenstrauss for arithmetic hyperbolic surfaces in the case when the orthonormal eigenfunctions ψ_n are also eigenfunctions of the Hecke operators. In this case the only limit is the normalized Lebesgue measure (see [15]).

A weak form of the quantum unique ergodicity conjecture, stating that any weak* limit is nonsingular, is closely related to the validity of Eq. 6 at the threshold 2. Indeed, Anantharaman and Nonnenmacher [2] considered both a general hyperbolic surface and more general eigenfunctions on an arithmetic surface and verified that any weak* limit is the projection of a φ-invariant measure with dimension at least 2. Rivière in turn showed that this property remains true on surfaces with variable negative curvature [24]. In [6] we showed that one cannot conclude from the results of [2] and [24] that a weak form of the quantum unique ergodicity conjecture holds by verifying the following result the proof of which we shortly sketch:

Theorem 1. *For any compact surface M whose curvature is everywhere negative, there exists an ergodic φ-invariant measure μ on T^1M such that* $\dim_H \Pi_*\mu = 2$ *and $\Pi_*\mu$ is singular with respect to the Lebesgue measure on M.*

Proof. According to Eq. 5, it is sufficient to construct an ergodic φ-invariant measure μ such that $\dim_H \mu = 2$ and μ is singular with respect to the Lebesgue measure.

Letting m be an ergodic φ-invariant measure on T^1M, we have

$$\dim_H m = 1 + 2\frac{h_m(\varphi)}{\lambda(m)},$$

where $h_m(\varphi)$ is the entropy and $\lambda(m)$ is the Lyapunov exponent [22]. This gives

$$\dim_H m = 2 \iff h_m(\varphi)/\lambda(m) = 1/2.$$

Hence, on any family for which the ratio $h_m(\varphi)/\lambda(m)$ varies continuously from 0 to 1, there will be measures having Hausdorff dimension 2. To construct such measures, we consider Markov measures in a symbolic coding of the geodesic flow. For the existence of the symbolic coding, see [23].

For any $n \times n$-matrix $A = A_{ij}$ with entries 0 or 1, define a subshift of finite type $\Sigma \subset \{1,\dots,n\}^{\mathbf{Z}}$ as the set of sequences $\underline{\omega} = (\omega_k)$ such that

$$A_{\omega_k \omega_{k+1}} = 1 \text{ for all } k.$$

Furthermore, letting σ be the left shift on Σ and letting r be a positive function on Σ, define the special flow $(\widetilde{\Sigma}_r, \tilde{\sigma}_t)$ by translation on the second coordinate where

$$\widetilde{\Sigma}_r = \{(\underline{\omega}, s) \mid \underline{\omega} \in \Sigma, 0 \leq s \leq r(\underline{\omega})\}/(\underline{\omega}, r(\underline{\omega})) \sim (\sigma(\underline{\omega}), 0),$$

that is, for $t \geq 0$, we have

$$\tilde{\sigma}_t(\underline{\omega}, s) = (\sigma^k(\underline{\omega}), u),$$

where $u = t + s - \sum_{j=0}^{k-1} r(\sigma^j(\underline{\omega}))$ and k is the unique natural number satisfying $0 \le u < r(\sigma^k(\underline{\omega}))$ and similarly for $t < 0$.

In what follows we select the function r in a special way. Indeed, due to [23], there is a mixing subshift of finite type (Σ, σ) and Hölder functions $r : \Sigma \to \mathbf{R}$ and $\pi : \widetilde{\Sigma}_r \to T^1M$ such that $\pi \circ \tilde{\sigma}_t = \varphi_t \circ \pi$. Choosing such r, it is enough to construct a measure on $\widetilde{\Sigma}_r$ with appropriate properties. Projecting this measure to T^1M by π gives the desired measure μ.

Let $P = P_{ij}$ be a Markov matrix such that $P_{ij} > 0$ if and only if $A_{ij} = 1$ and let μ_P be a σ-invariant Markov measure on Σ. We define a $\widetilde{\sigma}$-invariant probability measure $\widetilde{\mu}_P$ on $\widetilde{\Sigma}_r$ by

$$\widetilde{\mu}_P = \frac{\int_\Sigma \mathscr{L}|_{[0,r)} \, \mathrm{d}\mu_P}{\int_\Sigma r \, \mathrm{d}\mu_P}.$$

Now the measure $m_P = \pi_* \widetilde{\mu}_P$ is ergodic for φ, and using Abramov formula [1], we can calculate the ratio

$$R = \frac{h_{m_P}(\varphi)}{\lambda(m_P)}.$$

Using l-step Markov measures guarantees that there are many Markov matrices P on Σ such that $R = 1/2$ implying $\dim_H m_P = 2$. It remains to verify the singularity with respect to the Lebesgue measure.

The aim is to show that P can be chosen in such a way that the upper derivative of m_P with respect to the Lebesgue measure is typically infinite, that is,

$$\limsup_{\varepsilon \to 0} \frac{m_P(B((x,v),\varepsilon))}{\varepsilon^2} = \infty \tag{8}$$

for m_P-almost all $(x,v) \in T^1M$. Here $B((x,v),\varepsilon))$ is a closed ball with centre (x,v) and radius ε. By [18, Theorem 2.12], Eq. 8 implies singularity.

For the purpose of giving the main idea of the proof of Eq. 8, we consider fluctuations of measures of balls and define observables (X_n^u, Y_n^u) and (X_n^s, Y_n^s) such that X_n describes the mass of a ball and Y_n describes the radius on local unstable (u) and stable (s) manifolds. The vector-valued almost sure invariance principle [20] implies that (X_n^u, Y_n^u) and (X_n^s, Y_n^s) can be approximated by two-dimensional Brownian motions meaning that for $v \in \{u,s\}$ there exist $\lambda > 0$ and a probability space $(X, \mathbf{P})$ supporting a sequence of random variables $(\widetilde{X}_n^v, \widetilde{Y}_n^v)$ having the same distribution as (X_n^v, Y_n^v) and a two-dimensional Brownian motion W^v with a covariance matrix Q^v such that

$$|(\widetilde{X}_n^v, \widetilde{Y}_n^v) - W^v(n)| \ll n^{\frac{1}{2} - \lambda}$$

$\mathbf{P}$-almost surely for large n. From this we deduce that the fluctuations of the observables are large enough which implies the claim Eq. 8. The crucial steps are to show that the covariance matrix is nonsingular and to handle stable and unstable manifolds simultaneously. For details, see [6].

The measure constructed in theorem 1 is supported by the whole unit tangent bundle T^1M. In [7] it is shown that on a certain class of Riemann surfaces with constant negative curvature and with boundary, there exist natural two-dimensional measures invariant under the geodesic flow having two-dimensional supports such that their projections to the base manifold are two-dimensional, but the supports of the projections are Lebesgue negligible. In this case the singularity is due to the projection. In [7] the main tool is Besicovitch–Federer projection theorem for transversal families of mappings stated as follows (in theorem 2 we denote by $\mathscr{H}^m$ the m-dimensional Hausdorff measure):

Theorem 2. *Let $l, m,$ and n be integers with $m \leq l$ and $m < n$. Let $E \subset \mathbf{R}^n$ be $\mathscr{H}^m$-measurable with $\mathscr{H}^m(E) < \infty$. Assume that $\Lambda \subset \mathbf{R}^l$ is open and $\{P_\lambda : \mathbf{R}^n \to \mathbf{R}^m\}_{\lambda \in \Lambda}$ is a transversal family of maps. Then E is purely m-unrectifiable, if and only if $\mathscr{H}^m(P_\lambda(E)) = 0$ for $\mathscr{L}^l$-almost all $\lambda \in \Lambda$.*

Acknowledgements We acknowledge the support of the Centre of Excellence in Analysis and Dynamics Research funded by the Academy of Finland.

References

1. Abramov, L.: On the entropy of a flow (Russian). Dokl. Akad. Nauk SSSR **128**, 873–875 (1959)
2. Anantharaman, N., Nonnenmacher, S.: Half-delocalization of eigenfunctions for the Laplacian on an Anosov manifold. Ann. Inst. Fourier **57**, 2465–2723 (2007)
3. Falconer, K., Mattila, P.: The packing dimension of projections and sections of measures. Math. Proc. Cambridge Philos. Soc. **119**, 695–713 (1996)
4. Falconer, K., O'Neil, T.C.: Convolutions and the geometry of multifractal measures. Math. Nachr. **204**, 61–82 (1999)
5. Hovila, R.: The dimension spectrum of projected measures on Riemann manifolds. Ann. Acad. Sci. Fenn. Math. **35**, 595–608 (2010)
6. Hovila, R., Järvenpää, E., Järvenpää, M., Ledrappier, F.: Singularity of projections of 2-dimensional measures invariant under the geodesic flow. Comm. Math. Phys. **312**, 127–136 (2012)
7. Hovila, R., Järvenpää, E., Järvenpää, M., Ledrappier, F.: Besicovitch-Federer projection theorem and geodesic flows on Riemann surfaces. Geom. Dedicata. **161**, 51–61 (2012)
8. Hu, X., Taylor, J.: Fractal properties of products and projections of measures in $\mathbf{R}^n$. Math. Proc. Cambridge Philos. Soc. **115**, 527–544 (1994)
9. Hunt, B.R., Kaloshin, V.Yu.: How projections affect the dimension spectrum of fractal measures? Nonlinearity **10**, 1031–1046 (1997)
10. Hunt, B.R., Kaloshin, V.Yu.: Regularity of embeddings of infinite-dimensional fractal sets into finite-dimensional spaces. Nonlinearity **12**, 1263–1275 (1999)
11. Järvenpää, E., Järvenpää, M., Leikas, M.: (Non)regularity of projections of measures invariant under geodesic flow. Comm. Math. Phys. **254**, 695–717 (2005)
12. Kaufman, R.: On Hausdorff dimension of projections. Mathematika **15**, 153–155 (1968)
13. Ledrappier, F., Lindenstrauss E.: On the projections of measures invariant under the geodesic flow. Int. Math. Res. Not. **9**, 511–526 (2003)
14. Leikas, M.: Packing dimensions, transversal mappings and geodesic flows. Ann. Acad. Sci. Fenn. Math. **29**, 489–500 (2004)

15. Lindenstrauss, E.: Invariant measures and quantum unique ergodicity. Ann. Math. **163**, 165–219 (2006)
16. Marstrand, M.: Some fundamental geometrical properties of plane sets of fractional dimension. Proc. London Math. Soc. **4**, 257–302 (1954)
17. Mattila, P.: Orthogonal projections, Riesz capacities and Minkowski content. Indiana Univ. Math. J. **39**, 185–198 (1990)
18. Mattila, P.: Geometry of Sets and Measures in Euclidean Spaces: Fractals and Rectifiability. Cambridge University Press, Cambridge (1995)
19. Mattila, P.: Hausdorff dimension, projections, and Fourier transform. Publ. Mat. **48**, 3–48 (2004)
20. Melbourne, I., Nicol, M.: A vector-valued almost sure invariance principle for hyperbolic dynamical systems. Ann. Probab. **37**, 478–505 (2009)
21. Peres, Y., Schlag, W.: Smoothness of projections, Bernoulli convolutions, and the dimensions of exceptions. Duke Math. J. **102**, 193–251 (2000)
22. Pesin, Ya., Sadovskaya, V.: Multifractal analysis of conformal Axiom A flows. Comm. Math. Phys. **216**, 277–312 (2001)
23. Ratner, M.: Markov decomposition for an U-flow on a 3-dimensional manifold. Math. Notes **6**, 880–887 (1969)
24. Rivière, G.: Entropy of semiclassical measures in dimension 2. Duke Math. J. **155**, 271–335 (2010)
25. Sauer T.D., Yorke, J.A.: Are dimensions of a set and its image equal for typical smooth functions? Ergodic Theory Dynam. Syst. **17**, 941–956 (1997)

Multifractal Tubes

Lars Olsen

Abstract Tube formulas refer to study of volumes of r neighbourhoods of sets. For sets satisfying some (possible very weak) convexity conditions, this has a long history going back to Steiner in the early Nineteenth century. However, within the past 20 years, Lapidus has initiated and pioneered a systematic study of tube formulas for fractal sets. Following this line of investigation, it is natural to ask as to what extent it is possible to develop a theory of multifractal tubes. In this survey we will explain one approach to this problem based on Olsen (Multifractal tubes, Preprint, 2011). In particular, we will propose a general framework for studying tube formulas of multifractals and, as an example, we give a complete description of the asymptotic behaviour of the multifractal tube formulas for self-similar measures satisfying the Open Set Condition.

1 Fractal Tubes

Let E be a subset of $\mathbb{R}^d$ and $r > 0$. We now write $B(E,r)$ for the open r neighbourhood of E, i.e.

$$B(E,r) = \left\{ x \in \mathbb{R}^d \,\middle|\, \operatorname{dist}(x,E) < r \right\}.$$

Intuitively we will think of the set $B(E,r)$ as consisting of the E surrounded by a "tube" of width r. Our main interest is to compute the volume of the "tube" of width r surrounding E or equivalently computing the volume of the set $B(E,r)$ and

L. Olsen (✉)
Department of Mathematics, School of Mathematics and Statistics,
University of St. Andrews, St. Andrews, Fife KY16 9SS, Scotland, UK
e-mail: lo@st-and.ac.uk

J. Barral and S. Seuret (eds.), *Further Developments in Fractals and Related Fields*, Trends in Mathematics, DOI 10.1007/978-0-8176-8400-6_9,

subtract the volume of E. To make this formal, we define the Minkowski volume $V_r(E)$ of E by

$$V_r(E) = \frac{1}{r^d}\mathscr{L}^d(B(E,r));$$

here and below $\mathscr{L}^d$ denotes the Lebesgue measure in $\mathbb{R}^d$ and the normalizing factor $\frac{1}{r^d}$ is included to make the subsequent results simpler—we note that different authors use different normalizing factors. Tube formulas refer to formulas for computing the Minkowski volume $V_r(E)$ as a function of the width r of the "tube" surrounding E. In particular, one is typically interested in the following two types of results:

1. Asymptotic behaviour: finding a formula for the asymptotic behaviour of $V_r(E)$ as $r \searrow 0$.
2. Explicit formulas: finding an explicit formulas for $V_r(E)$ valid for all small r.

For convex sets E, this problem has a rich and fascinating history starting with the work of Steiner in the early Nineteenth century. This theory reached its mature form in the 1960s where Federer [13, 14] unified the tube formulas of Steiner for convex bodies and of Weyl for smooth submanifolds, as described in [2, 21, 50], and extended these results to sets of positive reach. Federer's tube formula has since been extended in various directions by a number of researchers in integral geometry and geometric measure theory, including [18, 19, 47–49, 52, 53] and most recently (and most generally) in [25]. The books [21, 35, 48] contain extensive endnotes with further information and many other references. While the above references investigate tube formulas for sets that satisfy some (possibly very weak) convexity conditions, very recently there has been significant interest in developing a theory of tube formulas for fractal sets and a number of exciting works have appeared. Indeed, in the early 1990s, Lapidus introduced the notion of "complex dimensions" and has during the past 20 years very successfully pioneered the use of "complex dimension" to obtain explicit tube formulas for certain classes of fractal sets; this exiting theory is described in detail in Lapidus and van Frankenhuysen's intriguing books [29, 30]. In a parallel development and building on earlier work by Lalley [26–28] and Gatzouras [20] (see also [11]), Winter [51] has initiated the systematic study of curvatures of fractal sets and applied this theory to study the asymptotic behaviour of the Minkowski volume $V_r(E)$ of fractal sets E using methods from renewal theory.

The Minkowski volume $V_r(E)$ is closely related to various notions from Fractal Geometry. Indeed, using the Minkowski volume $V_r(E)$, we define the lower and upper Minkowski dimension of E by

$$\underline{\dim}_{\mathsf{M}}(E) = \liminf_{r\searrow 0} \frac{\log V_r(E)}{-\log r},$$

$$\overline{\dim}_{\mathsf{M}}(E) = \limsup_{r\searrow 0} \frac{\log V_r(E)}{-\log r}.$$

The link with Fractal Geometry is now explained as follows. Namely, box dimensions play an important role in Fractal Geometry and it is not difficult to see that the lower Minkowski dimension equals the lower box dimension and that the upper Minkowski dimension equals the upper box dimension; for the definition of the box dimensions the reader is referred to Falconer's textbook [10].

It is clearly also of interest to analyse the behaviour of the Minkowski volume $V_r(E)$ itself as $r \searrow 0$. Indeed, if, for example, $a_1,\ldots,a_d,b_1,\ldots,b_d$ are real numbers with $a_i \le b_i$ for all i, and U denotes the rectangle $[a_1,b_1]\times\cdots\times[a_d,b_d]$ in $\mathbb{R}^d$, then it is clear that $\frac{1}{r^{-d}}V_r(U)\to(b_1-a_1)\cdots(b_d-a_d)=\mathscr{L}^d(U)$. This suggests that if t is a real number, then the limit $\lim_{r\searrow 0}\frac{1}{r^{-t}}V_r(E)$ (if it exists) may be interpreted as the t-dimensional volume of E. Motivated by this, for a real number t, we therefore define the lower and upper t-dimensional Minkowski content of E by

$$\begin{aligned}\underline{M}^t(E) &= \liminf_{r\searrow 0}\frac{1}{r^{-t}}V_r(E),\\ \overline{M}^t(E) &= \limsup_{r\searrow 0}\frac{1}{r^{-t}}V_r(E).\end{aligned}$$

If $\underline{M}^t(E)=\overline{M}^t(E)$, i.e. if the limit $\lim_{r\searrow 0}\frac{1}{r^{-t}}V_r(E)$ exists, then we say the E is t Minkowski measurable, and we will denote the common value of $\underline{M}^t(E)$ and $\overline{M}^t(E)$ by $M^t(E)$, i.e. we will write

$$M^t(E)=\underline{M}^t(E)=\overline{M}^t(E).$$

Of course, a set E may not be Minkowski measurable, i.e. the limit $\lim_{r\searrow 0}\frac{1}{r^{-t}}V_r(E)$ may not exist. In this case it is natural to study the limiting behaviour of "averages" of $\frac{1}{r^{-t}}V_r(E)$. We therefore define the lower and upper average t-dimensional Minkowski content of E by

$$\begin{aligned}\underline{M}^t_{\mathsf{ave}}(E) &= \liminf_{r\searrow 0}\frac{1}{-\log r}\int_r^1\frac{1}{s^{-t}}V_s(E)\,\frac{\mathrm{d}s}{s},\\ \overline{M}^t_{\mathsf{ave}}(E) &= \liminf_{r\searrow 0}\frac{1}{-\log r}\int_r^1\frac{1}{s^{-t}}V_s(E)\,\frac{\mathrm{d}s}{s}.\end{aligned}$$

If $\underline{M}^t_{\mathsf{ave}}(E)=\overline{M}^t_{\mathsf{ave}}(E)$, i.e. if the limit $\lim_{r\searrow 0}\frac{1}{-\log r}\int_r^1\frac{1}{s^{-t}}V_s(E)\,\frac{\mathrm{d}s}{s}$ exists, then we say the E is t average Minkowski measurable, and we will denote the common value of $\underline{M}^t_{\mathsf{ave}}(E)$ and $\overline{M}^t_{\mathsf{ave}}(E)$ by $M^t_{\mathsf{ave}}(E)$, i.e. we will write

$$M^t_{\mathsf{ave}}(E)=\underline{M}^t_{\mathsf{ave}}(E)=\overline{M}^t_{\mathsf{ave}}(E).$$

While the Minkowski dimensions in many cases can be computed rigorously relatively easy, it is a notoriously difficult problem to compute the Minkowski content. In fact, it is only within the past 15 years that the Minkowski content of non-trivial examples has been computed. Indeed, using techniques from complex analysis, Lapidus and collaborators [29, 30] have computed the Minkowski content of certain self-similar subsets of the real line, and using ideas from the theory of Mercerian theorems, Falconer [11] has obtained similar results.

It is our intention to extend the notion of Minkowski volume $V_r(E)$ to multifractals and investigate the asymptotic behaviour of the corresponding multifractal Minkowski volume as $r \searrow 0$ for self-similar multifractals. In order to motivate our definitions we will now explain what the term "multifractal analysis" covers.

2 Multifractals

2.1 Multifractal Spectra

Distributions with widely varying intensity occur often in the physical sciences, for example, the spatial–temporal distribution of rainfall, the spatial distribution of oil and gas in the underground, the distribution of galaxies in the universe, the dissipation of energy in a highly turbulent fluid flow and the occupation measure on strange attractors. Such distributions are called multifractals and have recently been the focus of much attention in the physics literature.

Figure 1 shows a typical multifractal, i.e. a measure with widely varying intensity. Dark regions have high concentration of mass and light regions have low concentration of mass. For a Borel measure μ on a $\mathbb{R}^d$ and a positive number α, let us consider the set $\Delta_\mu(\alpha)$ of those points x in $\mathbb{R}^d$ for which the measure $\mu(B(x,r))$ of the ball $B(x,r)$ with centre x and radius r behaves like r^α for small r, i.e. the set

Fig. 1 A typical multifractal, i.e. a measure with widely varying intensity. *Dark regions* have high concentration of mass and *light regions* have low concentration of mass

$$\Delta_\mu(\alpha) = \left\{ x \in \operatorname{supp}\mu \,\middle|\, \lim_{r \searrow 0} \frac{\log \mu(B(x,r))}{\log r} = \alpha \right\},$$

where $\operatorname{supp}\mu$ denotes the support of the measure. If the intensity of the measure μ varies very widely, it may happen that the sets $\Delta_\mu(\alpha)$ display a fractal-like character for a range of values of α. If this is the case, then the measure is called a multifractal measure or simply a multifractal, and it is natural to study the sizes of the sets $\Delta_\mu(\alpha)$ as α varies. There are two approaches to this. We may consider the measure $\mu(\Delta_\mu(\alpha))$ of the sets $\Delta_\mu(\alpha)$ as α varies. This approach was adopted by Cutler in a series of papers [5–7] and leads to a "decomposition" of the measure into its α-dimensional components. However, typically the sets $\Delta_\mu(\alpha)$ have zero μ measure except for a few exceptional values of α. Hence, the measure $\mu(\Delta_\mu(\alpha))$ does in general not allow us to distinguish between the sets $\Delta_\mu(\alpha)$. The other approach is to find the (fractal) dimension of $\Delta_\mu(\alpha)$. In most examples of interest the set $\Delta_\mu(\alpha)$ is dense in the support of μ for all values of α for which $\Delta_\mu(\alpha)$ is non-empty, and thus

$$\underline{\dim}_{\mathsf{B}}\Delta_\mu(\alpha) = \underline{\dim}_{\mathsf{B}}\overline{\Delta_\mu(\alpha)} = \underline{\dim}_{\mathsf{B}} \operatorname{supp}\mu$$

and

$$\overline{\dim}_{\mathsf{B}}\Delta_\mu(\alpha) = \overline{\dim}_{\mathsf{B}}\overline{\Delta_\mu(\alpha)} = \overline{\dim}_{\mathsf{B}} \operatorname{supp}\mu$$

for all values of α for which $\Delta_\mu(\alpha) \neq \varnothing$, where $\underline{\dim}_{\mathsf{B}}$ and $\overline{\dim}_{\mathsf{B}}$ denote the lower and upper box dimension, respectively. Box dimensions are thus in general of little use in discriminating between the size of the sets $\Delta_\mu(\alpha)$. It is therefore more natural to study the Hausdorff dimension,

$$f_\mu(\alpha) = \dim \Delta_\mu(\alpha), \tag{1}$$

of the sets $\Delta_\mu(\alpha)$ as a function of α where dim denotes the Hausdorff dimension. The function in Eq. (1) and similar functions are generically known as "the multifractal spectrum of μ", "the singularity spectrum of μ" or "the spectrum of scaling indices", and one of the main problems in multifractal analysis is to study these and related functions. The function $f_\mu(\alpha)$ was first explicitly defined by the physicists Halsey et al. in 1986 in their seminal paper [22]. The concepts underlying the above mentioned multifractal decompositions go back to two early papers by Mandelbrot [32,33] from 1972 and 1974, respectively. Mandelbrot [32,33] suggests that the bulk of intermittent dissipation of energy in a highly turbulent fluid flow occurs over a set of fractal dimension. The ideas introduced in [32, 33] were taken up by Frisch and Parisi [17] in 1985 and finally by Halsey et al. [22] in 1986. Of course, for many measures, the limit $\lim_{r \searrow 0} \frac{\log \mu(B(x,r))}{\log r}$ may fail to exist for all or many x, in which case we need to work with lower or upper limits as r tends to 0 and (perhaps) replace "$= \alpha$" in the definition of $\Delta_\mu(\alpha)$ with "$\leq \alpha$" or "$\geq \alpha$".

2.2 Renyi Dimensions

Based on a remarkable insight together with a clever heuristic argument Halsey et al. [22] suggest that the multifractal spectrum $f_\mu(\alpha)$ can be computed in the following way—known as the so-called "Multifractal Formalism" in the physics literature. The "Multifractal Formalism" involves the so-called Renyi dimensions which we will now define. Let μ be a Borel measure on $\mathbb{R}^d$. For $q \in \mathbb{R}$ and $r > 0$, we define the qth moment $I^q_{\mu,r}(E)$ of a subset E of $\mathbb{R}^d$ with respect to μ at scale r by

$$I^q_{\mu,r}(E) = \int_E \mu(B(x,r))^{q-1}\,\mathrm{d}\mu(x). \tag{2}$$

Next, the lower and upper Renyi dimensions of E with respect to μ are defined by

$$\underline{\dim}^q_{\mathsf{R},\mu}(E) = \liminf_{r\searrow 0} \frac{\log I^q_{\mu,r}(E)}{-\log r}, \tag{3}$$

$$\overline{\dim}^q_{\mathsf{R},\mu}(E) = \limsup_{r\searrow 0} \frac{\log I^q_{\mu,r}(E)}{-\log r}. \tag{4}$$

In particular, the Renyi dimensions of the support of μ play an important role in the statement of the "Multifractal Formalism". For this reason it is useful to denote these dimensions by separate notation, and we therefore define the lower and upper Renyi spectra $\underline{\tau}_\mu(q), \overline{\tau}_\mu(q) : \mathbb{R} \to [-\infty,\infty]$ of μ by

$$\underline{\tau}_\mu(q) = \underline{\dim}^q_{\mathsf{R},\mu}(\operatorname{supp}\mu) = \liminf_{r\searrow 0} \frac{\log I^q_{\mu,r}(\operatorname{supp}\mu)}{-\log r},$$

$$\overline{\tau}_\mu(q) = \overline{\dim}^q_{\mathsf{R},\mu}(\operatorname{supp}\mu) = \limsup_{r\searrow 0} \frac{\log I^q_{\mu,r}(\operatorname{supp}\mu)}{-\log r}.$$

2.3 The Multifractal Formalism

We can now state the "Multifractal Formalism". Loosely speaking the "Multifractal Formalism" says the the multifractal spectrum f_μ and the Renyi dimensions carry the same information. More precisely, the multifractal spectrum equals the Legendre transform of the Renyi dimensions. Before stating this formally, we remind the reader that if $\varphi : \mathbb{R} \to \mathbb{R}$ is a real-valued function, then the Legendre transform $\varphi^* : \mathbb{R} \to [-\infty,\infty]$ of φ is defined by

$$\varphi^*(x) = \inf_y (xy + \varphi(y)).$$

The Multifractal Formalism: A Physics Folklore Theorem. *The multifractal spectrum f_μ of μ equals the Legendre transforms,s $\underline{\tau}_\mu^*$ and $\overline{\tau}_\mu^*$, of the Renyi dimensions, i.e.*

$$f_\mu(\alpha) = \underline{\tau}_\mu^*(\alpha) = \overline{\tau}_\mu^*(\alpha)$$

for all $\alpha \geq 0$.

The "Multifractal Formalism" is a truly remarkable result: it states that the locally defined multifractal spectrum f_μ can be computed in terms of the Legendre transforms of the globally defined moment scaling functions $\underline{\tau}_\mu^*$ and $\overline{\tau}_\mu^*$. There is apriori no reason to expect that the Legendre transforms of the moment scaling functions $\underline{\tau}_\mu^*$ and $\overline{\tau}_\mu^*$ should provide any information about the fractal dimension of the set of points x such that $\mu(B(x,r)) \approx r^\alpha$ for $r \approx 0$. In some sense the "Multifractal Formalism" is a genuine mystery.

During the past 20 years there has been an enormous interest in verifying the Multifractal Formalism and computing the multifractal spectra of measures in the mathematical literature. In the mid-1990s Cawley and Mauldin [3] and Arbeiter and Patzschke [1] verified the Multifractal Formalism for self-similar measures satisfying the open set condition (OSC), and within the last 10 years the multifractal spectra of various classes of measures in Euclidean space $\mathbb{R}^d$ exhibiting some degree of self-similarity have been computed rigorously, cf. the textbooks [12, 42] and the references therein.

3 Multifractal Tubes

3.1 Multifractal Tubes

Motivated by Lapidus and van Frankenhuysen investigations [29, 30] of tube formulas for fractal sets, it is natural to develop a theory of multifractal tube formulas for multifractal measures. In this section we will present a framework attempting to do this. As an example, we will also give a complete description of the asymptotic behaviour of the multifractal tube formulas for self-similar measures satisfying the Open Set Condition.

Multifractal tube formulas are defined as follows. First note that if $r > 0$ and E is a subset of $\mathbb{R}^d$, then the Minkowski volume $V_r(E)$ is given by

$$V_r(E) = \frac{1}{r^d}\mathscr{L}^d(B(E,r)) = \frac{1}{r^d}\int\limits_{B(E,r)} \mathrm{d}\mathscr{L}^d(x),$$

where we have rewritten the Lebesgue measure $\mathscr{L}^d(B(E,r))$ of $B(E,r)$ as the integral $\int_{B(E,r)} \mathrm{d}\mathscr{L}^d(x)$. Motivated by the Renyi dimensions (i.e. Eqs. (2) and (4)) and the above expression for $V_r(E)$, we now define the multifractal Minkowski volume as follows. Namely, let $r > 0$ and E be a subset of $\mathbb{R}^d$. For real number q and a Borel measure μ on $\mathbb{R}^d$, we now define the multifractal q Minkowski volume $V^q_{\mu,r}(E)$ of E with respect to the measure μ by

$$V^q_{\mu,r}(E) = \frac{1}{r^d} \int\limits_{B(E,r)} \mu(B(x,r))^q \,\mathrm{d}\mathscr{L}^d(x)\,.$$

Note, that if $q = 0$, then the q multifractal Minkowski volume $V^q_{\mu,r}(E)$ reduces to the usual Minkowski volume, i.e.

$$V^0_{\mu,r}(E) = V_r(E)\,.$$

The importance of the Renyi dimensions in multifractal analysis together with the formal resemblance between the multifractal Minkowski volume $V^q_{\mu,r}(E)$ and the moments $I^q_{\mu,r}(E)$ used in the definition the Renyi dimensions may be seen as a justification for calling the quantity $V^q_{\mu,r}(E)$ for the *multifractal* Minkowski volume; a further justification for this terminology will be proved below.

Using the multifractal Minkowski volume we can define multifractal Minkowski dimensions. For real number q and a Borel measure μ on $\mathbb{R}^d$, we define the lower and upper multifractal q Minkowski dimension of E, by

$$\begin{aligned}\underline{\dim}^q_{\mathsf{M},\mu}(E) &= \liminf_{r\searrow 0} \tfrac{\log V^q_{\mu,r}(E)}{-\log r}\,,\\ \overline{\dim}^q_{\mathsf{M},\mu}(E) &= \limsup_{r\searrow 0} \tfrac{\log V^q_{\mu,r}(E)}{-\log r}\,.\end{aligned}$$

Again we note the close similarity between the multifractal Minkowski dimensions and the Renyi dimensions. Indeed, the next proposition shows that this similarity is not merely a formal resemblance. In fact, for $q \geq 0$, the multifractal Minkowski dimensions and the Renyi dimensions coincide. This clearly provides further justification for calling the quantity $V^q_{\mu,r}(E)$ for the *multifractal* Minkowski volume.

Proposition 1 ([38]). *Let μ be a Borel measure on $\mathbb{R}^d$ and $E \subseteq \mathbb{R}^d$. If $q \geq 0$, then*

$$\begin{aligned}\underline{\dim}^q_{\mathsf{M},\mu}(E) &= \underline{\dim}^q_{\mathsf{R},\mu}(E)\,,\\ \overline{\dim}^q_{\mathsf{M},\mu}(E) &= \overline{\dim}^q_{\mathsf{R},\mu}(E)\,.\end{aligned}$$

In particular, if $q \geq 0$, then

$$\underline{\dim}^q_{\mathsf{M},\mu}(\operatorname{supp}\mu) = \underline{\tau}_\mu(q)\,,$$

$$\overline{\dim}^q_{\mathsf{M},\mu}(\operatorname{supp}\mu) = \overline{\tau}_\mu(q).$$

Proof. This follows easily from the definitions. □

Having defined multifractal Minkowski dimensions, we also define multifractal Minkowski content and average multifractal Minkowski content. For real numbers q and t, we define the lower and upper (q,t)-dimensional multifractal Minkowski content of E with respect to μ by

$$\underline{M}^{q,t}_\mu(E) = \liminf_{r\searrow 0} \tfrac{1}{r^{-t}} V^q_{\mu,r}(E),$$
$$\overline{M}^{q,t}_\mu(E) = \limsup_{r\searrow 0} \tfrac{1}{r^{-t}} V^q_{\mu,r}(E).$$

If $\underline{M}^{q,t}_\mu(E) = \overline{M}^{q,t}_\mu(E)$, i.e. if the limit $\lim_{r\searrow 0} \frac{1}{r^{-t}} V^q_{\mu,r}(E)$ exists, then we say the E is (q,t) multifractal Minkowski measurable with respect to μ, and we will denote the common value of $\underline{M}^{q,t}_\mu(E)$ and $\overline{M}^{q,t}_\mu(E)$ by $M^{q,t}_\mu(E)$, i.e. we will write

$$M^{q,t}_\mu(E) = \underline{M}^{q,t}_\mu(E) = \overline{M}^{q,t}_\mu(E).$$

Of course, sets may not be multifractal Minkowski measurable, and it is therefore useful to introduce a suitable averaging procedure when computing the multifractal Minkowski content. Motivated by this we define the lower and upper (q,t)-dimensional average multifractal Minkowski content of E with respect to μ by

$$\underline{M}^{q,t}_{\mu,\mathsf{ave}}(E) = \liminf_{r\searrow 0} \tfrac{1}{-\log r} \int_r^1 \tfrac{1}{s^{-t}} V^q_{\mu,s}(E)\, \tfrac{\mathrm{d}s}{s},$$
$$\overline{M}^{q,t}_{\mu,\mathsf{ave}}(E) = \liminf_{r\searrow 0} \tfrac{1}{-\log r} \int_r^1 \tfrac{1}{s^{-t}} V^q_{\mu,s}(E)\, \tfrac{\mathrm{d}s}{s}.$$

If $\underline{M}^{q,t}_{\mu,\mathsf{ave}}(E) = \overline{M}^{q,t}_{\mu,\mathsf{ave}}(E)$, i.e. if the limit $\lim_{r\searrow 0} \frac{1}{-\log r} \int_r^1 \frac{1}{s^{-t}} V^q_{\mu,s}(E)\, \frac{\mathrm{d}s}{s}$ exists, then we say the E is (q,t) average multifractal Minkowski measurable with respect to μ, and we will denote the common value of $\underline{M}^{q,t}_{\mu,\mathsf{ave}}(E)$ and $\overline{M}^{q,t}_{\mu,\mathsf{ave}}(E)$ by $M^{q,t}_{\mu,\mathsf{ave}}(E)$, i.e. we will write

$$M^{q,t}_{\mu,\mathsf{ave}}(E) = \underline{M}^{q,t}_{\mu,\mathsf{ave}}(E) = \overline{M}^{q,t}_{\mu,\mathsf{ave}}(E).$$

3.2 Multifractal Tubes of Self-similar Measures

As an example, we will now compute the multifractal Minkowski content of self-similar measures. We first recall the definition of self-similar measures. Let $S_i : \mathbb{R}^d \to \mathbb{R}^d$ for $i = 1,\ldots,N$ be contracting similarities and let $(p_1,\ldots,p_N)$ be a probability vector. We denote the Lipschitz constant of S_i by $r_i \in (0,1)$. Let K and μ be the self-similar set associated with the list $(S_1,\ldots,S_N)$ and the self-similar measure associated with the list $(S_1,\dots,S_N,p_1,\dots,p_N)$, i.e. K is the unique non-empty compact subset of $\mathbb{R}^d$ such that

$$K=\bigcup_i S_i(K)\,, \tag{5}$$

and μ the unique Borel probability measure on $\mathbb{R}^d$ such that

$$\mu=\sum_i p_i\mu\circ S_i^{-1}\,, \tag{6}$$

cf. [24]. We note that it is well-known that $\operatorname{supp}\mu=K$.

We will frequently assume that the list $(S_1,\dots,S_N)$ satisfies certain "disjointness" conditions, viz the OSC or the strong separation condition (SSC) defined below.

The Open Set Condition: There exists an open non-empty and bounded subset U of $\mathbb{R}^d$ with $\cup_i S_iU\subseteq U$ and $S_iU\cap S_jU=\varnothing$ for all i,j with $i\neq j$.

The Strong Separation Condition: There exists an open non-empty and bounded subset U of $\mathbb{R}^d$ with $\cup_i S_iU\subseteq U$ and $\overline{S_iU}\cap\overline{S_jU}=\varnothing$ for all i,j with $i\neq j$.

Multifractal analysis of self-similar measures has attracted an enormous interest during the past 20 years. For example, using methods from ergodic theory, Peres and Solomyak [43] have recently shown that for any self-similar measure μ, the Renyi dimensions always exist, i.e. the limit $\lim_{r\searrow 0}\frac{\log I_{\mu,r}^q(K)}{-\log r}$ always exists regardless of whether or not the OSC is satisfied provided $q\geq 0$. If in addition the OSC is satisfied, an explicit expression for the two limits $\underline{\tau}_\mu(q)=\liminf_{r\searrow 0}\frac{\log I_{\mu,r}^q(K)}{-\log r}$ and $\overline{\tau}_\mu(q)=\limsup_{r\searrow 0}\frac{\log I_{\mu,r}^q(K)}{-\log r}$ can be obtained. Indeed, Arbeiter and Patzschke [1] and Cawley and Mauldin [3] proved that if the OSC is satisfied, then

$$\begin{aligned}\underline{\tau}_\mu(q)&=\liminf_{r\searrow 0}\frac{\log I_r^q(K)}{-\log r}\\&=\beta(q)\,,\\\overline{\tau}_\mu(q)&=\limsup_{r\searrow 0}\frac{\log I_r^q(K)}{-\log r}\\&=\beta(q)\,,\end{aligned} \tag{7}$$

for $q\in\mathbb{R}$, where $\beta(q)$ is defined by

$$\sum_i p_i^q r_i^{\beta(q)}=1\,. \tag{8}$$

Arbeiter and Patzschke [1] and Cawley and Mauldin [3] also verified the Multifractal Formalism for self-similar measures satisfying the OSC. Namely, in [1, 3], it is proved that if μ is a self-similar measure satisfying the OSC, then

$$f_\mu(\alpha) = \beta^*(\alpha)$$

for all $\alpha \geq 0$; recall, that the definition of the Legendre transform φ^* of a real-valued function $\varphi : \mathbb{R} \to \mathbb{R}$ is given in Sect. 2.3. We continue this line of investigation by computing the multifractal Minkowski dimensions and multifractal Minkowski content of self-similar measures satisfying various separation conditions. Firstly, we note that the multifractal Minkowski dimensions coincide with $\beta(q)$. This is not a deep fact and is included mainly for completeness.

Theorem 1 ([38]). *Let K and μ be given by Eqs. (5) and (6). Fix $q \in \mathbb{R}$ and assume that Condition (i) or Condition (ii) below is satisfied.*

(i) The OSC is satisfied and $0 \leq q$.
(ii) The SSC is satisfied.

Then we have

$$\underline{\dim}^q_{\mathsf{M},\mu}(K) = \overline{\dim}^q_{\mathsf{M},\mu}(K) = \beta(q)$$

for all $q \in \mathbb{R}$.

Proof. As noted above, this is not a deep fact and follows from the definitions using standard arguments similar to those in [1] or Falconer's textbook [12]. □

Next, we give a complete description of the asymptotic behaviour of the multifractal tube formulas for self-similar measures satisfying the Open Set Condition. In particular, we prove that if the set $\{\log r_1^{-1}, \ldots, \log r_N^{-1}\}$ is not contained in a discrete additive subgroup of $\mathbb{R}$, then K is $(q, \beta(q))$ multifractal Minkowski measurable with respect to μ, and if the set $\{\log r_1^{-1}, \ldots, \log r_N^{-1}\}$ is contained in a discrete additive subgroup of $\mathbb{R}$, then K is $(q, \beta(q))$ average multifractal Minkowski measurable with respect to μ. This is the content of Theorem 2. The proof of Theorem 2 is based on Renewal Theory and will be discussed after the statement of the theorem.

Theorem 2 ([38]). *Let K and μ be given by Eqs. (5) and (6). Fix $q \in \mathbb{R}$ and assume that Condition (i) or Condition (ii) below is satisfied.*

(i) The OSC is satisfied and $0 \leq q$.
(ii) The SSC is satisfied.

Define $\lambda_q : (0, \infty) \to \mathbb{R}$ by

$$\lambda_q(r) = V^q_{\mu,r}(K) - \sum_i p_i^q \mathbf{1}_{(0,r_i]}(r) V^q_{\mu, r_i^{-1} r}(K).$$

Then we have the following:

1. *If the set $\{\log r_1^{-1}, \ldots, \log r_N^{-1}\}$ is not contained in a discrete additive subgroup of $\mathbb{R}$, then*

$$\frac{1}{r^{-\beta(q)}} V^q_{\mu,r}(K) = c_q + \varepsilon_q(r),$$

where $c_q \in \mathbb{R}$ *is the constant given by*

$$c_q = \frac{1}{-\sum_i p_i^q r_i^{\beta(q)} \log r_i} \int_0^1 r^{\beta(q)} \lambda_q(r) \frac{\mathrm{d}r}{r}$$

and $\varepsilon_q(r) \to 0$ *as* $r \searrow 0$. *In addition,* K *is* $(q,\beta(q))$ *multifractal Minkowski measurable with respect to* μ*:*

$$M^{q,\beta(q)}_{\mu}(K) = \frac{1}{-\sum_i p_i^q r_i^{\beta(q)} \log r_i} \int_0^1 r^{\beta(q)} \lambda_q(r) \frac{\mathrm{d}r}{r}.$$

2. *If the set* $\{\log r_1^{-1}, \ldots, \log r_N^{-1}\}$ *is contained in a discrete additive subgroup of* $\mathbb{R}$ *and* $\langle \log r_1^{-1}, \ldots, \log r_N^{-1} \rangle = u\mathbb{Z}$ *with* $u > 0$*, then*

$$\frac{1}{r^{-\beta(q)}} V^q_{\mu,r}(K) = \pi_q(r) + \varepsilon_q(r),$$

where $\pi_q : (0,\infty) \to \mathbb{R}$ *is the multiplicatively periodic function with period equal to* e^u*, i.e.*

$$\pi_q(e^u r) = \pi_q(r)$$

for all $r \in (0,\infty)$*, given by*

$$\pi_q(r) = \frac{1}{-\sum_i p_i^q r_i^{\beta(q)} \log r_i} \sum_{n \in \mathbb{Z}, re^{nu} \le 1} (re^{un})^{\beta(q)} \lambda_q(re^{un})\, u$$

and $\varepsilon_q(r) \to 0$ *as* $r \searrow 0$. *In addition,* K *is* $(q,\beta(q))$ *average multifractal Minkowski measurable with respect to* μ *with*

$$M^{q,\beta(q)}_{\mu,\mathrm{ave}}(K) = \frac{1}{-\sum_i p_i^q r_i^{\beta(q)} \log r_i} \int_0^1 r^{\beta(q)} \lambda_q(r) \frac{\mathrm{d}r}{r}.$$

It is instructive to consider the special case $q = 0$. Indeed, since the multifractal Minkowski volume for $q = 0$ equals the usual Minkowski volume and since the (q,t)-dimensional multifractal Minkowski content for $q = 0$ equals the usual t-dimensional Minkowski content, the following corollary providing formulas for the asymptotic behaviour of the Minkowski volume of self-similar sets follows immediately from Theorem 2 by putting $q = 0$. This result was first obtained by Gatzouras [20] and later by Winter [51].

Corollary 1 ([20]). *Let* K *be given by Eqs. (5) and (6). Assume that the OSC is satisfied. Let* t *denote the common value of the box dimensions and the Hausdorff* *dimension of K, i.e. t is the unique number such that* $\sum_i r_i^t = 1$ *(see [12] or [24]). Define* $\lambda : (0,\infty) \to \mathbb{R}$ *by*

$$\lambda(r) = V_r(K) - \sum_i \mathbf{1}_{(0,r_i]}(r)\, V_{r_i^{-1}r}(K).$$

Then we have:

1. *If the set* $\{\log r_1^{-1}, \dots, \log r_N^{-1}\}$ *is not contained in a discrete additive subgroup of* $\mathbb{R}$*, then*
$$\frac{1}{r^{-t}} V_r(K) = c + \varepsilon(r),$$
where $c \in \mathbb{R}$ *is the constant given by*
$$c = \frac{1}{-\sum_i r_i^t \log r_i} \int_0^1 r^{\beta(q)} \lambda(r) \frac{\mathrm{d}r}{r}$$
and $\varepsilon(r) \to 0$ *as* $r \searrow 0$*. In addition, K is t Minkowski measurable with*
$$M^t(K) = \tfrac{1}{-\sum_i r_i^t \log r_i} \int_0^1 r^t \lambda(r) \tfrac{\mathrm{d}r}{r}.$$

2. *If the set* $\{\log r_1^{-1}, \dots, \log r_N^{-1}\}$ *is contained in a discrete additive subgroup of* $\mathbb{R}$ *and* $\langle \log r_1^{-1}, \dots, \log r_N^{-1} \rangle = u\mathbb{Z}$ *with* $u > 0$*, then*
$$\frac{1}{r^{-t}} V_r(K) = \pi(r) + \varepsilon(r),$$
where $\pi : (0,\infty) \to \mathbb{R}$ *is the multiplicatively periodic function with period equal to* e^u*, i.e.*
$$\pi(e^u r) = \pi(r)$$
for all $r \in (0,\infty)$*, given by*
$$\pi(r) = \frac{1}{-\sum_i r_i^t \log r_i} \sum_{n \in \mathbb{Z}, re^{nu} \le 1} (re^{un})^t \lambda(re^{un})\, u$$
and $\varepsilon(r) \to 0$ *as* $r \searrow 0$*. In addition, K is t average Minkowski measurable with*
$$M^t_{\mathsf{ave}}(K) = \tfrac{1}{-\sum_i r_i^t \log r_i} \int_0^1 r^t \lambda(r) \tfrac{\mathrm{d}r}{r}.$$

Proof. Since $\beta(0) = \underline{\dim}_{\mathsf{B}}(K) = \overline{\dim}_{\mathsf{B}}(K) = \dim(K) = t$ (see [12] or [24]) and $V^0_{\mu,r}(K) = V_r(K)$, this follows from Theorem 2 by putting $q = 0$. □

3.3 How Does One Prove Theorem 2 on the Asymptotic Behaviour of Multifractal Tubes of Self-similar Measures?

How does one prove Theorem 2? The proof is based on Renewal Theory and, in particular, on a very recent renewal theorem by Levitin and Vassiliev [31]. Below we state Levitin and Vassiliev's Renewal Theorem.

Theorem 3 (Levitin and Vassiliev's Renewal Theorem [31]). *Let $t_1,\dots,t_N > 0$ and $p_1,\dots,p_N > 0$ with $\sum_i p_i = 1$. Define the probability measure P by*

$$P = \sum_i p_i \delta_{t_i}\,.$$

Let $\lambda, \Lambda : \mathbb{R} \to \mathbb{R}$ be real-valued functions satisfying the following conditions:

1. *The function λ is piecewise continuous.*
2. *There are constants $c, k > 0$ such that*

   $$|\lambda(t)| \leq ce^{-k|t|}$$

   for all $t \in \mathbb{R}$.
3. *We have*

   $$\Lambda(t) \to 0 \text{ as } t \to -\infty.$$

4. *We have*

   $$\Lambda(t) = \int \Lambda(t-s)\,\mathrm{d}P(s) + \lambda(t)$$

   for all $t \in \mathbb{R}$.

Then the following holds:

1. *The non-arithmetic case: If $\{t_1,\dots,t_N\}$ is not contained in a discrete additive subgroup of $\mathbb{R}$, then*

   $$\Lambda(t) = c + \varepsilon(t)$$

   for all $t \in \mathbb{R}$ where

   $$c = \frac{1}{\int s\,\mathrm{d}P(s)} \int \lambda(s)\,\mathrm{d}s$$

   and $\varepsilon(t) \to 0$ as $t \to \infty$. In addition,

   $$\frac{1}{T}\int_0^T \Lambda(t)\,\mathrm{d}t \to c = \frac{1}{\int s\,\mathrm{d}P(s)} \int \lambda(s)\,\mathrm{d}s \text{ as } T \to \infty. \quad (9)$$

2. *The arithmetic case: If $\{t_1,\dots,t_N\}$ is contained in a discrete additive subgroup of $\mathbb{R}$ and $\langle t_1,\dots,t_N\rangle = u\mathbb{Z}$ with $u > 0$, then*

   $$\Lambda(t) = \pi(t) + \varepsilon(t)$$

for all $t \in \mathbb{R}$ where $\pi : \mathbb{R} \to \mathbb{R}$ is the periodic function with period equal to u, i.e.

$$\pi(t+u) = \pi(t)$$

for all $t \in \mathbb{R}$, given by

$$\pi(t) = \frac{1}{\int s\,\mathrm{d}P(s)} u \sum_{n\in\mathbb{Z}} \lambda(t+nu)$$

and $\varepsilon(t) \to 0$ as $t \to \infty$. In addition

$$\frac{1}{T}\int_0^T \Lambda(t)\,\mathrm{d}t \to c = \frac{1}{\int s\,\mathrm{d}P(s)} \int \lambda(s)\,\mathrm{d}s \text{ as } T \to \infty. \tag{10}$$

Proof. All statements, except Eqs. (9) and (10), follow [31]. Below we prove Eqs. (9) and (10). Indeed, Eq. (9) follows immediately and Eq. (10) is proved as follows. Namely, since π is periodic with period equal to u, we conclude that

$$\begin{aligned} \frac{1}{T}\int_0^T \Lambda(t)\,\mathrm{d}t &= \frac{1}{T}\int_0^T \pi(t)\,\mathrm{d}t + \frac{1}{T}\int_0^T \varepsilon(t)\,\mathrm{d}t \\ &\to \frac{1}{u}\int_0^u \pi(t)\,\mathrm{d}t \\ &= \frac{1}{\int t\,\mathrm{d}P(t)} \int_0^u \sum_{n\in\mathbb{Z}} \lambda(t+nu)\,\mathrm{d}t\,. \end{aligned} \tag{11}$$

Next, observe that since $|\lambda(t)| \le ce^{-k|t|}$ for all $t \in \mathbb{R}$ and $\int ce^{-k|t|}\,\mathrm{d}t < \infty$, it follows from two applications of Lebesgue's Dominated Convergence Theorem and the fact that π is periodic with period equal to u that

$$\begin{aligned} \int_0^u \sum_{n\in\mathbb{Z}} \lambda(t+nu)\,\mathrm{d}t &= \sum_{n\in\mathbb{Z}} \int_0^u \lambda(t+nu)\,\mathrm{d}t \\ &= \sum_{n\in\mathbb{Z}} \int_{nu}^{(n+1)u} \lambda(t)\,\mathrm{d}t \\ &= \sum_{n\in\mathbb{Z}} \int \mathbf{1}_{[nu,(n+1)u)}(t)\,\lambda(t)\,\mathrm{d}t \\ &= \int \sum_{n\in\mathbb{Z}} \mathbf{1}_{[nu,(n+1)u)}(t)\,\lambda(t)\,\mathrm{d}t \\ &= \int \lambda(t)\,\mathrm{d}t\,. \end{aligned} \tag{12}$$

Finally, combining Eqs. (11) and (12) shows that

$$\frac{1}{T}\int_0^T \Lambda(t)\,\mathrm{d}t \to \frac{1}{\int t\,\mathrm{d}P(t)}\int \lambda(t)\,\mathrm{d}t\,.$$

This completes the proof. □

The key difference between Levitin and Vassiliev's Renewal Theorem and the classical renewal theorem from Feller's books [15, 16] is the conclusion in the arithmetic case. While the assumptions in the classical renewal theorem are weaker, the conclusion in the arithmetic case is also weaker. More precisely, in the arithmetic case, Levitin and Vassiliev's Renewal Theorem says that the error-term $\varepsilon(t)$ tends to 0 as t tends to infinity, i.e.

$$\lim_{t\to\infty}\varepsilon(t)=0\,,$$

whereas the classical renewal theorem only allows us to conclude that the error-term $\varepsilon(t)$ tends to 0 as t tends to infinity through "steps" of length u, i.e.

$$\lim_{\substack{n\in\mathbb{N}\\ n\to\infty}}\varepsilon(nu+s)=0$$

for all $s\in\mathbb{R}$.

Using Levitin and Vassiliev's Renewal Theorem (Theorem 3) we can now prove Theorem 2. Below is a sketch of the proof.

Sketch of Proof of Theorem 2

In order to prove Theorem 2, we will apply Levitin and Vassiliev's Renewal Theorem to the probability measure $P=P_q$ and the functions $\lambda=\lambda_q^0$ and $\Lambda=\Lambda_q^0$ defined below. First recall that $\lambda_q:(0,\infty)\to\mathbb{R}$ is defined by

$$\lambda_q(r)=V_{\mu,r}^q(K)-\Sigma_i\,p_i^q\,\mathbf{1}_{(0,r_i]}(r)\,V^q_{\mu,r_i^{-1}r}(K)\,,$$

and define $\Lambda_q:(0,\infty)\to\mathbb{R}$ by

$$\Lambda_q(r)=V_{\mu,r}^q(K)\,.$$

We can now define the functions $\lambda_q^0,\Lambda_q^0:\mathbb{R}\to\mathbb{R}$. Namely, define $\lambda_q^0:\mathbb{R}\to\mathbb{R}$ by

$$\lambda_q^0(t)=\mathbf{1}_{[0,\infty)}(t)\,e^{-t\beta(q)}\lambda_q(e^{-t})\,,$$

and define $\Lambda_q^0:\mathbb{R}\to\mathbb{R}$ by

$$\Lambda_q^0(t)=\mathbf{1}_{[0,\infty)}(t)\,e^{-t\beta(q)}\Lambda_q(e^{-t})\,.$$

Finally, define the probability measure P_q by

$$P_q = \sum_i p_i^q r_i^{\beta(q)} \delta_{\log r_i^{-1}}.$$

The crux of the matter now is to show that the probability measure $P = P_q$ and the functions $\lambda = \lambda_q^0$ and $\Lambda = \Lambda_q^0$ satisfy conditions (1)–(4) in Levitin and Vassiliev's Renewal Theorem.

Condition (1) is satisfied. This is not difficult to show. Indeed, it follows by applying results from [34] that the function $f : (0,\infty) \to \mathbb{R}$ defined by $f(r) = \int_{B(K,r)} \mu(B(x,r))^q \, d\mathscr{L}^d(x)$ is continuous. This clearly implies that condition (1) is satisfied.

Condition (2) is satisfied. This is the difficult part of the proof and requires a number of very delicate estimates. In particular, the proof of condition (2) is based on the three key estimates below. The proofs of Key Estimate 2 and Key Estimate 3 are both highly technical and require a number very delicate estimates. Below we state the three key estimates. However, we will not prove the estimates. Instead the reader is referred to [38]. Before we can state the key estimates we need to introduce some notation. For $i \neq j$ and $r > 0$, let

$$Q_{i,j}^q(r) = \frac{1}{r^d} \int_{B(S_iK,r)\cap B(S_jK,r)} \mu(B(x,r))^q \, d\mathscr{L}^d(x).$$

Let $\Sigma = \{1,\ldots,N\}$ and write

$$\begin{aligned} \Sigma^m &= \{1,\ldots,N\}^m, \\ \Sigma^* &= \bigcup_m \Sigma_m, \end{aligned}$$

i.e. Σ^m is the family of all strings $\mathbf{i} = i_1 \ldots i_m$ of length m with $i_j \in \{1,\ldots,N\}$, and Σ^* is the family of all finite strings $\mathbf{i} = i_1 \ldots i_m$ with $i_j \in \{1,\ldots,N\}$. For $\mathbf{i} \in \Sigma^m$, we write $|\mathbf{i}| = m$ for the length of $\mathbf{i}$ and for a positive integer n with $n \leq m$, we write $\mathbf{i}|n = i_1 \ldots i_n$ for the truncation of $\mathbf{i}$ to the nth place. Also, for $\mathbf{i} = i_1 \ldots i_m, \mathbf{j} = j_1 \ldots j_n \in \Sigma^*$, let $\mathbf{ij} = i_1 \ldots i_m j_1 \ldots j_n$ denote the concatenation of $\mathbf{i}$ and $\mathbf{j}$. Next, if $\mathbf{i} = i_1 \ldots i_m \in \Sigma^*$, we will write

$$\begin{aligned} S_{\mathbf{i}} &= S_{i_1} \circ \cdots \circ S_{i_m}, \\ r_{\mathbf{i}} &= r_{i_1} \cdots r_{i_m}, \\ p_{\mathbf{i}} &= p_{i_1} \cdots p_{i_m}. \end{aligned} \tag{13}$$

Also for brevity, put $r_{\min} = \min_{i=1,\ldots,N} r_i$ and $r_{\max} = \max_{i=1,\ldots,N} r_i$.

For $\mathbf{i},\mathbf{h} \in \Sigma^*$, we write $\mathbf{i} \prec \mathbf{h}$ if and only if $\mathbf{i}$ is a substring of $\mathbf{h}$, i.e. if and only if there are strings $\mathbf{s},\mathbf{t} \in \Sigma^*$ such that $\mathbf{h} = \mathbf{sit}$. If $(S_1,\ldots,S_n)$ satisfies the OSC, then it follows from a result by Schief [46] that there exists an open, bounded and non-empty subset U of $\mathbb{R}^d$ with $\cup_i S_i U \subseteq U$, $S_i U \cap S_j U = \varnothing$ for all i, j with $i \neq j$, and $U \cap K \neq \varnothing$. In particular, since $U \cap K \neq \varnothing$, we can choose $\mathbf{l} \in \Sigma^*$ such that

$$S_{\mathbf{l}} K \subseteq U\,, \tag{14}$$

and the compactness of $S_{\mathbf{l}} K$ now implies that $d_0 = \operatorname{dist}(S_{\mathbf{l}} K, \mathbb{R}^d \setminus U) > 0$. For brevity write $D_0 = \mathrm{d}K$. Choose a positive integer M such that $\frac{1}{r_{\max}^{M-1}} \geq 2\frac{D_0}{d_0}$, and put $a = \frac{1}{D_0}\frac{r_{\min}}{r_{\max}^{M+1}}$ and $b = \frac{1}{D_0}\frac{1}{r_{\min}^{M+1}}$. Finally, define $Z^q : (0,\infty) \to \mathbb{R}$ by

$$Z^q(r) = \sum_{\mathbf{h} \in \Sigma^*, |\mathbf{h}| \geq |\mathbf{l}|, ar \leq r_{\mathbf{h}} \leq br, \mathbf{l} \not\prec \mathbf{h}} p_{\mathbf{h}}^q .$$

The three key estimates are now:

Key Estimate 1. $|\lambda_q(r)| \leq \sum_{i \neq j} Q_{i,j}^q(r)$ *for all* $0 < r < r_{\min}$.
Key Estimate 2. *There is a constant* $c > 0$ *such that*

$$\sum_{i \neq j} Q_{i,j}^q(r) \leq \begin{cases} c Z^q(\frac{1}{2}r) & \textit{for } q < 0 \textit{ and all } r > 0, \\ c Z^q(2r) & \textit{for } 0 \leq q \textit{ and all } r > 0. \end{cases}$$

Key Estimate 3. *There are constants* $k > 0$ *and* $\gamma(q) \in \mathbb{R}$ *with* $\gamma(q) < \beta(q)$ *such that*

$$Z^q(r) \leq k r^{-\gamma(q)} \quad \textit{for all } r > 0.$$

Combining the three key estimates we can now prove that condition (2) is satisfied. Indeed, choose $t_0 > 0$ such that $e^{-t} < r_{\min}$ for $t \geq t_0$. For $t \geq t_0$, we now have

$$|\lambda_q^0(t)| = \mathbf{1}_{[0,\infty)}(t)\, e^{-t\beta(q)}\, |\lambda_q(e^{-t})| \tag{15}$$

$$\leq e^{-t\beta(q)} \sum_{i \neq j} Q_{i,j}^q(e^{-t}) \qquad \text{[by Key Estimate 1]} \tag{16}$$

$$\leq \begin{cases} e^{-t\beta(q)} c Z^q(\frac{1}{2}e^{-t}) & \text{for } q < 0, \\ e^{-t\beta(q)} c Z^q(2e^{-t}) & \text{for } 0 \leq q \end{cases} \qquad \text{[by Key Estimate 2]} \tag{17}$$

$$\leq \begin{cases} e^{-t\beta(q)}\, ck\, (\frac{1}{2}e^{-t})^{-\gamma(q)} & \text{for } q < 0, \\ e^{-t\beta(q)}\, ck\, (2e^{-t})^{-\gamma(q)} & \text{for } 0 \leq q \end{cases} \qquad \text{[by Key Estimate 3]} \tag{18}$$

$$\tag{19}$$

$$= c_0 e^{-(\beta(q) - \gamma(q))t}, \tag{20}$$

where $c_0 = ck \max((\frac{1}{2})^{-\gamma(q)}, 2^{-\gamma(q)})$.

Next, since λ_q^0 is piecewise continuous (by condition (1)), we conclude that λ_q^0 is bounded on the compact interval $[0,t_0]$, and we therefore deduce that there is a constant M_0 such that $|\lambda_q^0(t)| \leq M_0$ for all $t \in [0,t_0]$. It follows from this and Eq. (20) that

$$|\lambda_q^0(t)| \leq \max\left(\frac{M_0}{e^{-(\beta(q)-\gamma(q))t_0}}\,,\, c_0\right) e^{-(\beta(q)-\gamma(q))t} \tag{21}$$

for all $t \geq 0$.

Inequality Eq. (21) and the fact that $\lambda_q^0(t) = 0$ for all $t < 0$ now prove that condition (2) is satisfied.

Condition (3) is satisfied. This follows trivially from the fact that $\Lambda_q^0(t) = 0$ for all $t < 0$.

Condition (4) is satisfied. Indeed, it follows immediately from the definitions of λ_q^0, Λ_q^0 and P_q that

$$\begin{aligned}
\Lambda_q^0(t) &= \mathbf{1}_{[0,\infty)}(t)\, e^{-t\beta(q)} \Lambda_q(e^{-t}) \\
&= \mathbf{1}_{[0,\infty)}(t)\, e^{-t\beta(q)} \left(\sum_i p_i^q\, \mathbf{1}_{(0,r_i]}(e^{-t})\, V^q_{\mu,r_i^{-1}e^{-t}}(K) + \lambda_q(e^{-t}) \right) \\
&= \sum_i p_i^q\, e^{-t\beta(q)}\, \mathbf{1}_{(0,r_i]}(e^{-t})\, \mathbf{1}_{[0,\infty)}(t)\, V^q_{\mu,r_i^{-1}e^{-t}}(K) + \lambda_q^0(t) \\
&= \sum_i p_i^q r_i^{\beta(q)}\, \mathbf{1}_{[0,\infty)}(t-\log r_i^{-1})\, \mathbf{1}_{[0,\infty)}(t)\, e^{-\beta(q)(t-\log r_i^{-1})} V^q_{\mu,e^{-(t-\log r_i^{-1})}}(K) + \lambda_q^0(t) \\
&= \sum_i p_i^q r_i^{\beta(q)}\, \mathbf{1}_{[0,\infty)}(t-\log r_i^{-1})\, e^{-\beta(q)(t-\log r_i^{-1})} V^q_{\mu,e^{-(t-\log r_i^{-1})}}(K) + \lambda_q^0(t) \\
&= \sum_i p_i^q r_i^{\beta(q)} \Lambda_q^0(t-\log r_i^{-1}) + \lambda_q^0(t) \\
&= \int \Lambda_q^0(t-s)\, \mathrm{d}P_q(s) + \lambda_q^0(t)
\end{aligned}$$

for all $t \in \mathbb{R}$. This proves that condition (4) is satisfied.

Since conditions (1)–(4) are satisfied, Levitin and Vassiliev's Renewal Theorem can now be applied to the probability measure $P = P_q$ and the functions $\lambda = \lambda_q^0$ and $\Lambda = \Lambda_q^0$. We divide the proof into two cases.

Case 1. If $\{\log r_1^{-1},\ldots,\log r_N^{-1}\}$ *is not contained in a discrete additive subgroup of* $\mathbb{R}$. If $\{\log r_1^{-1},\ldots,\log r_N^{-1}\}$ is not contained in a discrete additive subgroup of $\mathbb{R}$, then Levitin and Vassiliev's Renewal Theorem implies that

$$\Lambda_q^0(t) = c_q + \varepsilon_q^0(t),$$

where $c_q \in \mathbb{R}$ is the constant given by

$$\begin{aligned} c_q &= \frac{1}{\int s\,\mathrm{d}P_q(s)} \int \lambda_q^0(s)\,\mathrm{d}s \\ &= \frac{1}{-\sum_i p_i^q r_i^{\beta(q)} \log r_i} \int_0^\infty e^{-s\beta(q)} \lambda_q(e^{-s})\,\mathrm{d}s \\ &= \frac{1}{-\sum_i p_i^q r_i^{\beta(q)} \log r_i} \int_0^1 r^{\beta(q)} \lambda_q(r)\,\frac{\mathrm{d}r}{r} \end{aligned}$$

and

$$\varepsilon_q^0(t) \to 0 \text{ as } t \to \infty.$$

In particular, we have

$$r^{\beta(q)} V_{\mu,r}^q(K) = \Lambda_q^0(\log \tfrac{1}{r}) = c_q + \varepsilon_q(r), \tag{22}$$

where $\varepsilon_q(r) = \varepsilon_q^0(\log \frac{1}{r}) \to 0$ as $r \searrow 0$.

Finally, it follows from Eq. (22) that

$$r^{\beta(q)} V_{\mu,r}^q(K) \to c_q \text{ as } r \searrow 0.$$

This completes the proof of Theorem 2 in Case 1.

Case 2. If $\{\log r_1^{-1}, \dots, \log r_N^{-1}\}$ *is contained in a discrete additive subgroup of* $\mathbb{R}$. If $\{\log r_1^{-1}, \dots, \log r_N^{-1}\}$ is contained in a discrete additive subgroup of $\mathbb{R}$ and $\langle t_1, \dots, t_N \rangle = u\mathbb{Z}$ with $u > 0$, then Levitin and Vassiliev's Renewal Theorem implies that

$$\Lambda_q^0(t) = \pi_q^0(r) + \varepsilon_q^0(t),$$

where $\pi_q^0 : \mathbb{R} \to \mathbb{R}$ is the function given by

$$\begin{aligned} \pi_q^0(t) &= \frac{1}{\int s\,\mathrm{d}P_q(s)} u \sum_{n\in\mathbb{Z}} \lambda_q^0(t+nu) \\ &= \frac{1}{-\sum_i p_i^q r_i^{\beta(q)} \log r_i} u \sum_{n\in\mathbb{Z}} \lambda_q^0(t+nu) \end{aligned}$$

and

$$\varepsilon_q^0(t) \to 0 \text{ as } t \to \infty.$$

Moreover, we have

$$\pi_q^0(t+u) = \pi_q^0(t)$$

for all $t \in \mathbb{R}$, i.e. π_q^0 is additively periodic with period equal to u. In particular, we have

$$r^{\beta(q)} V_{\mu,r}^q(K) = \Lambda_q^0(\log \tfrac{1}{r}) = \pi_q(r) + \varepsilon_q(r),$$

where $\pi_q : \mathbb{R} \to \mathbb{R}$ is the function given by

$$\begin{aligned}
\pi_q(r) &= \pi_q^0(\log \tfrac{1}{r}) \\
&= \frac{1}{-\sum_i p_i^q r_i^{\beta(q)} \log r_i} u \sum_{n \in \mathbb{Z}} \lambda_q^0(\log \tfrac{1}{r} + nu) \\
&= \frac{1}{-\sum_i p_i^q r_i^{\beta(q)} \log r_i} u \sum_{n \in \mathbb{Z}} \mathbf{1}_{[0,\infty)}(\log \tfrac{1}{r} + nu)\, e^{-\beta(q)(\log \frac{1}{r} + nu)} \lambda_q(e^{-(\log \frac{1}{r} + nu)}) \\
&= \frac{1}{-\sum_i p_i^q r_i^{\beta(q)} \log r_i} u \sum_{n \in \mathbb{Z}, re^{nu} \le 1} (re^{nu})^{\beta(q)} \lambda_q(re^{nu})
\end{aligned}$$

and $\varepsilon_q(r) = \varepsilon_q^0(\log \frac{1}{r}) \to 0$ as $r \searrow 0$. Moreover, since π_q^0 is additively periodic with period equal to u, we have

$$\pi_q(e^u r) = \pi_q^0(\log \tfrac{1}{e^u r}) = \pi_q^0(\log \tfrac{1}{r} - u) = \pi_q^0(\log \tfrac{1}{r}) = \pi_q(r)$$

for all $r > 0$, i.e. π_q is multiplicatively periodic with period equal to e^u.

Finally it follows from Levitin and Vassiliev's Renewal Theorem that

$$\frac{1}{T} \int_0^T \Lambda_q^0(t)\, \mathrm{d}t \to c_q \text{ as } T \to \infty.$$

However, since

$$\begin{aligned}
\frac{1}{T} \int_0^T \Lambda_q^0(t)\, \mathrm{d}t &= \frac{1}{T} \int_0^T e^{-t\beta(q)} V_{\mu,e^{-t}}^q(K)\, \mathrm{d}t \\
&= \frac{1}{-\log e^{-T}} \int_{e^{-T}}^1 s^{\beta(q)} V_{\mu,s}^q(K) \frac{\mathrm{d}s}{s},
\end{aligned}$$

we now conclude that

$$\frac{1}{-\log r} \int_r^1 s^{\beta(q)} V_{\mu,s}^q(K) \frac{\mathrm{d}s}{s} \to c_q \text{ as } r \searrow 0.$$

This completes the proof of Theorem 2 in Case 2. □

4 Multifractal Tube Measures

4.1 Multifractal Tube Measures

The statement in Theorem 2 is a global one: it provides information about the limiting behaviour of the suitably normalized multifractal Minkowski volume

$$\frac{1}{r^{-\beta(q)}} V_{\mu,r}^{q}(K)$$

of the entire support K of μ as $r \searrow 0$. However, it is equally natural to ask for local versions of Theorem 2 describing the limiting behaviour of the normalized multifractal Minkowski volume

$$\frac{1}{r^{-\beta(q)}} V_{\mu,r}^{q}(E)$$

of (well-behaved) subsets E of the support of μ as $r \searrow 0$. In order to address this question, we now introduce multifractal tube measures. A further motivation for introducing multifractal tube measures comes from convex geometry and will be discussed below.

The multifractal tube measures are defined as follows. Fix a Borel measure μ on $\mathbb{R}^d$ and $r > 0$. For a real number q, we define the multifractal Minkowski tube measure $\mathscr{I}_{\mu,r}^{q}$ by

$$\mathscr{I}_{\mu,r}^{q}(E) = \frac{1}{r^d} \int_{E \cap B(\operatorname{supp}\mu, r)} \mu(B(x,r))^q \, \mathrm{d}\mathscr{L}^d(x)$$

for Borel subsets E of $\mathbb{R}^d$. Of course, the measures $\mathscr{I}_{\mu,r}^{q}$ will, in general, not converge weakly as $r \searrow 0$ (indeed, this is clear since Theorem 2 shows that, in general, $\mathscr{I}_{\mu,r}^{q}(\mathbb{R}^d) = V_{\mu,r}^{q}(K)$ does not converge as $r \searrow 0$). Hence in order to ensure weak convergence of $\mathscr{I}_{\mu,r}^{q}$ as $r \searrow 0$ it is necessary to normalize the measures $\mathscr{I}_{\mu,r}^{q}$. There are two natural ways to normalized. Firstly we can normalize by volume. More precisely, we define the volume normalized multifractal tube measure $\mathscr{V}_{\mu,r}^{q}$ by

$$\mathscr{V}_{\mu,r}^{q} = \frac{1}{\mathscr{I}_{\mu,r}^{q}(\mathbb{R}^d)} \mathscr{I}_{\mu,r}^{q} .$$

Secondly, we can normalize by scaling. More precisely, we defined the lower and upper scaling normalized multifractal tube measures $\underline{\mathscr{S}}_{\mu,r}^{q}$ and $\overline{\mathscr{S}}_{\mu,r}^{q}$ by

$$\underline{\mathscr{S}}_{\mu,r}^{q} = \frac{1}{r^{-\underline{\dim}_{\mathsf{M},\mu}^{q}(\operatorname{supp}\mu)}} \mathscr{I}_{\mu,r}^{q} ,$$

$$\overline{\mathscr{S}}_{\mu,r}^{q} = \frac{1}{r^{-\overline{\dim}_{\mathsf{M},\mu}^{q}(\operatorname{supp}\mu)}} \mathscr{I}_{\mu,r}^{q} .$$

It is instructive to consider the particular case $q = 0$. To discuss this case we first make the following definition. Namely, if U is a closed subset of $\mathbb{R}^d$ and $r > 0$, the parallel volume measure $\mathsf{V}_{U,r}$ of U is defined by

$$\mathsf{V}_{U,r}(E) = \frac{\mathscr{L}^d(E \cap B(U,r))}{\mathscr{L}^d(B(U,r))},$$

see, for example, the texts [21,35,48]. We now note that if $q = 0$ and μ is any Borel measure with $\operatorname{supp}\mu = U$, then the volume normalized multifractal tube measure $\mathscr{V}^q_{\mu,r}$ simplifies to

$$\begin{aligned}\mathscr{V}^0_{\mu,r}(E) &= \frac{\mathscr{L}^d(E \cap B(\operatorname{supp}\mu,r))}{\mathscr{L}^d(B(\operatorname{supp}\mu,r))} \\ &= \frac{\mathscr{L}^d(E \cap B(U,r))}{\mathscr{L}^d(B(U,r))} \\ &= \mathsf{V}_{U,r}(E). \qquad (23)\end{aligned}$$

This observation provides a further motivation for introducing multifractal tube measures. Namely, the measure $\mathscr{V}^0_{\mu,r}(E) = \mathsf{V}_{U,r}(E)$ is closely related to the notion of curvature measures in convex geometry. Curvature measures were introduced in the 1950s and are now recognized as a very powerful tool for analysing geometric properties of convex sets; see [21, 35, 48]. Indeed, if U is a closed convex subset of $\mathbb{R}^d$ with non-empty interior and $l = 0,1,2,\ldots,d$, then it is possible to define the lth order curvature measure V^l_U associated with U. Each curvature measure V^l_U is defined as the weak limit $\mathsf{V}^l_U = \lim_{r\searrow 0}\mathsf{V}^l_{U,r}$ of a certain family $(\mathsf{V}^l_{U,r})_{r>0}$ of measures. While we will not provide the reader with the definition of the measures $\mathsf{V}^l_{U,r}$ for a general integer $l = 0,1,2,\ldots,d$ (instead the interested reader can find the definition in previously mentioned texts [21,35,48]), we do note that if $l = d$, then $\mathsf{V}^d_{U,r} = \mathsf{V}_{U,r}$. In particular, the d-th order curvature measure V^d_U is defined by

$$\begin{aligned}\mathsf{V}^d_U &= \lim_{r\searrow 0}\mathsf{V}^d_{U,r} \\ &= \lim_{r\searrow 0}\mathsf{V}_{U,r},\end{aligned}$$

where lim denotes the limit with respect to the weak topology. This and the fact that $\mathscr{V}^0_{\mu,r} = \mathsf{V}_{U,r}$ show that the weak limit

$$\lim_{r\searrow 0}\mathscr{V}^q_{\mu,r}$$

(if it exists) may be viewed as a dth order multifractal curvature measure and the study of multifractal tube measures can therefore be seen as a first attempt to create a theory of multifractal curvatures.

It is, of course, also possible to define versions of the parallel volume measure analogous to $\underline{\mathscr{S}}^q_{\mu,r}$ and $\overline{\mathscr{S}}^q_{\mu,r}$. Indeed, if U is a closed subset of $\mathbb{R}^d$ and $r > 0$, we define the lower and upper scaling parallel volume measures $\underline{\mathsf{S}}_{U,r}$ and $\overline{\mathsf{S}}_{U,r}$ of U by

$$\underline{\mathsf{S}}_{U,r}(E) = \frac{1}{r^{-\underline{\dim}_{\mathsf{M}}(U)+d}} \mathscr{L}^d(E \cap B(U,r)),$$

$$\overline{\mathsf{S}}_{U,r}(E) = \frac{1}{r^{-\overline{\dim}_{\mathsf{M}}(U)+d}} \mathscr{L}^d(E \cap B(U,r));$$

recall that $\underline{\dim}_{\mathsf{M}}$ and $\overline{\dim}_{\mathsf{M}}$ denote the lower and upper Minkowski dimension, respectively. As above, we note that if $q = 0$ and μ is any probability measure with $\operatorname{supp}\mu = U$, then the scaling normalized multifractal tube measures $\underline{\mathscr{S}}^q_{\mu,r}$ and $\overline{\mathscr{S}}^q_{\mu,r}$ simplify to

$$\underline{\mathscr{S}}^0_{\mu,r}(E) = \underline{\mathsf{S}}_{U,r}(E), \tag{24}$$

$$\overline{\mathscr{S}}^0_{\mu,r}(E) = \overline{\mathsf{S}}_{U,r}(E). \tag{25}$$

4.2 Multifractal Tube Measures of Self-similar Measures

For self-similar measures μ satisfying the OSC, we will now investigate the existence of the weak limits of the multifractal tube measures $\mathscr{V}^q_{\mu,r}$, $\underline{\mathscr{S}}^q_{\mu,r}$ and $\overline{\mathscr{S}}^q_{\mu,r}$ as $r \searrow 0$. In fact, in many cases, these limits exist and equal (the suitably) normalized multifractal Hausdorff measure restricted to the support of μ.

We start by recalling the definition of the multifractal Hausdorff measure. In an attempt to develop a general theoretical framework for studying the multifractal structure of Borel measures, Olsen [36], Pesin [41] and Peyrière [44] introduced a family of measures $\{\mathscr{H}^{q,t}_\mu \mid q,t \in \mathbb{R}\}$ based on certain generalizations of the Hausdorff measure. The measures $\mathscr{H}^{q,t}_\mu$ have subsequently been investigated further by a large number of authors, including [4, 8, 9, 23, 37, 39, 40, 45]. Let $E \subseteq \mathbb{R}^d$ and $\delta > 0$. A countable family $\mathscr{B} = (B(x_i,r_i))_i$ of closed balls in $\mathbb{R}^d$ is called a centred δ-covering of E if $E \subseteq \cup_i B(x_i,r_i)$, $x_i \in E$ and $0 < r_i < \delta$ for all i. For $E \subseteq \mathbb{R}^d$, $q,t \in \mathbb{R}$ and $\delta > 0$ write

$$\begin{aligned}
\overline{\mathscr{H}}^{q,t}_{\mu,\delta}(E) &= \inf\left\{ \sum_i \mu(B(x_i,r_i))^q (2r_i)^t \,\middle|\, (B(x_i,r_i))_i \text{ is a centred } \delta\text{-covering of } E \right\}, \\
\overline{\mathscr{H}}^{q,t}_{\mu}(E) &= \sup_{\delta>0} \overline{\mathscr{H}}^{q,t}_{\mu,\delta}(E), \\
\mathscr{H}^{q,t}_{\mu}(E) &= \sup_{F\subseteq E} \overline{\mathscr{H}}^{q,t}_{\mu}(F).
\end{aligned}$$

It follows from [36] that $\mathscr{H}_\mu^{q,t}$ is a measure on the family of Borel subsets of $\mathbb{R}^d$. The measure $\mathscr{H}_\mu^{q,t}$ is, of course, a multifractal generalization of the centred Hausdorff measure. In fact, it is easily seen that if $t \geq 0$, then $2^{-t}\mathscr{H}_\mu^{0,t} \leq \mathscr{H}^t \leq \mathscr{H}_\mu^{0,t}$ where $\mathscr{H}^t$ denotes the t-dimensional Hausdorff measure. It is also easily seen that the measure $\mathscr{H}_\mu^{q,t}$ in the usual way assign a dimension to each subset E of $\mathbb{R}^d$ (cf. [36]): there exists a unique number $\dim_\mu^q(E) \in [-\infty,\infty]$ such that

$$\mathscr{H}_\mu^{q,t}(E) = \begin{cases} \infty & \text{for} \quad t < \dim_\mu^q(E) \\ 0 & \text{for} \quad \dim_\mu^q(E) < t \end{cases}.$$

The number $\dim_\mu^q(E)$ is an obvious multifractal analogue of the Hausdorff dimension $\dim(E)$ of E. In fact, it follows immediately from the definitions that $\dim(E) = \dim_\mu^0(E)$. One of the main importances of the multifractal Hausdorff measure $\mathscr{H}_\mu^{q,t}$ is its connection with the multifractal spectrum of μ. Indeed, if we define the dimension function $b_\mu : \mathbb{R} \to [-\infty,\infty]$ by

$$b_\mu(q) = \dim_\mu^q(\operatorname{supp}\mu),$$

then it follows from [36] that the multifractal spectrum f_μ of μ (cf. Eq. (1)) is bounded above by the Legendre transform b_μ^* of b_μ, i.e.

$$f_\mu(\alpha) \leq b_\mu^*(\alpha)$$

for all $\alpha \geq 0$, cf. [36]; recall, that the definition of the Legendre transform φ^* of a real-valued function $\varphi : \mathbb{R} \to \mathbb{R}$ is given in Sect. 2.3. This inequality may be viewed as a rigorous version of the "Multifractal Formalism". Furthermore, for many natural families of measure we have $f_\mu(\alpha) = b_\mu^*(\alpha)$ for all $\alpha \geq 0$, cf. [4, 8, 9, 36, 37].

We can now explicitly identify the weak limits of the multifractal tube measures $\mathscr{V}_{\mu,r}^q$, $\underline{\mathscr{S}}_{\mu,r}^q$ and $\overline{\mathscr{S}}_{\mu,r}^q$ as $r \searrow 0$ for self-similar measures μ. The first result shows that the weak limit of $\mathscr{V}_{\mu,r}^q$ (as $r \searrow 0$) always exists and equals the normalized multifractal Hausdorff measure.

Theorem 4 ([38]). *Let K and μ be given by Eqs. (5) and (6). Fix $q \in \mathbb{R}$ and assume that Condition (i) or Condition (ii) below is satisfied.*

(i) The OSC is satisfied and $0 \leq q$.
(ii) The SSC is satisfied.

Then we have

$$\mathscr{V}_{\mu,r}^q \to \frac{1}{\mathscr{H}_\mu^{q,\beta(q)}(K)} \mathscr{H}_\mu^{q,\beta(q)} \llcorner K \qquad \textit{weakly.}$$

Next, we study the limiting behaviour of $\underline{\mathscr{S}}^q_{\mu,r}$ and $\overline{\mathscr{S}}^q_{\mu,r}$ as $r \searrow 0$ for self-similar measures μ. Contrary to Theorem 4, the weak limits of $\underline{\mathscr{S}}^q_{\mu,r}$ and $\overline{\mathscr{S}}^q_{\mu,r}$ as $r \searrow 0$ may not exist. Indeed, if the set $\{\log r_1^{-1},\dots,\log r_N^{-1}\}$ is contained in a discrete additive subgroup of $\mathbb{R}$, then the weak limits of $\underline{\mathscr{S}}^q_{\mu,r}$ and $\overline{\mathscr{S}}^q_{\mu,r}$ as $r \searrow 0$ do not necessarily exist; however, the weak limits of certain averages of $\underline{\mathscr{S}}^q_{\mu,r}$ and $\overline{\mathscr{S}}^q_{\mu,r}$ exist and equal a multiple of the normalized multifractal Hausdorff measure. On the other hand, if the set $\{\log r_1^{-1},\dots,\log r_N^{-1}\}$ is not contained in a discrete additive subgroup of $\mathbb{R}$, then the weak limits of $\underline{\mathscr{S}}^q_{\mu,r}$ and $\overline{\mathscr{S}}^q_{\mu,r}$ as $r \searrow 0$ always exist and, as above, they equal a multiple of the normalized multifractal Hausdorff measure.

Theorem 5 ([38]). *Let K and μ be given by Eqs. (5) and (6). Fix $q \in \mathbb{R}$ and assume that Condition (i) or Condition (ii) below is satisfied.*

(i) The OSC is satisfied and $0 \le q$.
(ii) The SSC is satisfied.

Then the following holds:

(1) We have

$$\underline{\mathscr{S}}^q_{\mu,r} = \overline{\mathscr{S}}^q_{\mu,r} = \frac{1}{r^{-\beta(q)}}\,\mathscr{I}^q_{\mu,r}\,.$$

Write $\mathscr{S}^q_{\mu,r}$ for the common value of $\underline{\mathscr{S}}^q_{\mu,r}$ and $\overline{\mathscr{S}}^q_{\mu,r}$, i.e. write

$$\mathscr{S}^q_{\mu,r} = \frac{1}{r^{-\beta(q)}}\,\mathscr{I}^q_{\mu,r}\,.$$

Also, write

$$\mathscr{S}^q_{\mu,r,\mathsf{ave}} = \frac{1}{-\log r}\int_r^1 \frac{1}{s^{-\beta(q)}}\,\mathscr{I}^q_{\mu,s}\,\frac{\mathrm{d}s}{s}\,.$$

(2) If the set $\{\log r_1^{-1},\dots,\log r_N^{-1}\}$ is not contained in a discrete additive subgroup of $\mathbb{R}$, then

$$\mathscr{S}^q_{\mu,r} \to M^{q,\beta(q)}_{\mu}(K)\,\frac{1}{\mathscr{H}^{q,\beta(q)}_{\mu}(K)}\,\mathscr{H}^{q,\beta(q)}_{\mu}\llcorner K \qquad \textit{weakly,}$$

$$\mathscr{S}^q_{\mu,r,\mathsf{ave}} \to M^{q,\beta(q)}_{\mu,\mathsf{ave}}(K)\,\frac{1}{\mathscr{H}^{q,\beta(q)}_{\mu}(K)}\,\mathscr{H}^{q,\beta(q)}_{\mu}\llcorner K \qquad \textit{weakly;}$$

recall that K is $(q,\beta(q))$ multifractal Minkowski measurable with respect to μ and $(q,\beta(q))$ average multifractal Minkowski measurable with respect to μ by Theorem 2 and the multifractal Minkowski content $M^{q,\beta(q)}_{\mu}(K)$ and the average multifractal Minkowski content $M^{q,\beta(q)}_{\mu,\mathsf{ave}}(K)$ are therefore well defined.

(3) If the set $\{\log r_1^{-1},\ldots,\log r_N^{-1}\}$ is contained in a discrete additive subgroup of $\mathbb{R}$, then

$$\mathscr{S}^q_{\mu,r,\mathsf{ave}} \to M^{q,\beta(q)}_{\mu,\mathsf{ave}}(K)\,\frac{1}{\mathscr{H}^{q,\beta(q)}_\mu(K)}\,\mathscr{H}^{q,\beta(q)}_\mu \llcorner K \qquad \textit{weakly;}$$

recall that K is $(q,\beta(q))$ average multifractal Minkowski measurable with respect to μ by Theorem 2 and the average multifractal Minkowski content $M^{q,\beta(q)}_{\mu,\mathsf{ave}}(K)$ is therefore well defined.

As with Theorem 2, it is instructive to consider the special case $q = 0$. Indeed, we note (cf. Eq. (23)) that

$$\begin{aligned}\mathscr{V}^0_{\mu,r}(E) &= \frac{\mathscr{L}^d(E\cap B(K,r))}{\mathscr{L}^d(B(K,r))}\\ &= \mathsf{V}_{K,r}(E),\end{aligned}$$

i.e. $\mathscr{V}^0_{\mu,r}$ equals the normalized parallel body measure $\mathsf{V}_{K,r}$. Also, writing t for the common value of the box dimensions and Hausdorff dimension of K, we note [see Eq. (25)] that

$$\begin{aligned}\underline{\mathscr{S}}^0_{\mu,r}(E) = \overline{\mathscr{S}}^0_{\mu,r}(E) &= \frac{1}{r^{-t+d}}\,\mathscr{L}^d(E\cap B(K,r))\\ &= \underline{\mathsf{S}}_{K,r}(E) = \overline{\mathsf{S}}_{K,r}(E),\end{aligned}$$

i.e. $\underline{\mathscr{S}}^0_{\mu,r}$ and $\overline{\mathscr{S}}^0_{\mu,r}$ equal the scaling parallel body measures $\underline{\mathsf{S}}_{K,r}$ and $\overline{\mathsf{S}}_{K,r}$. The following corollaries therefore follow immediately from Theorem 2, Theorems 1 and 2 by putting $q = 0$. These results were first obtained by Winter in his doctoral dissertation [51].

Corollary 2 ([51]). *Let K be given by Eq. (5). Assume that the OSC is satisfied. Let t denote the common value of the box dimensions and the Hausdorff dimension of K, i.e. t is the unique number such that $\sum_i r_i^t = 1$. For $r > 0$, the normalized parallel body measure $\mathsf{V}_{K,r}$ is given by*

$$\mathsf{V}_{K,r}(E) = \frac{1}{\mathscr{L}^d(B(K,r))}\,\mathscr{L}^d(E\cap B(K,r)).$$

Then we have

$$\mathsf{V}_{K,r} \to \frac{1}{\mathscr{H}^t(K)}\,\mathscr{H}^t \llcorner K \qquad \textit{weakly.}$$

Proof. Since $\mathscr{V}^0_{\mu,r} = \mathsf{V}_{K,r}$, this follows from Theorem 4 by putting $q = 0$. □

Corollary 3 ([51]). *Let K be given by Eq. (5). Assume that the OSC is satisfied. Let t denote for the common value of the box dimensions and the Hausdorff dimension of K, i.e. t is the unique number such that $\sum_i r_i^t = 1$.*

(1) We have

$$\underline{\mathsf{S}}_{K,r}(E) = \overline{\mathsf{S}}_{K,r}(E) = \frac{1}{r^{-t+d}} \mathscr{L}^d(E \cap B(K,r)).$$

Write $\mathsf{S}_{K,r}$ for the common value of $\underline{\mathsf{S}}_{K,r}$ and $\overline{\mathsf{S}}_{K,r}$, i.e. write

$$\mathsf{S}_{K,r}(E) = \frac{1}{r^{-t+d}} \mathscr{L}^d(E \cap B(K,r)).$$

Also, write

$$\mathsf{S}_{K,r,\mathsf{ave}}(E) = \frac{1}{-\log r} \int_r^1 \frac{1}{s^{-t+d}} \mathscr{L}^d(E \cap B(K,s)) \frac{ds}{s}.$$

(2) If the set $\{\log r_1^{-1}, \ldots, \log r_N^{-1}\}$ is not contained in a discrete additive subgroup of $\mathbb{R}$, then

$$\mathsf{S}_{K,r} \to M^t(K) \frac{1}{\mathscr{H}^t(K)} \mathscr{H}^t \llcorner K \qquad \textit{weakly,}$$

$$\mathsf{S}_{K,r,\mathsf{ave}} \to M^t_{\mathsf{ave}}(K) \frac{1}{\mathscr{H}^t(K)} \mathscr{H}^t \llcorner K \qquad \textit{weakly;}$$

recall that K is t Minkowski measurable and t average Minkowski measurable by Corollary 1 and the Minkowski content $M^t(K)$ and the average Minkowski content $M^t_{\mathsf{ave}}(K)$ are therefore well defined.

(3) If the set $\{\log r_1^{-1}, \ldots, \log r_N^{-1}\}$ is contained in a discrete additive subgroup of $\mathbb{R}$ then

$$\mathsf{S}_{K,r,\mathsf{ave}} \to M^t_{\mathsf{ave}}(K) \frac{1}{\mathscr{H}^t(K)} \mathscr{H}^t \llcorner K \qquad \textit{weakly;}$$

recall that K is t average Minkowski measurable by Corollary 1 and the average multifractal Minkowski content $M^t_{\mathsf{ave}}(K)$ is therefore well defined.

Proof. Since $\underline{\mathscr{S}}^0_{\mu,r} = \overline{\mathscr{S}}^0_{\mu,r} = \mathsf{S}_{K,r}$, this follows from Theorem 5 by putting $q = 0$. □

In Sect. 4.1 it was suggested that one motivation for introducing the multifractal tube measures $\mathscr{V}^q_{\mu,r}$ is that the limiting behaviour of $\mathscr{V}^q_{\mu,r}$ may be viewed as providing a local version of Theorem 2. Namely, Theorem 2 describes the limiting behaviour of $\frac{1}{r^{-\beta(q)}} V^q_{\mu,r}(K)$ as $r \searrow 0$, whereas Theorem 4 provides information about the the limiting behaviour of $\frac{1}{r^{-\beta(q)}} V^q_{\mu,r}(E)$ as $r \searrow 0$ for "well-behaved" subsets E of K. The viewpoint is made precise in the next corollary. Below we use the following notation, namely, if X is a metric space and $E \subseteq X$, then we will denote the the boundary of E in X by ∂E.

Corollary 4. *Let K and μ be given by Eqs. (5) and (6). Fix $q \in \mathbb{R}$ and assume that Condition (i) or Condition (ii) below is satisfied.*

(i) The OSC is satisfied and $0 < q$.
(ii) The SSC is satisfied.

Let $E \subseteq \mathbb{R}^d$ be a Borel set with:

1. $\mathscr{H}_{\mu}^{q,\beta(q)}(E \cap K) > 0$
2. $\mathscr{H}_{\mu}^{q,\beta(q)}(\partial E \cap K) = 0$
3. $E \cap B(K,r) = B(E \cap K, r)$ *for r small enough*

(Observe that, for example, the set $E = \mathbb{R}^d$ satisfies the above conditions, and if $K = L \cup M$ with $\operatorname{dist}(L,M) > 0$ and $\mathscr{H}_{\mu}^{q,\beta(q)}(L) > 0$ and $0 < \delta < \operatorname{dist}(L,M)$, then the set $E = B(L,\delta)$ satisfies the above conditions.)

Then we have the following:

1. *If the set $\{\log r_1^{-1}, \ldots, \log r_N^{-1}\}$ is not contained in a discrete additive subgroup of $\mathbb{R}$, then $E \cap K$ is $(q,\beta(q))$ multifractal Minkowski measurable with respect to μ with*

$$M_{\mu}^{q,\beta(q)}(E \cap K) = M_{\mu}^{q,\beta(q)}(K)\, \frac{\mathscr{H}_{\mu}^{q,\beta(q)}(E \cap K)}{\mathscr{H}_{\mu}^{q,\beta(q)}(K)};$$

recall that K is $(q,\beta(q))$ average multifractal Minkowski measurable with respect to μ by Theorem 2 and the multifractal Minkowski content $M_{\mu}^{q,\beta(q)}(K)$ is therefore well defined.

2. *If the set $\{\log r_1^{-1}, \ldots, \log r_N^{-1}\}$ is contained in a discrete additive subgroup of $\mathbb{R}$, then $E \cap K$ is $(q,\beta(q))$ average multifractal Minkowski measurable with respect to μ with*

$$M_{\mu,\mathsf{ave}}^{q,\beta(q)}(E \cap K) = M_{\mu,\mathsf{ave}}^{q,\beta(q)}(K)\, \frac{\mathscr{H}_{\mu}^{q,\beta(q)}(E \cap K)}{\mathscr{H}_{\mu}^{q,\beta(q)}(K)};$$

recall that K is $(q,\beta(q))$ average multifractal Minkowski measurable with respect to μ by Theorem 2 and the average multifractal Minkowski content $M_{\mu,\mathsf{ave}}^{q,\beta(q)}(K)$ is therefore well defined.

Proof. This follows immediately from Theorem 5 since the condition $E \cap B(K,r) = B(E \cap K, r)$ implies that

$$\begin{aligned} \mathscr{I}_{\mu,r}^{q}(E) &= \frac{1}{r^d} \int_{E \cap B(K,r)} \mu(B(x,r))^q \, \mathrm{d}\mathscr{L}^d(x) = \frac{1}{r^d} \int_{B(E \cap K,r)} \mu(B(x,r))^q \, \mathrm{d}\mathscr{L}^d(x) \\ &= V_{\mu,r}^{q}(E \cap K). \end{aligned}$$

□

Note that Corollary 4 is a genuine extension of Theorem 2: namely, if we let $E = K$ in Corollary 4, then Corollary 4 simplifies to Theorem 2.

References

1. Arbeiter, M., Patzschke, N.: Random self-similar multifractals. Math. Nachr. **181**, 5–42 (1996)
2. Berger, M., Gostiaux, B.: Differential Geometry: Manifolds, Curves and Surfaces. Springer, Berlin (1988)
3. Cawley, R., Mauldin, R.D.: Multifractal decomposition of Moran fractals. Adv. Math. **92**, 196–236 (1992)
4. Cole, J.: The Geometry of Graph Directed Self-Conformal Multifractals. Ph.D. Thesis, University of St. Andrews, Scotland (1998)
5. Cutler, C.D.: The Hausdorff distribution of finite measures in Euclidean space. Can. J. Math. **38**, 1459–1484 (1986)
6. Cutler, C.D.: Connecting ergodicity and dimension in dynamical systems. Ergod. Theor. Dyn. Syst. **10**, 451–462 (1990)
7. Cutler, C.D.: Measure disintegrations with respect to σ-stable monotone indices and pointwise representation of packing dimension. Proceedings of the 1990 Measure Theory Conference at Oberwolfach. Supplemento Ai Rendiconti del Circolo Mathematico di Palermo, Ser. II, No. 28, pp. 319–340 (1992)
8. Das, M.: Local properties of self-similar measures. Illinois Math. J. **42**, 313–332 (1998)
9. Das, M.: Hausdorff measures, dimensions and mutual singularity. Trans. Amer. Math. Soc. **357**, 4249–4268 (2005)
10. Falconer, K.J.: Fractal Geometry — Mathematical Foundations and Applications. Wiley, Chichester (1990)
11. Falconer, K.J.: On the Minkowski measurability of fractals. Proc. Am. Math. Soc. **123**, 1115–1124 (1995)
12. Falconer, K.J.: Techniques in Fractal Geometry. Wiley, Chichester (1997)
13. Federer, H.: Curvature measures. Trans. Amer. Math. Soc. **93**, 418–491 (1959)
14. Federer, H.: Geometric Measure Theory. Springer, New York (1969)
15. Feller, W.: An Introduction to Probability Theory and Its Applications, vol. 2, 2nd edn. Wiley, New York (1971)
16. Feller, W.: An Introduction to Probability Theory and its Applications, vol. 1, 3rd edn. Wiley, New York (1986)
17. Frisch, U., Parisi, G.: On the singularity structure of fully developed turbulence. Appendix to U. Frisch, Fully developed turbulence and intermittency, Turbulence and Predictability in Geophysical Fluid Dynamics and Climate Dynamics. Proceedings of the International School of Physics "Enrico Fermi", Course LXXXVIII, pp. 84–88. North Holland, Amsterdam (1985)
18. Fu, J.H.G.: Tubular neighborhoods in Euclidean spaces. Duke Math. J. **52**, 1025–1046 (1985)
19. Fu, J.H.G.: Curvature measures of subanalytic sets. Amer. Math. J. **116**, 819–880 (1994)
20. Gatzouras, D.: Lacunarity of self-similar and stochastically self-similar sets. Trans. Amer. Math. Soc. **352**, 1953–1983 (2000)
21. Gray, A.: Tubes. Progress in Mathematics, vol. 221. Birkhäuser, Boston (2004)
22. Halsey, T.C., Jensen, M.H., Kadanoff, L.P., Procaccia, I., Shraiman, B.J.: Fractal measures and their singularities: the characterization of strange sets. Phys. Rev. A **33**, 1141–1151 (1986)
23. Hofbauer, F., Raith, P., Steinberger, T.: Multifractal dimensions for invariant subsets of piecewise monotonic interval maps Fund. Math. **176**, 209–232 (2003)
24. Hutchinson, J.: Fractals and self-similarity. Indiana Univ. Math. J. **30**, 713–747 (1981)
25. Hug, D., Last, G., Weil, W.: A local Steiner-type formula for general closed sets and applications. Math. Z. **246**, 237–272 (2004)
26. Lalley, S.: The packing and covering functions of some self-similar fractals. Indiana Univ. Math. J. **37**, 699–710 (1988)
27. Lalley, S.: Renewal theorems in symbolic dynamics, with applications to geodesic flows, non-Euclidean tessellations and their fractal limits. Acta Math. **163**, 1–55 (1989)
28. Lalley, S.: Probabilistic methods in certain counting problems of ergodic theory. Ergodic theory, symbolic dynamics, and hyperbolic spaces. In: Bedford, T., Keane, M., Caroline Series

(eds.) Papers from the Workshop on Hyperbolic Geometry and Ergodic Theory held in Trieste, April 17–28, 1989, pp. 223–257. Oxford Science Publications. The Clarendon Press, Oxford University Press, New York (1991)
29. Lapidus, M.L., van Frankenhuysen, M.: Fractal Geometry and Number Theory: Complex Dimensions of Fractal Strings and Zeros of Zeta Functions. Birkhäuser, Boston (2000)
30. Lapidus, M.L., van Frankenhuysen, M.: Fractal geometry, complex dimensions and zeta functions: geometry and spectra of fractal strings. Springer Monographs in Mathematics. Springer, New York (2006)
31. Levitin, M., Vassiliev, D.: Spectral asymptotics, renewal theorem, and the Berry conjecture for a class of fractals. Proc. London Math. Soc. **72**, 188–214 (1996)
32. Mandelbrot, B.: Possible refinement of the log-normal hypothesis concerning the distribution of energy dissipation in intermittent turbulence. Statistical Models and Turbulence. Lecture Notes in Physics, vol. 12. Springer, New York (1972)
33. Mandelbrot, B.: Intermittent turbulence in self-similar cascades: divergence of high moments and dimension of the carrier. Fluid. J. Mech. **62**, 331–358 (1974)
34. Mattila, P.: On the structure of self-similar fractals. Ann. Acad. Sci. Fenn. Ser. A I Math. **7**, 189–195 (1982)
35. Morvan, J.-M.: Generalized Curvatures. Springer, Berlin (2008)
36. Olsen, L.: A multifractal formalism. Adv. Math. **116**, 82–196 (1995)
37. Olsen, L.: Self-affine multifractal Sierpinski sponges in $\mathbb{R}^d$ Pacific J. Math. **183**, 143–199 (1998)
38. Olsen, L.: Multifractal tubes. Preprint (2011)
39. O'Neil, T.: The multifractal spectrum of quasi self-similar measures. J. Math. Anal. Appl. **211**, 233–257 (1997)
40. O'Neil, T.: The multifractal spectra of projected measures in Euclidean spaces. Chaos Solitons Fract. **11**, 901–921 (2000)
41. Pesin, Y.: Dimension type characteristics for invariant sets of dynamical systems. Russian Math. Surveys **43**, 111–151 (1988)
42. Pesin, Y.: Dimension theory in dynamical systems. Contemporary Views and Applications. The University of Chicago Press, Chicago (1997)
43. Peres, Y., Solomyak, B.: Existence of L^q dimensions and entropy dimension for self-conformal measures Indiana Univ. Math. J. **49**, 1603–1621 (2000)
44. Peyrière, J.: Multifractal measures. In: Proceedings of the NATO Advanced Study Institute on Probabilistic and Stochastic Methods in Analysis with Applications, Il Ciocco, pp. 175–186. NATO ASI Series, Series C: Mathematical and Physical Sciences, vol. 372, Kluwer, Dordrecht (1992)
45. Schechter, A.: On the centred Hausdorff measure. Bull. Lond. Math. Soc. **62**, 843–851 (2000)
46. Schief, A.: Separation properties for self-similar sets. Proc. Amer. Math. Soc. **122**, 111–115 (1994)
47. Schneider, R.: Curvature measures of convex bodies. Ann. Mat. Pura Appl. IV, **116**, 101–134 (1978)
48. Schneider, R.: Convex Bodies: The Brunn-Minkowski Theory. Cambridge University Press, Cambridge (1993)
49. Stacho, L.L.: On curvature measures. Acta Sci. Math. **41**, 191–207 (1979)
50. Weyl, H.: On the volume of tubes. Amer. Math, J. **61**, 461–472 (1939)
51. Winter, S.: Curvature measures and fractals. Diss. Math. **453**, 1–66 (2008)
52. Zähle, M.: Integral and current representation of Federer's curvature measures. Arch. Math. **46**, 557–567 (1986)
53. Zähle, M.: Curvatures and currents for unions of sets with positive reach. Geom. Dedicata **123**, 155–171 (1987)

The Multiplicative Golden Mean Shift Has Infinite Hausdorff Measure

Yuval Peres and Boris Solomyak

Abstract In an earlier work, joint with R. Kenyon, we computed the Hausdorff dimension of the "multiplicative golden mean shift" defined as the set of all reals in $[0,1]$ whose binary expansion (x_k) satisfies $x_k x_{2k} = 0$ for all $k \geq 1$. Here we show that this set has infinite Hausdorff measure in its dimension. A more precise result in terms of gauges in which the Hausdorff measure is infinite is also obtained.

1 Introduction

Consider the set

$$\Xi_G := \left\{ x = \sum_{k=1}^{\infty} x_k 2^{-k} : \; x_k \in \{0,1\}, \; x_k x_{2k} = 0 \text{ for all } k \right\}$$

which we call the "multiplicative golden mean shift." The reason for this term is that the set of binary sequences corresponding to the points of Ξ_G is invariant under the action of the semigroup of multiplicative positive integers $\mathbb{N}^*$: $M_r(x_k) = (x_{rk})$ for $r \in \mathbb{N}$. Fan, Liao, and Ma [3] showed that $\dim_M(\Xi_G) = \sum_{k=1}^{\infty} 2^{-k-1} \log_2 F_{k+1} = 0.82429\ldots$, where F_k is the k-th Fibonacci number: $F_1 = 1$, $F_2 = 2$, $F_{k+1} = F_{k-1} + F_k$, and raised the question of computing the Hausdorff dimension of Ξ_G.

Y. Peres
Microsoft Research, One Microsoft Way, Redmond, WA 98052, USA
e-mail: peres@microsoft.com

B. Solomyak (✉)
Department of Mathematics, University of Washington, Box 354350, UW, Seattle, WA 98195, USA
e-mail: solomyak@math.washington.edu

J. Barral and S. Seuret (eds.), *Further Developments in Fractals and Related Fields*, Trends in Mathematics, DOI 10.1007/978-0-8176-8400-6_10,

Theorem 1 ([5,6]). *We have* $\dim_H(\Xi_G) < \dim_M(\Xi_G)$. *In fact,*

$$\dim_H(\Xi_G) = -\log_2 p = 0.81137\ldots, \quad \text{where } p^3 = (1-p)^2, \quad 0 < p < 1. \tag{1}$$

Here we prove

Theorem 2. *(i) The set* Ξ_G *has infinite (not* σ*-finite) Hausdorff measure in its dimension. Moreover, let* $s = \dim_H(\Xi_G)$. *Then* $\mathscr{H}^\phi(\Xi_G) = \infty$ *for*

$$\phi(t) = t^s \exp\left[-c\frac{|\log t|}{(\log|\log t|)^2}\right] \tag{2}$$

provided that $c > 0$ *is sufficiently small, and furthermore,* Ξ_G *is not* σ*-finite with respect to* $\mathscr{H}^\phi$.

(ii) On the other hand, we have $\mathscr{H}^{\psi_\theta}(\Xi_G) = 0$ *for*

$$\psi_\theta(t) = t^s \exp\left[-\frac{|\log t|}{(\log|\log t|)^\theta}\right], \tag{3}$$

provided that $\theta < 2$.

Remarks. 1. In [6] we have pointed out a remarkable analogy between dimension properties of multiplicative shifts of finite type and self-affine carpets of Bedford and McMullen, see ([1,8]), although we are not aware of any direct connection. The stated theorem provides further evidence of this: it exactly corresponds to Theorem 3 from the paper by the first-named author [9]. We should point out, however, that our proof requires many new elements; in particular, the recurrence relation from Lemma 3 below has no parallels in [9].

2. For self-affine carpets with non-uniform horizontal fibers, there is an elegant "soft" argument showing that the Hausdorff measure of the set in its dimension cannot be positive and finite [7], and more generally, this holds for any gauge [9]. It would be interesting to find a similar argument for the multiplicative golden mean shift as well.

3. We expect that similar results hold for other multiplicative shifts of finite type considered in [6]. Since the proofs are quite technical, we decided to focus on the most basic example of Ξ_G.

2 Preliminaries and the Scheme of the Proof

It is more convenient to work in the symbolic space $\Sigma_2 = \{0,1\}^{\mathbb{N}}$, with the metric

$$\rho((x_k),(y_k)) = 2^{-\min\{n:\, x_n \neq y_n\}}.$$

It is well known that the dimensions of a compact subset of $[0,1]$ and the corresponding set of binary digit sequences in Σ_2 are equal (this is equivalent to replacing the covers by arbitrary interval with those by dyadic intervals), and the Hausdorff measures in the gauges that we are considering are comparable, up to a multiplicative constant. Thus, it suffices to work with the set X_G—the collection of all binary sequences (x_k) such that $x_k x_{2k} = 0$ for all k. Observe that

$$X_G = \{\omega = (x_k)_{k=1}^{\infty} \in \Sigma_2 : \ (x_{i2^r})_{r=0}^{\infty} \in \Sigma_G \ \text{ for all } i \text{ odd}\}, \tag{4}$$

where Σ_G is the usual (additive) golden mean shift:

$$\Sigma_G := \{(x_k)_{k=1}^{\infty} \in \Sigma_2, \ x_k x_{k+1} = 0, \ \text{ for all } k \geq 1\}.$$

We will use the Rogers–Taylor density theorem from [11]. We state it in the symbolic space Σ_2 where $[u]$ denotes the cylinder set of sequences starting with a finite "word" u and $x_1^n = x_1 \dots x_n$. Given a continuous increasing function ϕ on $[0,\infty)$, with $\phi(0) = 0$, we consider the generalized Hausdorff measure with the gauge ϕ, denoted by $\mathscr{H}^{\phi}$; see, e.g., [2, p.33] or [10, p.50] for the definition and basic properties.

Theorem 3 (Rogers and Taylor). *Let $\mathbb{P}$ be a finite Borel measure on Σ_2 and let Λ be a Borel set in Σ_2 such that $\mathbb{P}(\Lambda) > 0$. Let ϕ be any gauge function. If for all $x \in \Lambda$,*

$$\beta_1 \leq \liminf_{n\to\infty} \frac{\phi(2^{-n})}{\mathbb{P}[x_1^n]} \leq \beta_2 \tag{5}$$

(where β_1, β_2 may be zero or infinity), then

$$c_1 \beta_1 \mathbb{P}(\Lambda) \leq \mathscr{H}^{\phi}(\Lambda) \leq c_2 \beta_2 \mathbb{P}(\Lambda),$$

where c_1 and c_2 are positive and finite.

Corollary 1. *Let $\mathbb{P}$ be a finite Borel measure on Σ_2 and let Λ be a Borel set in Σ_2 such that $\mathbb{P}(\Lambda) > 0$. Let ϕ be any gauge function.*

(i) If for $\mathbb{P}$-a.e. $x \in \Lambda$

$$\lim_{n\to\infty} (\log_2 \mathbb{P}[x_1^n] - \log_2 \phi(2^{-n})) = -\infty,$$

then $\mathscr{H}^{\phi}(\Lambda) = \infty$.

(ii) If for all $x \in \Lambda$

$$\lim_{n\to\infty} (\log_2 \mathbb{P}[x_1^n] - \log_2 \phi(2^{-n})) = +\infty,$$

then $\mathscr{H}^{\phi}(\Lambda) = 0$.

For an odd i denote by $J(i) = \{2^r i\}_{r=0}^{\infty}$ the geometric progression with ratio 2 starting at i. Equation (4) says that $x \in X_G$ if and only if the "restriction" of x to every $J(i)$ belongs to Σ_G. We can define a measure on X_G by taking an infinite product of probability measures on each "copy" of Σ_G.

In order to compute $\dim_H(X_G)$, it was enough to take the same measure μ on each copy see [5]. Given a probability measure μ on Σ_G, we define a probability measure on X_G by

$$\mathbb{P}_\mu[u] := \prod_{i \le n,\, i \text{ odd}} \mu\left[u|_{J(i)}\right], \tag{6}$$

where $u|_{J(i)}$ denotes the "restriction" of the word u to the subsequence $J(i)$. It was proved in [5, 6] that there is a unique probability measure μ on Σ_G such that $\dim_H(\mathbb{P}_\mu) = \dim_H(X_G)$. Denote by $\mu(r)$ the Markov (nonstationary) measure on Σ_G, with initial probabilities $(r, 1-r)$ and the matrix of transition probabilities $P = (P(i,j))_{i,j=0,1} = \begin{pmatrix} r & 1-r \\ 1 & 0 \end{pmatrix}$. Then $\mu = \mu(p)$, where $p^3 = (1-p)^2$. The measure $\mu(r)$ on cylinder sets can be explicitly written as follows:

$$\mu(r)([u_1 \dots u_k]) = (1-r)^{N_1(u_1 \dots u_k)} r^{N_0(u_1 \dots u_k) - N_1(u_1 \dots u_{k-1})}, \tag{7}$$

where $u \in \{0,1\}^k$ is a word admissible in Σ_G, i.e., if $u_j = 1$, then $u_{j+1} = 0$ for $j \le k-1$, and $N_i(u)$ denotes the number of symbols i in the word u. To verify Eq. (7), note that the probability of a 1 is always $1-r$ (including the first position), and the probability of a 0 is r, except when it follows a 1, in which case its probability equals one.

For the lower bound, i.e., part (i) of Theorem 2, we have to "fine-tune" the measure $\mathbb{P}_\mu$ by taking a product of measures $\mu(p_k)$ on subsequences $J(i)$ with odd i such that $2^k \le i < 2^{k+1}$. It is clear that we must have $\lim_{k\to\infty} p_k = p$; in fact, we will take $p_k = p + \frac{\delta}{k}$. More precisely, let

$$\mu_k = \mu(p_k), \text{ where } p_k = p + \frac{\delta}{k},\ k \ge 1,\ p_0 = p, \tag{8}$$

and $\delta > 0$ is sufficiently small, so that $p_1 = p + \delta < 1$. Next, we define for $u \in \{0,1\}^n$, with $2^{\ell-1} < n \le 2^\ell$,

$$\mathbb{P}_\delta[u] := \prod_{k=1}^{\ell} \prod_{\frac{n}{2^k} < i \le \frac{n}{2^{k-1}},\, i \text{ odd}} \mu_{\ell-k}\left[u|_{J(i)}\right], \tag{9}$$

where $u|_{J(i)} = u_i \dots u_{2^{k-1}i}$ is a word of length k. It is easy to see that $\mathbb{P}_\delta$ is a probability measure on X_G.

Without loss of generality we can (and will) use logarithms base 2 in Eqs. (2) and (3). Theorem 2(i) immediately follows from Corollary 1(i) and the following proposition.

Proposition 1. *There exist constants $\delta > 0$ and $c > 0$ such that the measure $\mathbb{P}_\delta$ defined by Eq. (9) satisfies*

$$\lim_{n\to\infty}\left(\log_2 \mathbb{P}_\delta[x_1^n] - \log_2 \phi(2^{-n})\right) = -\infty$$

for $\mathbb{P}_\delta$-a.e. $x \in X_G$, where ϕ is the gauge function from Eq. (2). Equivalently,

$$\lim_{n\to\infty}\left(\log_2 \mathbb{P}_\delta[x_1^n] + ns + \frac{cn}{(\log_2 n)^2}\right) = -\infty \tag{10}$$

for $\mathbb{P}_\delta$-a.e. $x \in X_G$, where $s = -\log_2 p = \dim_H(X_G)$.

For the upper bound of the Hausdorff measure, i.e., part (ii) of Theorem 2, it is enough to take the same measure $\mu = \mu(p)$ as in [5,6]; however, the proof is rather delicate; it follows the scheme of [9, Theorem 3(ii)], but with many modifications.

We will need a classical large deviation inequality, which we state in the generality needed for us.

Lemma 1 (Hoeffding's inequality [4]). *Let $\{X_i\}_{i\geq 1}$ be a sequence of independent random variables with expectation zero, such that $|X_i| \leq C$, and let $S_n = \sum_{i=1}^n X_i$. Then*

$$\mathbb{P}\left(S_n \geq tn\right) \leq \exp\left(-\frac{t^2 n}{2C^2}\right) \tag{11}$$

for all $t > 0$ and $n \geq 1$.

3 Lower Estimates of Hausdorff Measure

Here we prove Proposition 1. We start with a reduction.

Lemma 2. *If Eq. (10) holds for positive integers n satisfying*

$$n = 2^{\lfloor \ell/2 \rfloor} d, \ \text{where}\ 2^{\ell-1} < n \leq 2^\ell,\ d \in \mathbb{N}, \tag{12}$$

with a constant $c > 0$, then Eq. (10) holds for all n with c replaced by $c/2$.

Proof. For a large integer $n \in (2^{\ell-1}, 2^\ell]$, let

$$d := \lfloor 2^{-\lfloor \ell/2 \rfloor} n \rfloor, \quad m := 2^{\lfloor \ell/2 \rfloor} d.$$

Then

$$n - \sqrt{2n} \leq n - 2^{\lfloor \ell/2 \rfloor} < m \leq n.$$

It is clear that m satisfies Eq. (12) (possibly with a different ℓ). Observe that

$$\log_2 \mathbb{P}_\delta[x_1^n] + ns + \frac{(c/2)n}{(\log_2 n)^2} \le \log_2 \mathbb{P}_\delta[x_1^m] + ms + \frac{cm}{(\log_2 m)^2}$$
$$+ s(n-m) + c\left[\frac{n/2}{(\log_2 n)^2} - \frac{m}{(\log_2 m)^2}\right].$$

Since

$$s(n-m) + c\left[\frac{n/2}{(\log_2 n)^2} - \frac{m}{(\log_2 m)^2}\right] \le s\sqrt{2n} + c\left[\frac{n/2}{(\log_2 n)^2} - \frac{n-\sqrt{2n}}{(\log_2 n)^2}\right] < 0$$

for large enough n, the claim follows. □

For $k \ge 1$ let α_k be the partition of Σ_G into cylinders of length k. For a measure μ on Σ_2 and a finite partition α, denote by $H^\mu(\alpha)$ the μ-entropy of the partition, with base 2 logarithms:

$$H^\mu(\alpha) = -\sum_{A\in\alpha} \mu(A)\log_2 \mu(A).$$

Let n be such that Eq. (12) holds. In view of Eq. (9),

$$\log_2 \mathbb{P}_\delta[x_1^n] \le \sum_{k=1}^{\lfloor \ell/2 \rfloor} \sum_{\frac{n}{2^k} < i \le \frac{n}{2^{k-1}},\ i \text{ odd}} \log_2 \mu_{\ell-k}\left[x_1^n|_{J(i)}\right]. \tag{13}$$

Note that $x_1^n|_{J(i)}$ is a word of length k for $i \in (n/2^k, n/2^{k-1}]$, with i odd, which is a beginning of a sequence in Σ_G. Thus, $[x_1^n|_{J(i)}]$ is an element of the partition α_k. The random variables $x \mapsto \log_2 \mu_{\ell-k}[x_1^n|_{J(i)}]$ are i.i.d for $i \in (n/2^k, n/2^{k-1}]$, with i odd, and their expectation equals $-H^{\mu_{\ell-k}}(\alpha_k)$, by the definition of entropy. Note that there are $n/2^{k+1}$ odds in $(n/2^k, n/2^{k-1}]$. It is easy to see from Eqs. (7) and (8) that

$$\left|\log_2 \mu_{\ell-k}[x_1^n|_{J(i)}]\right| \le Ck, \tag{14}$$

for $i \in (n/2^k, n/2^{k-1}]$, with some $C > 0$, independent of n and k. Let

$$S_{n/2^{k+1}} := \sum_{\frac{n}{2^k} < i \le \frac{n}{2^{k-1}},\ i \text{ odd}} \log_2 \mu_{\ell-k}[x_1^n|_{J(i)}]$$

and $S^*_{n/2^{k+1}} := S_{n/2^{k+1}} + \frac{n}{2^{k+1}} H^{\mu_{\ell-k}}(\alpha_k)$, be the corresponding sum of centered (zero expectation) random variables. Then we have, for $k = 1, \ldots \lfloor \ell/2 \rfloor$, and any $\varepsilon \in (0, \frac{1}{2})$, using Eq. (14) in Hoeffding's inequality Eq. (11):

$$\mathbb{P}_\delta\left(x:\ S_{n/2^{k+1}} > \frac{-n}{2^{k+1}} H^{\mu_{\ell-k}}(\alpha_k) + \left(\frac{n}{2^{k+1}}\right)^{1-\varepsilon}\right)$$

$$= \mathbb{P}_\delta\left(x:\ S^*_{n/2^{k+1}} > \left(\frac{n}{2^{k+1}}\right)^{1-\varepsilon}\right)$$

$$\leq \exp\left[-\frac{(n/2^{k+1})^{1-2\varepsilon}}{2C^2k^2}\right].$$

Denote $b_\varepsilon = \sum_{k=1}^{\infty} 2^{-(k+1)\varepsilon}$. Now it follows from Eq. (13) that

$$\mathbb{P}_\delta\left(x:\ \log_2 \mathbb{P}_\delta[x_1^n] > -n \sum_{k=1}^{\lfloor \ell/2 \rfloor} \frac{H^{\mu_{\ell-k}}(\alpha_k)}{2^{k+1}} + b_\varepsilon n^{1-\varepsilon}\right) \tag{15}$$

$$\leq \mathbb{P}_\delta\left(x:\ \sum_{k=1}^{\lfloor \ell/2 \rfloor} S^*_{n/2^{k+1}} > \sum_{k=1}^{\lfloor \ell/2 \rfloor} \left(\frac{n}{2^{k+1}}\right)^{1-\varepsilon}\right)$$

$$\leq \sum_{k=1}^{\lfloor \ell/2 \rfloor} \mathbb{P}_\delta\left(x:\ S^*_{n/2^{k+1}} > \left(\frac{n}{2^{k+1}}\right)^{1-\varepsilon}\right)$$

$$\leq \sum_{k=1}^{\lfloor \ell/2 \rfloor} \exp\left[-\frac{n^{1-2\varepsilon}}{2C^2 2^{(k+1)(1-2\varepsilon)} k^2}\right]$$

$$\leq \ell \exp\left[-\frac{C'}{\ell^2}\left(\frac{n}{2^{\lfloor \ell/2 \rfloor + 1}}\right)^{1-2\varepsilon}\right]$$

$$\leq \log_2(2n) \exp\left[-\frac{C'}{\log_2^2(2n)}\left(\frac{n}{8}\right)^{\frac{1}{2}-\varepsilon}\right], \tag{16}$$

where we used that $\sqrt{8n} > 2^{1+\lfloor \ell/2 \rfloor}$ and $\ell \leq \log_2(2n)$ by Eq. (12) in the last step. Since the last expression is summable in n, it follows from Borel–Cantelli that for $\mathbb{P}_\delta$-a.e. $x \in X_G$, the event in parentheses in Eq. (15) holds only for finitely many n. This is the set of full $\mathbb{P}_\delta$ measure for which we will prove Eq. (10), for n satisfying Eq. (12).

Below we let $H(r) = -r\log_2 r - (1-r)\log_2(1-r)$.

Lemma 3. *We have, for any $r \in (0,1)$ and the measure $\mu(r)$ defined by Eq. (7),*

$$H^{\mu(r)}(\alpha_k) = H(r)F_{k-1}(r),\ k \geq 1, \tag{17}$$

where $F_0(x) = 1,\ F_1(x) = 1 + x$, and

$$F_k(x) = 1 + xF_{k-1}(x) + (1-x)F_{k-2}(x),\ k \geq 2. \tag{18}$$

Moreover, the polynomials $F_k(x)$ can be expressed as follows:

$$F_k(x) = \frac{(x-1)^{k+2} - (k+2)x + (2k+3)}{(x-2)^2}, \quad k \geq 0. \tag{19}$$

Proof. For $k=1$ the formula Eq. (17) is trivially true. For $k \geq 2$ we have

$$H^{\mu(r)}(\alpha_k) = H^{\mu(r)}(\alpha_1) + H^{\mu(r)}(\alpha_k|\alpha_1) = H(r) + H^{\mu(r)}(\alpha_k|\alpha_1).$$

By the definition of conditional entropy and the properties of Σ_G, we have

$$H^{\mu(r)}(\alpha_k|\alpha_1) = rH^{\mu(r)}(\alpha_{k-1}) + (1-r)H^{\mu(r)}(\alpha_{k-2}).$$

(We set $H^{\mu(r)}(\alpha_0) = 0$ here.) Indeed, 0 in Σ_G can be followed by an arbitrary element of Σ_G, and 1 is followed by 0 and then by an arbitrary element of Σ_G. Now Eqs. (17) and (18) are easily checked by induction. The explicit formula for $F_k(x)$ was found using that

$$F_k(x) - F_{k-1}(x) = 1 - (1-x)(F_{k-1}(x) - F_{k-2}(x)),$$

and can also be checked by induction. □

Since $\mu_{\ell-k} = \mu(p_{\ell-k})$, we have by Eq. (17)

$$\sum_{k=1}^{\lfloor \ell/2 \rfloor} \frac{H^{\mu_{\ell-k}}(\alpha_k)}{2^{k+1}} = \sum_{k=1}^{\lfloor \ell/2 \rfloor} \frac{H(p_{\ell-k})F_{k-1}(p_{\ell-k})}{2^{k+1}}. \tag{20}$$

Recall that $p_{\ell-k} = p + \frac{\delta}{\ell-k}$. Next we write the Taylor estimate at p, such that $p^3 = (1-p)^2$. We have $p \approx 0.56984 > (1/2)$, so it suffices to consider $x \in (\frac{1}{2}, 1)$. Below C_i denote positive absolute constants. It follows from Eq. (19) that

$$|F_k(x)| \leq C_1 k, \;\; |F'(x)| \leq C_2 k, \;\; |F''(x)| \leq C_3 k, \; x \in (1/2, 1), \; k \geq 1. \tag{21}$$

Therefore,

$$\left| \sum_{k=1}^{\lfloor \ell/2 \rfloor} \frac{H(p_{\ell-k})F_{k-1}(p_{\ell-k})}{2^{k+1}} - \sum_{k=1}^{\lfloor \ell/2 \rfloor} \frac{H(p)F_{k-1}(p)}{2^{k+1}} - \sum_{k=1}^{\lfloor \ell/2 \rfloor} \frac{(HF_{k-1})'(p)}{2^{k+1}} \cdot \frac{\delta}{\ell-k} \right|$$

$$\leq C_4 \sum_{k=1}^{\lfloor \ell/2 \rfloor} \frac{k}{2^{k+1}} \cdot \left(\frac{\delta}{\ell-k}\right)^2 \leq C_5 \frac{\delta^2}{\ell^2}. \tag{22}$$

Lemma 4. *We have*

$$\sum_{k=1}^{\infty} \frac{H(p)F_{k-1}(p)}{2^{k+1}} = s = -\log_2 p \tag{23}$$

and

$$\sum_{k=1}^{\infty} \frac{(HF_{k-1})'(p)}{2^{k+1}} = 0. \tag{24}$$

Proof. One can verify directly that $A(r) := H(r)\sum_{k=1}^{\infty} \frac{F_{k-1}(r)}{2^{k+1}} = \frac{2H(r)}{3-r}$, and this function achieves its maximum at p. Alternatively, this follows from [5], since $A(r)$ equals what was denoted $s(\mu)$ in [5], for $\mu = \mu(r)$. □

In view of Eq. (21), we have $\left|\sum_{k=\lfloor \ell/2\rfloor+1}^{\infty} \frac{H(p)F_{k-1}(p)}{2^{k+1}}\right| \le C_6 \ell \cdot 2^{-\ell/2}$; hence Eq. (23) implies

$$\left|\sum_{k=1}^{\lfloor \ell/2\rfloor} \frac{H(p)F_{k-1}(p)}{2^{k+1}} - s\right| \le C_6 \ell \cdot 2^{-\ell/2}. \tag{25}$$

Next, writing $\frac{1}{\ell-k} = \frac{1}{\ell} + \frac{k}{\ell^2} + \frac{k^2}{\ell^2(\ell-k)}$, we obtain

$$\begin{aligned}
&\sum_{k=1}^{\lfloor \ell/2\rfloor} \frac{(HF_{k-1})'(p)}{2^{k+1}} \cdot \frac{\delta}{\ell-k} \\
&= \frac{\delta}{\ell}\sum_{k=1}^{\lfloor \ell/2\rfloor} \frac{(HF_{k-1})'(p)}{2^{k+1}} + \frac{\delta}{\ell^2}\sum_{k=1}^{\lfloor \ell/2\rfloor} \frac{k(HF_{k-1})'(p)}{2^{k+1}} + \frac{\delta}{\ell^2}\sum_{k=1}^{\lfloor \ell/2\rfloor} \frac{k^2(HF_{k-1})'(p)}{2^{k+1}(\ell-k)} \\
&=: S_1 + S_2 + S_3.
\end{aligned}$$

Using Eq. (21), by Eq. (24) we have

$$|S_1| \le \frac{\delta}{\ell}\left|\sum_{k=\lfloor \ell/2\rfloor+1}^{\infty} \frac{(HF_{k-1})'(p)}{2^{k+1}}\right| \le C_7 \frac{\delta}{\ell}\cdot\frac{\ell}{2^{\ell/2}} = \frac{C_7\delta}{2^{\ell/2}}, \tag{26}$$

and

$$|S_3| \le C_8 \frac{\delta}{\ell^3}. \tag{27}$$

Finally,

$$\left|S_2 - \frac{\delta}{\ell^2}\sum_{k=1}^{\infty} \frac{k(HF_{k-1})'(p)}{2^{k+1}}\right| \le C_9 \frac{\delta}{\ell^2}\cdot\frac{\ell^2}{2^{\ell/2}} = \frac{C_9\delta}{2^{\ell/2}}. \tag{28}$$

Lemma 5. *We have*

$$\tau := \sum_{k=1}^{\infty} \frac{k(HF_{k-1})'(p)}{2^{k+1}} > 0.$$

The proof uses a (rigorous) numerical calculation, and we postpone it to the end of the section. Combining Eqs. (20), (22), (25), (26), (27), and (28), we obtain

$$\left|\sum_{k=1}^{\lfloor \ell/2\rfloor} \frac{H^{\mu_{\ell-k}}(\alpha_k)}{2^{k+1}} - s - \frac{\tau\delta}{\ell^2}\right| \le \frac{C_5\delta^2}{\ell^2} + \frac{C_6\ell}{2^{\ell/2}} + \frac{(C_7+C_9)\delta}{2^{\ell/2}} + \frac{C_8\delta}{\ell^3}. \tag{29}$$

Now we can conclude the proof of the proposition. Let $x \in X_G$ be such that for all n sufficiently large, satisfying Eq. (12), we have

$$\log_2 \mathbb{P}_\delta[x_1^n] \le -n \sum_{k=1}^{\lfloor \ell/2\rfloor} \frac{H^{\mu_{\ell-k}}(\alpha_k)}{2^{k+1}} + b_\varepsilon n^{1-\varepsilon}.$$

Recall that this holds for $\mathbb{P}_\delta$-a.e. x by Eq. (16) and Borel–Cantelli Lemma. Then from Eq. (29) we obtain, keeping in mind that $n \in (2^{\ell-1}, 2^\ell]$:

$$\begin{aligned}\mathscr{S}_n(x) &:= \log_2 \mathbb{P}_\delta[x_1^n] + ns + \frac{cn}{(\log_2 n)^2} \\ &\le \frac{cn}{(\log_2 n)^2} - \tau\delta\frac{n}{(\log_2 n)^2} + b_\varepsilon n^{1-\varepsilon} + \frac{C_5\delta^2 n}{(\log_2 n)^2} \\ &\quad + C_6\sqrt{n}\log_2 n + (C_7+C_9)\delta\sqrt{n} + \frac{C_8\delta n}{(\log_2 n)^3}.\end{aligned}$$

Now we choose a positive $\delta < \frac{\tau}{3C_5}$, which is possible by Lemma 5, so that $C_5\frac{\delta^2 n}{(\log_2 n)^2} < \frac{1}{3}\frac{\tau\delta n}{(\log_2 n)^2}$, and then choose $c \in (0, \tau\delta/3)$, whence

$$\frac{cn}{(\log_2 n)^2} < \frac{1}{3}\cdot\tau\delta\frac{n}{(\log_2 n)^2}.$$

Then

$$\mathscr{S}_n(x) \le -\frac{1}{3}\frac{\tau\delta n}{(\log_2 n)^2} + b_\varepsilon n^{1-\varepsilon} + C_6\sqrt{n}\log_2 n + (C_7+C_9)\delta\sqrt{n} + \frac{C_8\delta n}{(\log_2 n)^3} \to -\infty,$$

as $n \to \infty$, and Eq. (10) follows. Proposition 1 is now proved completely.

Proof of Theorem 2(i). As already mentioned, $\mathscr{H}^\phi(\Xi_G) = \mathscr{H}^\phi(X_G) = \infty$ follows from the Rogers–Taylor density theorem (more precisely, from Corollary 1(i)). If $\mathscr{H}^\phi|_{\Xi_G}$ was σ-finite for some $c > 0$, we would have $\mathscr{H}^\phi(\Xi_G) = 0$ for all larger values of c, which is a contradiction. □

Remark. It is clear, without any calculation, that there exists $\gamma > 0$, arbitrarily small, such that

$$\tau_\gamma := \sum_{k=1}^{\infty} \frac{k^{1+\gamma}(HF_{k-1})'(p)}{2^{k+1}} \ne 0.$$

This implies, by a minor modification of the argument, that $\mathscr{H}^{\phi_\gamma}(X_G) = \infty$ for the gauge function

$$\phi_\gamma(t) = t^s \exp\Big[-c\frac{|\log t|}{(\log|\log t|)^{2+\gamma}}\Big].$$

To this end, we need to take $p_k = p \pm \frac{\delta}{k^{1+\gamma}}$ in Eq. (8), where the sign is that of τ_γ. The details are left to the reader.

Proof (Proof of Lemma 5). A numerical calculation (we used *Mathematica*) showed that

$$\sum_{k=1}^{12} \frac{k(HF_{k-1})'(p)}{2^{k+1}} \approx 0.187469.$$

Thus, we only need to estimate the remainder.

We have $(HF_{k-1})'(p) = H(p)F'_{k-1}(p) + H'(p)F_{k-1}(p)$. Recall that $p \approx 0.56984$, and a calculation gives

$$H(p) \approx 0.68336 < 0.7, \quad H'(p) \approx -0.281198; \text{ hence } |H'(p)| < 0.3.$$

Recall Eq. (19) that $F_n(x) = (x-2)^{-2}[(x-1)^{n+2} - (n+2)x + (2n+3)]$, whence

$$0 < F_n(p) < 2^{-(n+2)} - (n+2)/2 + (2n+3) < 3 + 3n/2.$$

Further,

$$F'_n(p) = \frac{2((p-1)^{n+2} - (n+2)p + (2n+3))}{(p-2)^3} + \frac{(n+2)(p-1)^{n+1} - (n+2)}{(p-2)^2}.$$

Note that in the expression for $F'_n(p)$ the 1st term is positive and the 2nd term is negative. The first term, in absolute value, is less than $2(3+3n/2) = 3n+6$, and the second term, in absolute value, is less than $n+3$ for $n \geq 1$. Thus,

$$|F'_n(p)| < 3n+6.$$

It follows (using a crude estimate) that

$$\left|\sum_{k=13}^{\infty} \frac{k(HF_{k-1})'(p)}{2^{k+1}}\right| < \sum_{k=13}^{\infty} \frac{(0.7(3k+3) + 0.3(3k+3)/2)k}{2^{k+1}} < \sum_{k=13}^{\infty} \frac{3k(k+1)}{2^{k+1}}.$$

Finally,

$$\sum_{k=13}^{\infty} \frac{3k(k+1)}{2^{k+1}} = (3/4)[(1-x)^{-1}x^{15}]''|_{x=1/2}$$
$$= (3/4)[2^{-11} + 30 \cdot 2^{-12} + 15 \cdot 14 \cdot 2^{-12}] < 0.1 < 0.187469,$$

completing the proof of the claim that $\sum_{k=1}^{\infty} \frac{k(HF_{k-1})'(p)}{2^{k+1}} > 0$. $\square$

4 Upper Bound for Hausdorff Measure

First we give a short proof of a weaker result: $\mathscr{H}^{\psi}(X_G) = 0$ where

$$\psi(t) = t^s \exp\left[-\frac{|\log_2 t|}{g(\log_2|\log_2 t|)}\right] \tag{30}$$

where g is increasing and $\int^{\infty} \frac{dt}{g(t)} = \infty$; in particular, this includes $\psi = \psi_\theta$ from Eq. (3) with $\theta = 1$.

Proof. We use the measure $\mathbb{P}_\mu$ from Eq. (6), where $\mu = \mu(p)$, $p^3 = (1-p)^2$, as in [5]. Consider **any** point $x \in X_G$. Then we obtain from Eq. (9), as in [5], for n even:

$$\begin{aligned}\mathbb{P}_\mu[x_1^n] &= (1-p)^{N_1(x_1^n)} p^{N_0(x_1^n) - N_1\left(x_1^{n/2}\right)} \\ &= p^n p^{N_0\left(x_1^{n/2}\right) - N_0(x_1^n)/2},\end{aligned} \tag{31}$$

in view of $1 - p = p^{3/2}$, $N_1(x_1^n) = n - N_0(x_1^n)$. Note that $\log_2 \psi(2^{-n}) = -ns - \frac{n}{(\ln 2) g(\log_2 n)}$. In view of $s = -\log_2 p$, we have

$$\frac{\log_2 \mathbb{P}_\mu[x_1^n] - \log_2 \psi(2^{-n})}{n} = \frac{s}{2}\left(\frac{N_0\left(x_1^{n/2}\right)}{n/2} - \frac{N_0(x_1^n)}{n}\right) + \frac{1}{(\ln 2) g(\log_2 n)}. \tag{32}$$

Denote

$$b_j := \frac{\log_2 \mathbb{P}_\mu\left[x_1^{2^j}\right] - \log_2 \psi\left(2^{-2^j}\right)}{2^j} = \frac{s}{2}\left(\frac{N_0\left(x_1^{2^{j-1}}\right)}{2^{j-1}} - \frac{N_0\left(x_1^{2^j}\right)}{2^j}\right) + \frac{1}{(\ln 2) g(j)}.$$

Then

$$b_1 + \cdots + b_\ell = \frac{s}{2}\left(N_0(x_1^1) - \frac{N_0\left(x_1^{2^\ell}\right)}{2^\ell}\right) + \sum_{j=1}^{\ell} \frac{1}{(\ln 2) g(j)} \to +\infty,\ \ell \to \infty,$$

by the assumption on the function g. It follows that $\limsup 2^j b_j = +\infty$; hence

$$\limsup_{n\to\infty}(\log_2 \mathbb{P}_\mu[x_1^n] - \log_2 \psi(2^{-n})) = +\infty,$$

and we obtain $\mathscr{H}^{\psi}(X_G) = 0$ by Corollary 1(ii). $\square$

Obtaining the same result for ψ_θ from Eq. (3) with $1 < \theta < 2$ is more delicate. Our proof follows the scheme of the proof of [9, Theorem 3(ii)], but we have to make a number of modifications. The following lemma is a version of [9, Lemma 5] in the form convenient for us.

Lemma 6. *(i) Let $1 < \eta < 2$. Suppose that $\{\gamma(n)\}_{n=1}^{\infty}$ is a real sequence such that*

$$C_1 := \sup_n |\gamma(n) - \gamma(n-1)| < \infty \tag{33}$$

and for all $n \geq n_0$

$$\gamma(n) \geq \frac{\gamma(2n)}{2} + \frac{n}{(\log_2(2n))^\eta}. \tag{34}$$

Then either there exists $c > 0$ such that for all $n \geq n_0$

$$\gamma(2n) \geq c\frac{2n}{(\log_2(2n))^{\eta-1}}, \tag{35}$$

or there exists $\varepsilon > 0$ such that for infinitely many n

$$\gamma(2n) \leq -\varepsilon n \quad \text{and} \quad \gamma(n) - \frac{\gamma(2n)}{2} \leq \frac{n}{\log_2(2n)}. \tag{36}$$

(ii) For any real sequence $\{\gamma(n)\}_{n=1}^{\infty}$ satisfying Eq. (33),

$$\gamma(n) - \frac{\gamma(2n)}{2} < \frac{n}{\log_2(2n)} \tag{37}$$

for infinitely many n.

Proof. (i) Iterating Eq. (34) we obtain for $n \geq n_0$ and $m \geq 1$:

$$\gamma(n) \geq \frac{\gamma(2^m n)}{2^m} + n \cdot \sum_{j=1}^{m} \frac{1}{(j + \log_2 n)^\eta}. \tag{38}$$

Case 1: $\gamma(n) \geq 0$ for all $n \geq n_0$. Then Eq. (38) implies for $n \geq n_0$:

$$\gamma(n) \geq n \sum_{j=1}^{\infty} \frac{1}{(j + \log_2 n)^\eta} \geq c\frac{n}{(\log_2 n)^{\eta-1}},$$

whence Eq. (35) holds.

Case 2: there exists $n_1 \geq n_0$ such that $\gamma(n_1) < -\varepsilon < 0$. Then Eq. (34) implies $\gamma(2n_1) \leq 2\gamma(n_1) < -2\varepsilon$, and inductively, $\gamma(2^m n_1) \leq -2^m \varepsilon$ for all $m \geq 1$. Moreover, for infinitely many m, we have

$$\gamma\left(2^{m-1} n_1\right) - \frac{\gamma(2^m n_1)}{2} \leq \frac{2^{m-1} n_1}{\log_2(2^m n_1)},$$

since otherwise,

$$\frac{\gamma(2^{m-1}n_1)}{2^{m-1}} - \frac{\gamma(2^m n_1)}{2^m} > \frac{n_1}{m+\log n_1}, \quad m \geq m_0+1,$$

and then taking the sum over m from m_0+1 to ℓ yields

$$\frac{\gamma(2^{m_0}n_1)}{2^{m_0}} - \frac{\gamma(2^{m_0+\ell}n_1)}{2^{m_0+\ell}} \to \infty, \quad \ell \to \infty,$$

which is a contradiction, since $|\gamma(i)| \leq C_1 i$ by Eq. (33). Thus, Eq. (36) holds for infinitely many $n = 2^m n_1$, as desired.

(ii) If the claim is not true, then Eq. (34) holds for $n \geq n_0$ with $\eta = 1$, for some $n_0 \in \mathbb{N}$. Then we obtain Eq. (38) with $\eta = 1$. But $\gamma(2^m n) \geq -C_1 2^m n$ by Eq. (33), and we get a contradiction letting $m \to \infty$. □

We still use the measure $\mathbb{P}_\mu$ from Eq. (6), as in [5], so by Eq. (31), keeping in mind that $s = -\log_2 p$, we have

$$\log_2 \mathbb{P}_\mu \left[x_1^{2n}\right] + s(2n) = \left[N_0(x_1^{2n})/2 - N_0(x_1^n)\right] s. \tag{39}$$

Observe that

$$N_0(x_1^n) = \sum_{k=1}^{\ell+1} \sum_{\frac{n}{2^k} < i \leq \frac{n}{2^{k-1}},\ i \text{ odd}} N_0(x_1^n|_{J_i}), \quad 2^{\ell-1} < n \leq 2^\ell. \tag{40}$$

By the definition of the measure $\mathbb{P}_\mu$, the random variables $N_0(x_1^n|_{J_i})$ are i.i.d. for odd $i \in \left(\frac{n}{2^k}, \frac{n}{2^{k-1}}\right]$. Note that $|x_1^n|_{J_i}| = k$ for such i, and the distribution of these random variables is the distribution of $N_0(u)$, $|u| = k$, where $\{u_i\}$ is the Markov chain corresponding to μ. By the definition of $\mu = \mu(p)$,

$$\mathbb{E}\left[N_0[u]\right] = \sum_{j=0}^{k-1} (\mathbf{p}P^j)_0, \quad |u| = k,$$

where $\mathbf{p} = (p, 1-p)$ and $P = \begin{pmatrix} p & 1-p \\ 1 & 0 \end{pmatrix}$. Since P has left eigenvectors $\pi = \left(\frac{1}{2-p}, \frac{1-p}{2-p}\right)$ and $\tau = (1,-1)$ corresponding to the eigenvalues 1 and $p-1$, respectively, we have

$$(\mathbf{p}P^j)_0 = \frac{1}{2-p}[1-(p-1)^{j+2}], \ j \geq 0;$$

hence

$$\mathbb{E}\left[N_0[u]\right] = \frac{k}{2-p} - \frac{1}{2-p}\sum_{j=0}^{k-1}(p-1)^{j+2} = \frac{k}{2-p} - \frac{(1-(p-1)^k)(p-1)^2}{(2-p)^2} =: L_k. \tag{41}$$

Lemma 7. *We have*

$$\left|\frac{\mathbb{E}\left[N_0(x_1^{2n})\right]}{2}-\mathbb{E}\left[N_0(x_1^n)\right]\right|<C(\log_2 n)^2,\quad n\in\mathbb{N},$$

for some $C>0$*, where* x *has the law of* $\mathbb{P}_\mu$.

Proof. Denote by $\mathbf{Z}_{\text{odd}}(a,b]$ the set of odd integers in the interval $(a,b]$, where $a<b$ are reals. We have from Eqs. (40) and (41)

$$\mathbb{E}\left[N_0(x_1^n)\right]=\sum_{k=1}^{\ell+1}\#\mathbf{Z}_{\text{odd}}\left(\frac{n}{2^k},\frac{n}{2^{k-1}}\right]\cdot L_k.$$

Note that $\mathbf{Z}_{\text{odd}}(\frac{n}{2^{\ell+1}},\frac{n}{2^\ell}]=\{1\}$ if $n=2^\ell$, and it is empty otherwise. It follows that

$$\frac{\mathbb{E}\left[N_0(x_1^{2n})\right]}{2}-\mathbb{E}\left[N_0(x_1^n)\right]=\sum_{k=1}^{\ell+1}\left(\frac{\#\mathbf{Z}_{\text{odd}}\left(\frac{n}{2^{k-1}},\frac{n}{2^{k-2}}\right]}{2}-\#\mathbf{Z}_{\text{odd}}\left(\frac{n}{2^k},\frac{n}{2^{k-1}}\right]\right)\cdot L_k+d\cdot L_{\ell+2},\tag{42}$$

where $d\in\{0,1/2\}$. It is easy to see that

$$\left|\#\mathbf{Z}_{\text{odd}}(a,b]-\left(\frac{b-a}{2}\right)\right|\leq 1,\quad\text{for }\ 0<a<b,\tag{43}$$

hence, taking Eq. (41) into account,

$$\left|\frac{\mathbb{E}\left[N_0(x_1^{2n})\right]}{2}-\mathbb{E}\left[N_0(x_1^n)\right]\right|\leq 2\sum_{k=1}^{\ell+2}L_k\leq C'\sum_{k=1}^{\ell+2}k\leq C''\ell^2\leq C(\log_2 n)^2.$$

□

Proof (Proof of Theorem 2(ii)). In order to show that $\mathscr{H}^{\psi_\theta}(X_G)=0$, we cover X_G by three subsets: B, L, and Λ, defined as follows. Let

$$B:=\left\{x\in X_G:\exists\eta>\theta,\ \frac{N_0(x_1^{2n})}{2}-N_0(x_1^n)>\frac{-2n}{(\log_2(2n))^\eta}\text{ for infinitely many }n\right\}.\tag{44}$$

Denote

$$N_0^*(x_1^n):=N_0(x_1^n)-\mathbb{E}\left[N_0(x_1^n)\right]$$

and let

$$L:=\left\{x\in X_G:\ \exists\varepsilon>0,\ \frac{N_0(x_1^{2n})}{2}-N_0(x_1^n)\geq\frac{-2n}{\log_2(2n)}\ \text{ and }\ N_0^*(x_1^{2n})\leq-\varepsilon n\ \text{ for infinitely many }n\right\}.\tag{45}$$

Finally, let $\Lambda = X_G \setminus (L \cup B)$. It suffices to verify that each of the three sets B, L, Λ has zero $\mathscr{H}^{\psi_\theta}$-measure (indeed, L and Λ even have zero $\mathscr{H}^s$-measure).

Step 1: $\mathscr{H}^{\psi_\theta}(B) = 0$. Let B_η be the set of $x \in X_G$ such that the condition in Eq. (44) holds for a fixed η. Thus, $B = \bigcup_{\eta > \theta} B_\eta = \bigcup_{\eta \in \mathbf{Q},\, \eta > \theta} B_\eta$, and it is enough to show that $\mathscr{H}^{\psi_\theta}(B_\eta) = 0$. We have from Eq. (39) and the definition of ψ_θ for all $x \in B_\eta$:

$$\log_2 \mathbb{P}_\mu[x_1^{2n}] - \log_2 \psi_\theta(2^{-2n}) = s\left(\frac{N_0(x_1^{2n})}{2} - N_0(x_1^n)\right) + \frac{2n}{(\ln 2)(\log_2(2n))^\theta}$$
$$> \frac{-2ns}{(\log_2(2n))^\eta} + \frac{2n}{(\ln 2)(\log_2(2n))^\theta}$$

for infinitely many n. Since $\eta > \theta$, it follows that

$$\limsup_{n\to\infty} \left(\log_2 \mathbb{P}_\mu \left[x_1^{2n}\right] - \log_2 \psi_\theta \left(2^{-2n}\right)\right) = +\infty;$$

hence $\mathscr{H}^{\psi_\theta}(B_\eta) = 0$ by Theorem 3.

Step 2: $\mathscr{H}^{\psi_\theta}(L) = 0$. Denote by $L(\varepsilon)$ the set of points $x \in X_G$ which satisfy the condition in Eq. (45) for a given $\varepsilon > 0$. For $\varepsilon > 0$ and $n \in \mathbb{N}$ let $\mathscr{L}_n(\varepsilon)$ be the set of words u of length $2n$ for which the condition in Eq. (45) holds. (Note that this condition depends only on the first $2n$ symbols of x; thus, Eq. (45) holds for all $x \in [u]$.) If $u \in \mathscr{L}_n(\varepsilon)$ then by Eqs. (39) and (45)

$$\log_2 \left(\mathbb{P}_\mu[u]\, 2^{2ns}\right) \geq \frac{-2sn}{\log_2(2n)};$$

hence

$$2^{-2ns} \leq \exp\left(\frac{2sn(\ln 2)}{\log_2(2n)}\right) \mathbb{P}_\mu[u]. \tag{46}$$

By the definition of $\mathscr{L}_n(\varepsilon)$ we have

$$\sum_{u \in \mathscr{L}_n(\varepsilon)} \mathbb{P}_\mu[u] \leq \mathbb{P}_\mu\left(x :\ N_0^*(x_1^{2n}) \leq -\varepsilon n\right). \tag{47}$$

The following lemma is a consequence of large deviation estimates; it will be used in the last step of the proof as well.

Lemma 8. *There exist $c_2, c_3 > 0$ such that for all $t > 0$ and $n \in \mathbb{N}$,*

$$\mathbb{P}_\mu\left(x :\ |N_0^*(x_1^{2n})| \geq tn\right) \leq c_2 \exp(-c_3 t^2 n).$$

Proof. We have by Eq. (40),

$$N_0^*(x_1^{2n}) = \sum_{k=1}^{\ell+1} S_{A_k}^*, \text{ where } A_k = \#\mathbf{Z}_{\text{odd}}\left(\frac{n}{2^k}, \frac{n}{2^{k-1}}\right],$$
$$S_{A_k}^* := \sum_{\frac{n}{2^k} < i \le \frac{n}{2^{k-1}},\ i \text{ odd}} N_0^*(x_1^n|_{J_i}),$$

and

$$N_0^*(x_1^n|_{J_i}) = N_0(x_1^n|_{J_i}) - \mathbb{E}[N_0(u)] \text{ for } |u| = k \text{ and } i \in \mathbf{Z}_{\text{odd}}\left(\frac{n}{2^k}, \frac{n}{2^{k-1}}\right].$$

Now,

$$\mathbb{P}_\mu\left(\left|\sum_{k=1}^{\ell+1} S_{A_k}^*\right| \ge tn\right) \le \sum_{k=1}^{\ell+1} \mathbb{P}_\mu\left(\left|S_{A_k}^*\right| \ge \frac{tn}{k(k+1)}\right),$$

since $\sum_{k=1}^{\infty} \frac{1}{k(k+1)} = 1$. Note that $S_{A_k}^*$ is a sum of A_k independent random variables, which are bounded by k in modulus; hence by Hoeffding's inequality Eq. (11)

$$\mathbb{P}_\mu\left(\left|S_{A_k}^*\right| \ge \frac{tn}{k(k+1)}\right) = \mathbb{P}_\mu\left(\left|S_{A_k}^*\right| \ge A_k \cdot \frac{tn}{k(k+1)A_k}\right)$$
$$\le 2\exp\left[-\frac{t^2 n^2}{2k^4(k+1)^2 A_k}\right].$$

Observe that $A_k \le \frac{n}{2^{k+1}} + 1 \le \frac{n}{2^k}$ by Eq. (43); hence

$$\mathbb{P}_\mu\left(\left|S_{A_k}^*\right| \ge \frac{tn}{k(k+1)}\right) \le 2\exp\left[-\frac{t^2 n \cdot 2^k}{2k^4(k+1)^2}\right],$$

and, therefore,

$$\mathbb{P}_\mu\big(x : |N_0^*(x_1^{2n})| \ge tn\big) \le 2\sum_{k=1}^{\infty} \exp\left[-\frac{t^2 n \cdot 2^k}{2k^4(k+1)^2}\right] \le c_2 \exp(-c_3 t^2 n),$$

for some positive c_2, c_3, as desired. □

Combining Eqs. (46), (47) and Lemma 8, with $t = \varepsilon$, yields

$$\sum_{u \in \mathcal{L}_n(\varepsilon)} (2^{-2n})^s \le c_2 \exp\left[\frac{2sn(\ln 2)}{\log_2(2n)} - c_3 \varepsilon^2 n\right].$$

The right-hand side of this inequality is summable in n, so by choosing large n_0, we can make the sum

$$\sum_{n \ge n_0} \sum_{u \in \mathcal{L}_n(\varepsilon)} (2^{-2n})^s = \sum_{n \ge n_0} \sum_{u \in \mathcal{L}_n(\varepsilon)} (d[u])^s$$

arbitrarily small. But for any n_0, the union $\bigcup_{n=n_0}^{\infty}\bigcup_{u\in\mathscr{L}_n(\varepsilon)}[u]$ forms a cover of $L(\varepsilon)$, proving that $\mathscr{H}^s(L(\varepsilon))=0$. Finally, $L=\bigcup_{\varepsilon\in\mathbf{Q}}L(\varepsilon)$, so we obtain that $\mathscr{H}^s(L)=0$ and certainly $\mathscr{H}^{\psi_\theta}(L)=0$.

Step 3. For $\eta\in(1,2)$, $\varepsilon\in(0,\eta)$, and $c>0$, let $\Lambda(\eta,\varepsilon,c)$ be the set of $x\in X_G$ such that for n sufficiently large we have

$$\frac{N_0(x_1^{2n})}{2}-N_0(x_1^n)\le\frac{-2n}{(\log_2(2n))^{\eta}} \tag{48}$$

and

$$N_0^*(x_1^{2n})\ge c\frac{2n}{(\log_2(2n))^{\eta-1}}, \tag{49}$$

but for infinitely many n,

$$\frac{N_0(x_1^{2n})}{2}-N_0(x_1^n)>\frac{-2n}{(\log_2(2n))^{\eta-\varepsilon}}. \tag{50}$$

By Lemma 6(ii), applied to $\{N_0(x_1^n)\}_{n\ge1}$, Eq. (50) certainly holds for $\varepsilon=\eta-1$.

We claim that

$$X_G\setminus(B\cup L)\subset\bigcup_{\eta\in(1,2)}\bigcup_{c>0}\bigcup_{\varepsilon\in(0,2-\eta)}\Lambda(\eta,\varepsilon,c). \tag{51}$$

Indeed, for $x\in X_G\setminus B$ let η^* be the infimum of η for which Eq. (48) holds for n sufficiently large (note that $x\notin B$ means such η exists). Then $\eta^*\in[1,2)$ by Lemma 6(ii), and Eq. (48) holds with $\eta=\eta^*+\frac{2-\eta^*}{3}$ for n sufficiently large, whereas Eq. (50) holds for $\varepsilon\in(\eta-\eta^*,2-\eta)=(\frac{2-\eta^*}{3},\frac{2(2-\eta^*)}{3})$. Let

$$\gamma(n):=N_0^*(x_1^n),\ \ n\ge1.$$

It is clear that

$$|\gamma(n+1)-\gamma(n)|\le2,\ \ n\ge1.$$

It follows from Eq. (48) and Lemma 7 that

$$\gamma(n)-\frac{\gamma(2n)}{2}\ge\frac{2n}{(\log_2(2n))^{\eta}}-C(\log_2 n)^2\ge\frac{n}{(\log_2(2n))^{\eta}}$$

for n sufficiently large. Thus, the sequence $\{\gamma(n)\}_{n\ge1}$ satisfies the assumptions of Lemma 6(i). By Lemma 6(i), either there exists $c>0$ such that for all n sufficiently large

$$N_0^*(x_1^{2n})\ge c\frac{2n}{(\log_2(2n))^{\eta-1}},$$

which together with the above yields that $x \in \Lambda(\eta,\varepsilon,c)$, or else there exists $\varepsilon > 0$ such that for infinitely many n,

$$N_0^*(x_1^{2n}) \le -\varepsilon n \quad \text{and} \quad N_0^*(x_1^n) - \frac{N_0^*(x_1^{2n})}{2} \le \frac{n}{\log_2(2n)};$$

hence by Lemma 7,

$$N_0(x_1^n) - \frac{N_0(x_1^{2n})}{2} \le \frac{n}{\log_2(2n)} + c(\log_2 n)^2 \le \frac{2n}{\log_2(2n)}$$

for infinitely many n, so that $x \in L$, proving the claim. Since the union in Eq. (51) can be taken over rational η, c, ε, it suffices to show that $\mathscr{H}^s(\Lambda(\eta,\varepsilon,c)) = 0$ for $\varepsilon \in (0, 2-\eta)$.

Let $\Gamma_n(\eta,\varepsilon,c)$ be the collection of words u of length $2n$ for which Eqs. (48)–(50) hold (as before, this is well defined). If $u \in \Gamma_n(\eta,\varepsilon,c)$, then by Eqs. (39) and (50)

$$\log_2(\mathbb{P}_\mu[u]\,2^{2ns}) \ge \frac{-2ns}{(\log_2(2n))^{\eta-\varepsilon}};$$

hence

$$2^{-2ns} \le \exp\left(\frac{2ns(\ln 2)}{\log_2(2n)^{\eta-\varepsilon}}\right)\mathbb{P}_\mu[u]. \tag{52}$$

By the definition of $\Gamma_n(\eta,\varepsilon,c)$ and Lemma 8, with $t = \frac{2c}{(\log_2(2n))^{\eta-1}}$,

$$\begin{aligned}\sum_{u\in\Gamma_n(\eta,\varepsilon,c)} \mathbb{P}_\mu[u] &\le \mathbb{P}_\mu\left(x:\ N_1^*(x_1^{2n}) \ge c\frac{2n}{(\log_2(2n))^{\eta-1}}\right)\\ &\le c_2\exp\left(-\frac{\tilde{c}n}{(\log_2(2n))^{2\eta-2}}\right),\end{aligned}$$

with $\tilde{c} = 4c_3c^2$. Combining this with Eq. (52) yields

$$\sum_{u\in\Gamma_n(\eta,\varepsilon,c)} 2^{-2ns} \le \exp c_2\left[\frac{2ns(\ln 2)}{\log_2(2n)^{\eta-\varepsilon}} - \frac{\tilde{c}n}{(\log_2(2n))^{2\eta-2}}\right].$$

Recall that $\varepsilon < 2-\eta$, and therefore $\eta-\varepsilon > 2\eta-2$ and the right-hand side of the last inequality is summable in n. It follows that, by taking n_1 sufficiently large, the sum

$$\sum_{n\ge n_1}\ \sum_{u\in\Gamma_n(\eta,\varepsilon,c)} 2^{-2ns}$$

can be made arbitrarily small. Since for any n_1, the union

$$\bigcup_{n \geq n_1} \bigcup_{u \in \Gamma_n(\eta,\varepsilon,c)} [u]$$

covers $\Lambda(\eta,\varepsilon,c)$, this implies $\mathscr{H}^s(\Lambda(\eta,\varepsilon,c)) = 0$ (and hence $\mathscr{H}^{\psi_\theta}(\Lambda(\eta,\varepsilon,c)) = 0$), completing the proof. □

Acknowledgements The research of B. S. was supported in part by the NSF grant DMS-0968879. He is grateful to the Microsoft Research Theory Group for hospitality during 2010–2011. He would also like to thank the organizers of the conference "Fractals and Related Fields II" for the excellent meeting and stimulating atmosphere. The authors are grateful to the referee for careful reading of the manuscript and many helpful comments.

References

1. Bedford, T.: Crinkly curves, Markov partitions and box dimension in self-similar sets. Ph.D. Thesis, University of Warwick, (1984)
2. Falconer, K.: Fractal Geometry. Mathematical Foundations and Applications. Wiley, Chichester (1990)
3. Fan, A., Liao, L., Ma, J.: Level sets of multiple ergodic averages. Preprint (2011) arXiv:1105.3032. Monatsh. Math. **168**(1), 17–26 (2012)
4. Hoeffding, W.: Probability inequalities for sums of bounded random variables. J. Am. Statist. Assoc. **58**(302), 13–30 (1963)
5. Kenyon, R., Peres, Y., Solomyak, B.: Hausdorff dimension of the multiplicative golden mean shift. C. R. Math. Acad. Sci. Paris **349**, 625–628 (2011)
6. Kenyon, R., Peres, Y., Solomyak, B.: Hausdorff dimension for fractals invariant under the multiplicative integers. Preprint (2011) arXiv 1102.5136. Ergodic Th. Dynam. Sys. **32**(5), 1567–1584 (2012)
7. Lalley, S., Gatzouras, D.: Hausdorff and box dimensions of certain self-affine fractals. Indiana Univ. Math. J. **41**(2), 533–568 (1992)
8. McMullen, C.: The Hausdorff dimension of general Sierpinski carpets. Nagoya Math. J. **96**, 1–9 (1984)
9. Peres, Y.: The self-affine carpets of McMullen and Bedford have infinite Hausdorff measure. Math. Proc. Camb. Philos. Soc. **116**, 513–526 (1994)
10. Rogers, C.A.: Hausdorff Measures. Cambridge University Press, Cambridge (1970).
11. Rogers, C.A., Taylor, S.J.: Functions continuous and singular with respect to a Hausdorff measure. Mathematika **8**, 1–31 (1961)

The Law of Iterated Logarithm and Equilibrium Measures Versus Hausdorff Measures for Dynamically Semi-regular Meromorphic Functions

Bartłomiej Skorulski and Mariusz Urbański

Abstract The Law of Iterated Logarithm for dynamically semi-regular meromorphic mappings and loosely tame observables is established. The equilibrium states of tame potentials are compared with an appropriate one-parameter family of generalized Hausdorff measures. The singularity/absolute continuity dichotomy is established. Both results utilize the concept of nice sets and the theory of infinite conformal iterated function systems.

1 Introduction

One of the central questions in ergodic theory is to find out how mixing and how random is a dynamical system which preserves a probability measure. There is an enormous literature on the subject establishing fast (desirably exponential) decay of correlations, the Central Limit Theorem, and the Law of Iterated Logarithm. The classical results concern Bernoulli shifts, Markov chains (see [2] or any standard book on probability theory), and Gibbs states of Hölder continuous potentials for dynamical systems exhibiting some sort of hyperbolic behavior (see, e.g., [10]. Strong stochastic laws such as exponential decay of correlations and the Central Limit Theorem were established in [9] for the class of dynamically semi-regular meromorphic functions (we refer the reader to Sect. 2 for the definition).

As it was shown in [8, 9], this is a large class of functions and its ergodic theory and thermodynamic formalism were well developed and understood. What was

B. Skorulski
Departamento de Matemáticas, Universidad Católica del Norte,
Avenida Angamos 0610, Antofagasta, Chile
e-mail: bskorulski@ucn.cl

M. Urbański (✉)
Department of Mathemsatics, University of North Texas, Denton, TX 76203-1430, USA
e-mail: urbanski@unt.edu web: www.math.unt.edu/~urbanski

J. Barral and S. Seuret (eds.), *Further Developments in Fractals and Related Fields*,
Trends in Mathematics, DOI 10.1007/978-0-8176-8400-6_11,

missing there was the Law of Iterated Logarithm. Some research to fill this gap was undertaken in [1] where the Law of Iterated Logarithm was established, provided that the dynamics of a semi-regular map f was conjugated (on a subset of the Julia set of f) to the shift map. In the present chapter we establish this law without these kinds of restrictions. In Sect. 4 we prove the following theorem (see Theorem 7).

Theorem 1. *Let $f:\mathbb{C}\to\hat{\mathbb{C}}$ be a dynamically semi-regular meromorphic function and fix $t>\rho/\alpha$. Let $\psi_t:J_f\to\mathbb{R}$ be a loosely tame function. Then the asymptotic variance $\sigma_f^2(\psi_t)$ exists and, if $\sigma_f^2(\psi_t)>0$, equivalently if $\psi_t:J_f\to\mathbb{R}$ is not cohomologous to a constant in the class of Hölder continuous functions on J_f, then the function $\psi_t:J_f\to\mathbb{R}$ satisfies the Law of Iterated Logarithm with respect to the dynamical system (f,μ_t) with $A_{\psi_t}=\sqrt{2}\sigma_{\hat{f}}(\hat{\psi}_t)>0$.*

Here, the constants ρ and α come from the definition of semi-regularity (see Sect. 2) and the class of loosely tame observables (see Sect. 2) include all bounded Hölder continuous functions. The measure μ_t is a Borel probability invariant measure which is an equilibrium state for ψ and is unique in the sense of Theorem 4. The existence of this kind of measure for semi-regular meromorphic function was proved in [9] and unfortunately is unknown in general. However, if one can prove their existence in general it seems possible to generalize results obtained here.

Our approach is based on the one hand on our observation that under relatively mild conditions the Law of Iterated Logarithm for an induced (first return) map entails this law for the original system (see Theorem 5), and, on the other hand, on the fact (see [4, 13, 15]) that each dynamically semi-regular function, as a matter of fact each tame meromorphic function, admits first return maps that form a very well understood class of conformal iterated function systems (see [7]). For this class of system all above-mentioned stochastic laws are known [7].

Sticking to the realm of dynamically semi-regular meromorphic functions, the second theme of our chapter is the issue of comparing the equilibrium states of tame potentials with an appropriate one-parameter family of generalized Hausdorff measures. This circle of investigations goes back to the fundamental work [5] of Makarov in potential theory (harmonic measure) and its dynamical counterpart [12]. The dichotomy phenomenon of singularity/absolute continuity observed in [12] has been afterward also detected in the context of parabolic Jordan curves [3] and conformal iterated function systems (see [16], comp [7]). In this chapter we exhibit it in the realm of meromorphic functions. In Sect. 5 we show the following theorem (see Theorem 9).

Theorem 2. *Let $f:\mathbb{C}\to\hat{\mathbb{C}}$ be a dynamically semi-regular meromorphic function and for every $t>\rho/\alpha$ let $\psi_t=-t\log|f'|_\tau+k$ be a loosely tame function. Suppose that $\sigma^2(\hat{\psi}_t)>0$ (this is in particular true if $t\neq \mathrm{HD}(J_{r,f})$, more particularly if $t\geq 2$) and that $h:(a,+\infty)\to(0,+\infty)$ is a slowly growing function. Then:*

(a) If h belongs to the upper class, then the measures μ_t and $\mathrm{H}_{\tilde{h}}|_{J_f}$ are mutually singular.
(b) If h belongs to the lower class, then μ_t is absolutely continuous with respect to $\mathrm{H}_{\tilde{h}}$.

As for the Law of Iterated Logarithm our approach here utilizes the concept of nice sets that generate infinite conformal iterated function systems in the sense of [7]. For them, as already mentioned, the dichotomy is known (see [16], comp [7]). It is then an easy observation that it also holds for original meromorphic functions. The key technical issues in here are to conclude that the asymptotic variance of an appropriate function related to the induced system (IFS) is positive and that these functions have finite moments of all orders.

2 Preliminaries

Let $f:\mathbb{C}\to\hat{\mathbb{C}}$ be a meromorphic function. The Fatou set of f consists of all points $z\in\mathbb{C}$ that admit an open neighborhood U_z such that all the forward iterates f^n, $n\geq 0$, of f are well-defined on U_z and the family of maps $\{f^n|_{U_z}:U_z\to\mathbb{C}\}_{n=0}^{\infty}$ is normal. The Julia set of f, denoted by J_f, is then defined as the complement of the Fatou set of f in $\mathbb{C}$. By $\mathrm{Sing}(f^{-1})$ we denote the set of singularities of f^{-1}. We define *the postsingular set* of $f:\mathbb{C}\to\hat{\mathbb{C}}$ as

$$\mathrm{PS}(f)=\overline{\bigcup_{n=0}^{\infty} f^n(\mathrm{Sing}(f^{-1}))}.$$

Given a set $F\subset\hat{\mathbb{C}}$ and $n\geq 0$, by $\mathrm{Comp}(f^{-n}(F))$ we denote the collection of all connected components of the inverse image $f^{-n}(F)$. A meromorphic function $f:\mathbb{C}\to\hat{\mathbb{C}}$ is called *tame* if its postsingular set does not contain its Julia set. This is the primary object of our interest in this chapter.

We make heavy use of the concept of *a nice set* which Rivera–Letelier introduced in [13] in the realm of the dynamics of rational maps of the Riemann sphere. In Neil Dobbs [4] proved their existence for tame meromorphic functions from $\mathbb{C}$ to $\hat{\mathbb{C}}$. The following theorem follows directly from Lemma 11 from [4].

Theorem 3. *Let $f:\mathbb{C}\to\hat{\mathbb{C}}$ be a tame meromorphic function. Fix $z\in\mathscr{J}(f)\setminus\mathscr{P}(f)$, $\kappa>1$, and $K>1$. Then there exists $L>1$ such that for all $r>0$ sufficiently small, there exists an open connected set $U=U(z,r)\subset\mathbb{C}\setminus\mathscr{P}(f)$ such that:*

(a) If $V\in Comp(f^{-n}(U))$ and $V\cap U\neq\emptyset$, then $V\subset U$.
(b) If $V\in Comp(f^{-n}(U))$ and $V\cap U\neq\emptyset$, then, for all $w,w'\in V$,

$$|(f^n)'(w)|\geq L \ \ and \ \ \frac{|(f^n)'(w)|}{|(f^n)'(w')|}\leq K.$$

(c) $\overline{B(z,r)}\subset U\subset B(z,\kappa r)\subset\mathbb{C}\setminus\mathscr{P}(f)$.

Let $\mathscr{U}$ be the collection of all nice sets of $f:\mathbb{C}\to\hat{\mathbb{C}}$, i.e., all the sets $U=U_z$ satisfying the above proposition with some $z\in J_f\setminus\mathrm{PS}(f)$, $\kappa>1$, $K>1$ $L>1$, and some $r > 0$. Note that if $U \in \mathscr{U}$ and $V \in \mathrm{Comp}(f^{-n}(U))$ satisfies the requirements (a), (b), and (c) from Theorem 3 then there exists a unique holomorphic inverse branch

$$f_V^{-n} : B(z, \kappa r) \to \mathbb{C} \text{ such that } f_V^{-n}(U) = V.$$

As noted in [15], the collection

$$\mathscr{S} = \mathscr{S}_U = \{\phi_e\}_{e \in E}$$

of all such inverse branches all restricted to the set $X = \overline{U}$, bijectively parametrized by some arbitrary countable set E, forms an *iterated function system* in the sense of [6, 7]. In particular, it clearly satisfies the Open Set Condition. We have just mentioned [6, 7]. In what concerns iterated function systems we try our concepts and notation to be compatible with that of [7]. We now recall that given $n \in \mathbb{N}$ and $\omega = \omega_1 \omega_2 \ldots \omega_n \in E^n$, we put

$$\phi_\omega = \phi_{\omega_1} \circ \phi_{\omega_2} \circ \ldots \phi_{\omega_n} : X \to X.$$

Given $\omega \in E^{\mathbb{N}}$, for every $n \in \mathbb{N}$, we put $\omega|_n := \omega_1 \omega_2 \ldots \omega_n \in E^n$. Then the sets

$$\{\phi_{\omega|_n}(X)\}_{n \geq 1}$$

form a descending sequence of nonempty compact sets and therefore

$$\bigcap_{n \geq 1} \phi_{\omega|_n}(X) \neq \emptyset.$$

Making use of the Poincaré (hyperbolic) metric on $B(z, \kappa r)$, a standard argument shows that Euclidean diameters of these sets decrease to zero uniformly exponentially fast. Thus, the intersection

$$\bigcap_{n \geq 1} \phi_{\omega|_n}\left(X_{t(\omega_n)}\right)$$

is a singleton and we denote its only element by $\pi(\omega)$. In this way we get the coding map π, also frequently called a projection, from the symbol space $E^{\mathbb{N}}$ to X:

$$\pi : E^{\mathbb{N}} \to X.$$

We put

$$J_{\mathscr{S}} := \pi(E^{\mathbb{N}}),$$

and call this set the *limit set* of the iterated function system $\mathscr{S}$. Given $\omega, \tau \in E^{\mathbb{N}}$, we define $\omega \wedge \tau \in E^{\mathbb{N}} \cup E^*$ to be the longest initial block common for both ω and τ. For each $\kappa > 0$, we define a metric, d_κ, on I^∞, by setting

$$d_\kappa(\omega, \tau) = \mathrm{e}^{-\kappa|\omega \wedge \tau|}.$$

These metrics are all equivalent and induce the same topology and Borel sets. A function is uniformly continuous with respect to one of these metrics if and only if it is uniformly continuous with respect to all. Also, a function is Hölder continuous with respect to one of these metrics if and only if it is Hölder continuous with respect to all of them. The projection $\pi : E^{\mathbb{N}} \to X$ is Hölder continuous.

In Sect. 4 we will make use of the thermodynamic formalism built out of such systems (see [7]). We now recall the basic concepts and theorems resulting forming this formalism. Let $\sigma : E^{\mathbb{N}} \to E^{\mathbb{N}}$ be the shift map, i.e., $\sigma(\omega_1\omega_2\omega_3\ldots) = \omega_2\omega_3\ldots$. Let

$$G = \{g^{(e)} : X \to \mathbb{R} : e \in E\}$$

be a family of real-valued functions. For every $n \geq 1$ and $\beta > 0$ let

$$V_n(G) = \sup_{\omega \in E^n} \sup_{x,y \in X_{t(\omega)}} \{|g^{(\omega_1)}(\phi_{\sigma(\omega)}(x)) - g^{(\omega_1)}(\phi_{\sigma(\omega)}(y))|\}e^{\beta(n-1)}.$$

We have made the conventions that the empty word $\emptyset$ is the only word of length 0 and $\phi_\emptyset = \mathrm{Id}_X$. Thus, $V_1(G) < \infty$ simply means the diameters of the sets $g^i(X)$ are uniformly bounded. The collection G is called a Hölder family of functions (of order β) if

$$V_\beta(G) = \sup_{n \geq 1}\{V_n(G)\} < \infty. \tag{1}$$

We call the Hölder family G summable Hölder (of order β) if (1) is satisfied and

$$\sum_{e \in E} e^{\sup(g^{(e)})}) < \infty. \tag{2}$$

The following limit exists and it is called topological pressure of the family G:

$$\mathrm{P}(G) = \lim_{n \to \infty} \frac{1}{n} \log \sum_{|\omega| = n} \exp\left(\sup\left(\sum_{j=1}^{n} g^{(\omega_j)} \circ \phi_{\sigma^j\omega}\right)\right).$$

A Borel probability measure m is said to be G-conformal, provided it is supported on the limit set $J_{\mathscr{S}}$ and the following two conditions are satisfied: for every $e \in E$ and for every Borel set $A \subset X$,

$$m(\phi_e(A)) = \int_A \exp\big(g^{(e)} - \mathrm{P}(G)\big)\, dm \tag{3}$$

and

$$m\big(\phi_a(X) \cap \phi_b(X)\big) = 0 \tag{4}$$

for all $a, b \in E$, $a \neq b$. It is proved in [7] that all Hölder summable families G of functions admit unique G-conformal measures m_G and their unique invariant versions, i.e., Borel probability measures μ_G on $J_{\mathscr{S}}$ that are equivalent to m_G with bounded (and separated from zero) Hölder continuous Radon–Nikodym derivatives with the property that

$$\mu_G(A) = \sum_{e\in E} \mu_G(\phi_e(A))$$

for every Borel set $A \subset X$.

We now continue with meromorphic functions alone. So, keep $f : \mathbb{C} \to \hat{\mathbb{C}}$ a meromorphic function. The function f is called *topologically hyperbolic* if

$$\mathrm{dist}_{\mathrm{Euclid}}(J_f, \mathrm{PS}(f)) > 0,$$

and it is called *expanding* if there exist $c > 0$ and $\lambda > 1$ such that

$$|(f^n)'(z)| \geq c\lambda^n$$

for all integers $n \geq 1$ and all points $z \in J_f \setminus f^{-n}(\infty)$. Note that every topologically hyperbolic meromorphic function is tame. A meromorphic function that is both topologically hyperbolic and expanding is called *hyperbolic*. The meromorphic function $f : \mathbb{C} \to \hat{\mathbb{C}}$ is called dynamically *semi-regular* if it is of finite order, denoted in this chapter by ρ, and satisfies the following rapid growth condition for its derivative:

$$|f'(z)| \geq \kappa^{-1}(1+|z|)^{\alpha_1}(1+|f(z)|)^{\alpha_2}, \quad z \in J_f, \tag{5}$$

with some constant $\kappa > 0$ and α_1, α_2 such that $\alpha_2 > \max\{-\alpha_1, 0\}$. Set $\alpha := \alpha_1 + \alpha_2$.

Let $k : J_f \to \mathbb{R}$ be a weakly Hölder continuous function in the sense of [9]. The definition introduced in [9] is somewhat technical and we will not provide it in the current chapter. What is important is that each bounded, uniformly locally Hölder function $k : J_f \to \mathbb{R}$ is weakly Hölder. Fix $\tau > \alpha_2$ as required in [9]. For $t \in \mathbb{R}$, let

$$\psi_t = -t\log|f'|_\tau + k, \tag{6}$$

where $|f'(z)|_\tau$ is the norm, or, equivalently, the scaling factor, of the derivative of f evaluated at a point $z \in J_f$ with respect to the Riemannian metric $|\mathrm{d}\tau(z)| = (1+|z|)^{-\tau}|\mathrm{d}z|$. Following [9] functions of the form (6) (frequently referred to as potentials) are called *loosely tame*. Let $\mathscr{L}_t : C_b(J_f) \to C_b(J_f)$ be the corresponding *Perron–Frobenius operator* given by the formula

$$\mathscr{L}_t g(z) = \sum_{w\in f^{-1}(z)} g(w)\mathrm{e}^{\psi_t(w)}.$$

It was shown in [9] that, for every $z \in J_f$ and for the function $\mathbf{1} : z \mapsto 1$, the limit

$$\lim_{n\to\infty} \frac{1}{n} \log \mathscr{L}_t \mathbf{1}(z)$$

exists and takes on the same common value, which we denote by $\mathrm{P}(t)$ and call *the topological pressure* of the potential ψ_t. The following theorem was proved in [9].

Theorem 4. *If $f : \mathbb{C} \to \hat{\mathbb{C}}$ is a dynamically semi-regular meromorphic function and $k : J_f \to \mathbb{R}$ is a weakly Hölder continuous potential, then for every $t > \rho/\alpha$ there exist uniquely determined Borel probability measures m_t and μ_t on J_f with the following properties:*

(a) $\mathcal{L}_t^* m_t = m_t$.
(b) $\mathrm{P}(t) = \sup\{\mathrm{h}_\mu(f) + \int \psi_t \, d\mu : \mu \circ f^{-1} = \mu \text{ and } \int \psi_t \, d\mu > -\infty\}$.
(c) $\mu_t \circ f^{-1} = \mu_t$, $\int \psi_t \, d\mu_t > -\infty$, *and* $\mathrm{h}_{\mu_t}(f) + \int \psi_t \, d\mu_t = \mathrm{P}(t)$.
(d) *The measures μ_t and m_t are equivalent and the Radon–Nikodym derivative $\frac{d\mu_t}{dm_t}$ has a nowhere-vanishing Hölder continuous version which is bounded above.*

3 The Law of Iterated Logarithm: Abstract Setting

In this section we deal with issues related to the Law of Iterated Logarithm in the setting of general measure-preserving transformations. Let (X, μ) be a probability space and let $T : X \to X$ be a measurable map preserving measure μ. Let $g : Y \to \mathbb{R}$ be a square integrable function and $S_n g = \sum_{k=0}^{n-1} g \circ T^k$. We put

$$\overline{\sigma}_T^2(g) := \limsup_{n\to\infty} \frac{1}{n} \int_X (S_n(g) - n\mu(g))^2 \, d\mu$$

and

$$\underline{\sigma}_T^2(g) := \liminf_{n\to\infty} \frac{1}{n} \int_X (S_n(g) - n\mu(g))^2 \, d\mu.$$

In the case when these two numbers are equal, we denote by $\sigma_T^2(g)$ their common value and call it the asymptotic variance of g.

Let us now briefly recall *the Rokhlin's natural extension* of the dynamical system (T, μ). The phase space is

$$\tilde{X} = \{(x_n)_{n\le 0} : T(x_n) = x_{n+1} \ \forall n \le -1\}.$$

The transformation $\tilde{T} : \tilde{X} \to \tilde{X}$ is determined by the property that

$$\left(\tilde{T}\left((x_n)_{n\le 0}\right)\right)_k = T(x_k).$$

Let $\pi_0 : \tilde{X} \to X$ be the canonical projection onto the zeroth coordinate, i.e,

$$\pi_0\left((x_n)_{n\le 0}\right) = x_0.$$

It is well known (see, e.g., [11]) that there exists a unique probability $\tilde{T}$-invariant measure $\tilde{\mu}$ on $\tilde{X}$ such that

$$\tilde{\mu} \circ \pi_0^{-1} = \mu.$$

The dynamical system $(\tilde{T},\tilde{\mu})$ is a measure-preserving automorphism and

$$\pi_0 \circ \tilde{T} = T \circ \pi_0.$$

This system is referred to as *the Rokhlin's natural extension of* (T,μ).

We say that two functions $g_1 : X \to \mathbb{R}$ and $g_2 : X \to \mathbb{R}$ are *cohomologous* in a class C of function from X to $\mathbb{R}$ if there exists a function $u \in C$ such that

$$g_2 - g_1 = u - u \circ T.$$

Any function cohomologous to the zero function is called a *coboundary*.

We shall prove the following generalization and extension of Lemma 53 in [17].

Lemma 1. *Let (X,μ) be a probability space and let $T : X \to X$ be a measurable map preserving measure μ. Fix A, a measurable subset of X with positive measure μ. Let $\tau : A \to \mathbb{N}$ be the first return time to A and let $T_A = T^{\tau_A} : A \to A$ be the corresponding first return map. Assume that*

$$\mu(\tau_A^{-1}([n,+\infty))) \le \text{const} \cdot n^{-\alpha},$$

for some $\alpha > 8$ and all $n \ge 1$. For every function $g : X \to \mathbb{R}$ let $\hat{g} : A \to \mathbb{R}$ be defined by the formula

$$\hat{g}(x) = \sum_{j=0}^{\tau_A(x)-1} g \circ T^j(x). \tag{7}$$

If $g \in L_4(\mu)$, $\overline{\sigma}_T^2(g) > 0$, and $\mu(g) = 0$, then $\hat{g} : A \to \mathbb{R}$ is not a coboundary in the class of bounded measurable functions on A.

Proof. Seeking contradiction suppose that $\hat{g} : A \to \mathbb{R}$ is such a coboundary, i.e.,

$$\hat{g} = u - u \circ T_A \tag{8}$$

with some bounded measurable function $u : A \to \mathbb{R}$. We may assume without loss of generality that the dynamical system (T,μ) is a measure-preserving automorphism. In fact, let $(\tilde{T},\tilde{\mu})$ be Rokhlin's natural extension of (T,μ), $\tilde{A} := \pi_0^{-1}(A)$, $\tilde{g} := \hat{g} \circ \pi_0$, and $\tilde{u} := u \circ \pi_0$. Then (8) implies that

$$\tilde{g} = \tilde{u} - \tilde{u} \circ \tilde{T}_{\tilde{A}}$$

and the set $\tau_A^{-1}(n)$ (the set of points with first return time n) will be mapped to the sets $\pi_0\left(\tau_A^{-1}(n)\right)$. In particular they will have the same measures, respectively, μ and $\tilde{\mu}$.

For every $n \ge 0$ let

$$A_n = \{x \in A : \tau_A(x) \ge n\}.$$

Fix an integer $n \geq 1$. For all $x \in A$ let

$$i = i(x) := \min\{0 \leq l \leq n \cdot T^l(x) \in A\}.$$

If no such l exists, set $i = n$. Let

$$j = j(x) := \max\{0 \leq l \leq n : T^l(x) \in A\}.$$

If no such l exists, set $j = 0$. We have

$$0 \leq i \leq j \leq n$$

and there exists a unique integer $0 \leq k \leq j - i$ such that

$$T^{j-i}(T^i(x)) = T_A^k(T^i(x)).$$

Hence we can write

$$S_n g(x) = S_i g(x) + S_k^{T_A}(\hat{g})(T^i(x)) + S_{n-j} g(T^j(x)) = a(x) + b(x) + c(x),$$

where $a(x) = S_i g(x)$, $b(x) = S_k^{T_A}(\hat{g})(T^i(x))$, and $c(x) = S_{n-j} g(T^j(x))$. In order to show that $\sigma_T^2(g) = 0$, we shall estimate

$$\left(\int (S_n(g))^2 \mathrm{d}\mu\right)^{\frac{1}{2}} = ||(S_n(g))||_2 \leq ||a||_2 + ||b||_2 + ||c||_2. \tag{9}$$

We shall deal with each of these three L_2 norms separately. Since $b(x) = S_k^{T_A}(\hat{g})(T^i(x))$ and $|S_k^{T_A}(\hat{g})(T^i(x))| \leq 2||u||_\infty$, we get immediately that

$$||b||_2 \leq 2||u||_\infty. \tag{10}$$

Next, we estimate $||a||_2$. We have

$$a(x) = \sum_{l=0}^{n} \mathbf{1}_{i^{-1}(l)}(x) S_l g(x).$$

Applying Cauchy–Schwarz inequality, we therefore get

$$\begin{aligned}
||a||_2 &\leq \sum_{l=0}^{n} ||\mathbf{1}_{i^{-1}(l)} S_l g||_2 = \sum_{l=0}^{n} \left(\int \mathbf{1}_{i^{-1}(l)} (S_l g)^2 \mathrm{d}\mu\right)^{\frac{1}{2}} \\
&\leq \sum_{l=0}^{n} \left(\int \mathbf{1}_{i^{-1}(l)} \mathrm{d}\mu\right)^{\frac{1}{4}} \left(\int (S_l g)^4 \mathrm{d}\mu\right)^{\frac{1}{4}}
\end{aligned}$$

$$
\begin{aligned}
&= \sum_{l=0}^{n} (\mu(i^{-1}(l)))^{\frac{1}{4}} \left(\int (S_l g)^4 \mathrm{d}\mu \right)^{\frac{1}{4}} \\
&= \sum_{l=0}^{n} (\mu(i^{-1}(l)))^{\frac{1}{4}} ||S_l g||_4 \\
&= \sum_{l=0}^{n} (\mu(i^{-1}(l)))^{\frac{1}{4}} \left|\left| \sum_{s=0}^{l-1} g \circ T^s \right|\right|_4 \\
&\leq \sum_{l=0}^{n} (\mu(i^{-1}(l)))^{\frac{1}{4}} \sum_{s=0}^{l-1} ||g \circ T^s||_4 \\
&= \sum_{l=0}^{n} (\mu(i^{-1}(l)))^{\frac{1}{4}} \sum_{s=0}^{l-1} ||g||_4 \\
&= ||g||_4 \sum_{l=1}^{n} l(\mu(i^{-1}(l)))^{\frac{1}{4}}. \qquad (11)
\end{aligned}
$$

Now notice that for μ almost every $x \in X$, there exist $x' \in A$ and an integer $k \geq 1$ such that $T^k(x') = x$ and $\tau_A(x') \geq k + i(x)$ (the strict inequality can hold only if $i(x) = n$ and $T^n(x) \notin A$). Hence,

$$
i^{-1}(l) \subset \bigcup_{k=1}^{\infty} T^k(\tau_A^{-1}(k+l)).
$$

Thus, as $T : X \to X$ is an automorphism, we get that

$$
\mu(i^{-1}(l)) \leq \sum_{k=1}^{\infty} \mu\big(T^k(\tau_A^{-1}(k+l))\big) = \sum_{k=1}^{\infty} \mu\big(\tau_A^{-1}(k+l)\big) = \mu(A_{l+1}).
$$

Inserting this to (11), we obtain

$$
\begin{aligned}
||a||_2 &\leq ||g||_4 \sum_{l=0}^{n} l(\mu(A_{l+1}))^{\frac{1}{4}} \leq \mathrm{const}||g||_4 \sum_{l=1}^{n} l(l+1)^{-\alpha/4} \\
&\leq \mathrm{const}||g||_4 \sum_{l=1}^{\infty} l^{1-\frac{\alpha}{4}} < +\infty. \qquad (12)
\end{aligned}
$$

The last inequality holds, since $\alpha > 8$.

The upper estimate of $||c||_2$ can be done similarly. Indeed, exactly as (11), we obtain the following.

$$
||c||_2 = ||S_{n-j} g \circ (T^j)||_2 = ||S_{n-j} g||_2 \leq ||g||_4 \sum_{l=0}^{n} (n-l)(\mu(j^{-1}(l))^{\frac{1}{4}}. \qquad (13)
$$

Now notice that if $T^{i(x)}(x) \notin A$, then $c(x) = 0$ and otherwise $T^{j(x)}(x) \in A$ and $\tau_A(T^{j(x)}(x)) > n - j(x)$. So,

$$j^{-1}(l) \subset T^{-l}(A_{n-l+1}) \subset T^{-l}(A_{n-l}).$$

Inserting this to (13), we thus get

$$\begin{aligned}
||c||_2 &\leq ||g||_4 \sum_{l=0}^{n} (n-l)(\mu(A_{n-l}))^{\frac{1}{4}} v = ||g||_4 \sum_{l=0}^{n-1} (n-l)(\mu(A_{n-l}))^{\frac{1}{4}} \\
&= ||g||_4 \sum_{l=1}^{n} l(\mu(A_l))^{\frac{1}{4}} \leq ||g||_4 \sum_{l=1}^{n} l \mathrm{const} l^{-\alpha/4} \\
&= \mathrm{const}||g||_4 \sum_{l=1}^{\infty} l^{1-\frac{\alpha}{4}} \\
&< +\infty. \qquad (14)
\end{aligned}$$

Again, the last inequality holds, since $\alpha > 8$.

Combining this, (10) and (12) and inserting them to (9), we see that the integrals $\int (S_n g)^2 \mathrm{d}\mu$ remain uniformly bounded as $n \to \infty$. This obviously implies that $\sigma_T^2(g) = 0$. This contradiction finishes the proof. □

We shall now show that under mild conditions if a first return map satisfies the Law of Iterated Logarithm, then so does the original map. Precisely, we say that a μ-integrable function $g : X \to \mathbb{R}$ satisfies the Law of Iterated Logarithm if there exists a positive constant A_g such that

$$\limsup_{n\to\infty} \frac{S_n g - n \int g \mathrm{d}\mu}{\sqrt{n \log\log n}} = A_g.$$

From now on we assume without loss of generality that

$$\mu(g) = \int g \mathrm{d}\mu = 0.$$

Keep a measurable set $A \subset X$ with $\mu(A) > 0$. Given a point $x \in A$, the sequence $(\tau_n(x))_{n=1}^{\infty}$ is then defined as follows:

$$\tau_1(x) := \tau_A(x) \quad \text{and} \quad \tau_n(x) = \tau_{n-1}(x) + \tau(T^{\tau_{n-1}(x)}(x)).$$

Finally we can prove the following theorem. Its proof can be also found in [17].

Theorem 5. *Let $T : X \to X$ be a measurable dynamical system preserving a probability measure μ on X. Assume that the dynamical system (T, μ) is ergodic. Fix A, a measurable subset of X having a positive measure μ. Let $g : X \to \mathbb{R}$ be a* *measurable function such that the function $\hat{g} : A \to \mathbb{R}$ satisfies the Law of Iterated Logarithm with respect to the dynamical system (T_A, μ_A). If, in addition,*

$$\int |\hat{g}|^{2+\gamma} d\mu < \infty \tag{15}$$

for some $\gamma > 0$, then the function $g : X \to \mathbb{R}$ satisfies the Law of Iterated Logarithm with respect to the original dynamical system (T, μ) and $A_g = \sqrt{\mu(A)}$.

Proof. Since the Law of Iterated Logarithm holds for a point $x \in X$ if and only if it holds for $T(x)$, in virtue of ergodicity of T, it suffices to prove our theorem for almost all points in A. By our assumptions there exists a positive constant $A_{\hat{g}}$ such that

$$\limsup_{n\to\infty} \frac{S_{\tau_n} g(x)}{\sqrt{n \log\log n}} = A_{\hat{g}}$$

for μ_A-a.e. $x \in A$. Since, by Kac's Lemma,

$$\lim_{n\to\infty} \frac{\tau_n}{n} = \frac{1}{\mu(A)} \int_A \tau d\mu = \frac{1}{\mu(A)}, \tag{16}$$

μ_A-a.e. on A, we thus have

$$\limsup_{n\to\infty} \frac{S_{\tau_n} g}{\sqrt{\tau_n \log\log \tau_n}} = \limsup_{n\to\infty} \frac{S_{\tau_n} g}{\sqrt{n \log\log n}} = \sqrt{\mu(A)} \tag{17}$$

μ_A-a.e. on A. Now, for every $n \in \mathbb{N}$ and (almost) every $x \in A$, let $k = k(x,n)$ be the positive integer uniquely determined by the condition that

$$\tau_k(x) \le n < \tau_{k+1}(x).$$

Since

$$S_n g(x) = S_{\tau_k(x)} g(x) + S_{n-\tau_k(x)} g(T^{\tau_k(x)}(x)),$$

we have that

$$\frac{S_n g(x)}{\sqrt{n \log\log n}} = \frac{S_{\tau_k(x)} g(x)}{\sqrt{n \log\log n}} + \frac{S_{n-\tau_k(x)} g(x)}{\sqrt{n \log\log n}}. \tag{18}$$

Since by (16)

$$\lim_{n\to\infty} \frac{\tau_{k+1}(x)}{\tau_k(x)} = 1,$$

we get from (17) that

$$\limsup_{n\to\infty} \frac{S_{\tau_k} g(x)}{\sqrt{n \log\log n}} = \limsup_{n\to\infty} \frac{S_{\tau_k} g(x)}{\sqrt{\tau_k \log\log \tau_k}} = A_{\hat{g}}.$$

Because of this and because of (18), we are only left to show that

$$\lim_{n\to\infty} \frac{S_{n-\tau_{k(n)}} g(x)}{\sqrt{n \log\log n}} = 0 \tag{19}$$

μ_A-a.e. on A. To do this, note first that

$$\frac{S_{\tau_{k+1}-\tau_k}|g|(T^{\tau_k}(x))}{\sqrt{k\log\log k}} = \frac{|\hat{g}|(T_A^k(x))}{\sqrt{k\log\log k}}.$$

Take an arbitrary $\varepsilon \in (0,\gamma)$. Since

$$\begin{aligned}
\mu\big(\{x\in A : |\hat{g}|(T_A^k(x)) \geq \varepsilon\sqrt{k\log\log k}\}\big) \\
&= \mu\big(\{x\in A : |\hat{g}|(x) \geq \varepsilon\sqrt{k\log\log k}\}\big) \\
&= \mu\big(\{x\in A : |\hat{g}|^{2+\varepsilon}(x) \geq \varepsilon^{2+\varepsilon}(k\log\log k)^{1+\varepsilon/2}\}\big) \\
&\leq \frac{\int |\hat{g}|^{2+\varepsilon} \mathrm{d}\mu}{\varepsilon^{2+\varepsilon}(k\log\log k)^{1+\varepsilon/2}},
\end{aligned} \tag{20}$$

using (15), we conclude that

$$\sum_{k=1}^{\infty} \mu\big(\{x\in A : |\hat{g}|(x) \geq \varepsilon\sqrt{k\log\log k}\}\big) < \infty.$$

So, applying Borel–Cantelli lemma, (19) follows. This finishes the proof. □

4 The Law of Iterated Logarithm: Meromorphic Functions

Let $f : \mathbb{C} \to \hat{\mathbb{C}}$ be a dynamically semi-regular meromorphic function and $t > \rho/\alpha$. Let $\mathscr{S} = \{\phi_e\}_{e\in E}$ be the iterated function system induced by some nice set U for f. Our first technical result, ultimately aiming at the Law of Iterated Logarithm, is this.

Lemma 2. *If $f : \mathbb{C} \to \hat{\mathbb{C}}$ is a dynamically semi-regular meromorphic function and $t > \rho/\alpha$, then*

$$\limsup_{n\to\infty} \frac{1}{n} \log \mu_t\Big(\bigcup_{|\omega|\geq n} \phi_\omega(U)\Big) < 0.$$

Proof. Noting that $U \cap \bigcup_{k=1}^{n-1} f^k(\phi_\omega(U)) = \emptyset$ and repeating the proof of Proposition 6.3 from [15], we show that

$$\mathrm{P}_c(t) := \limsup_{n\to\infty} \frac{1}{n} \log \sum_{|\omega|=n} \exp\big(\sup\big(S_{n-1}(\psi_t \circ f \circ \phi_\omega)\big)\big) < \mathrm{P}(t).$$

Hence,

$$\begin{aligned}\limsup_{n\to\infty}\frac{1}{n}\log m_t\Big(\bigcup_{|\omega|=n} f(\phi_\omega(U))\Big)\\ &\le \limsup_{n\to\infty}\frac{1}{n}\log\sum_{|\omega|=n}\exp\big(\sup\big(S_{n-1}(\psi_t\circ f\circ\phi_\omega)\big)-\mathrm{P}(t)(n-1)\big)\\ &= \mathrm{P}_c(t)-\mathrm{P}(t)<0.\end{aligned}$$

Since, see Theorem 4(d), the Radon–Nikodym derivative $\frac{\mathrm{d}\mu_t}{\mathrm{d}m_t}$ is uniformly bounded above, we thus get that

$$\limsup_{n\to\infty}\frac{1}{n}\log\mu_t\Bigg(f\Big(\bigcup_{|\omega|=n}\phi_\omega(U)\Big)\Bigg)\le \mathrm{P}_c(t)-\mathrm{P}(t).$$

Since the probability measure μ_t is f-invariant, we have

$$\mu_t\Bigg(f\Big(\bigcup_{|\omega|=n}\phi_\omega(U)\Big)\Bigg)\ge\mu_t\Big(\bigcup_{|\omega|=n}\phi_\omega(U)\Big),$$

and therefore,

$$\limsup_{n\to\infty}\frac{1}{n}\log\mu_t\Big(\bigcup_{|\omega|=n}\phi_\omega(U)\Big)\le \mathrm{P}_c(t)-\mathrm{P}(t).$$

So, finally,

$$\limsup_{n\to\infty}\frac{1}{n}\log\mu_t\Big(\bigcup_{|\omega|\ge n}\phi_\omega(U)\Big)\le \mathrm{P}_c(t)-\mathrm{P}(t)<0.$$

The proof is complete. □

Because of Lemma 2 we obviously have some constant $C>0$ such that

$$\mu_t\Big(\bigcup_{|\omega|\ge n}\phi_\omega(U)\Big)\le Cn^{-9} \tag{21}$$

for all $n\ge 1$. For every $e\in E$ let $N_e\ge 1$ be the unique integer determined by the property that $f^{N_e}\circ\phi_e=\mathrm{Id}$. Let $\hat{f}:J_{\mathscr{S}}\to J_{\mathscr{S}}$ be the first return map on $J_{\mathscr{S}}$, i.e., $\hat{f}$ is defined by the formula

$$\hat{f}(\phi_e(z))=f^{N_e}(\phi_e(z))=z$$

for all $e \in E$ and all $z \in J_{\mathscr{S}}$. N_e is then the first return time to $J_{\mathscr{S}}$. Recall from the previous section that given $g : J_f \to \mathbb{R}$, the function $\hat{g} : J_{\mathscr{S}} : \mathbb{R}$ is given by the following formula:

$$\hat{g}(\phi_e(z)) = \sum_{j=0}^{N_e-1} g \circ f^j(\phi_e(z))$$

for all $e \in E$ and all $z \in J_{\mathscr{S}}$. Let $\hat{m}_t$ and $\hat{\mu}_t$ be the probability conditional measures on $J_{\mathscr{S}}$ respectively of m_t and μ_t. The measure $\hat{\mu}_t$ is then $\hat{f}$-invariant. Let

$$\psi_t^{(e)}(z) = \hat{\psi}_t(\phi_e(z)) - \mathrm{P}(t)N_e.$$

Then, a straightforward formal calculation shows that $\hat{m}_t$ is the F_t-conformal measure for the Hölder summable family $F_t = \{\psi_t^{(e)}\}_{e\in E}$ and that $\hat{\mu}_t$ is the corresponding invariant version of $\hat{m}_t$. Therefore all the results proved in [7] for conformal and invariant measures of summable Hölder families apply to measures $\hat{m}_t$ and $\hat{\mu}_t$. We will need them at the end of the section. At the moment, as an immediate consequence of Theorem 5, we get the following.

Theorem 6. *Let $f : \mathbb{C} \to \hat{\mathbb{C}}$ be a dynamically semi-regular meromorphic function and fix $t > \rho/\alpha$. Let $g : J_f \to \mathbb{R}$ be a measurable function such that the function $\hat{g} : J_{\mathscr{S}} \to \mathbb{R}$ satisfies the Law of Iterated Logarithm with respect to the dynamical system $(\hat{f}, \hat{\mu}_t)$. If, in addition,*

$$\int_{J_f} |\hat{g}|^{2+\gamma} d\mu_t < \infty \tag{22}$$

for some $\gamma > 0$, then the function $g : J_f \to \mathbb{R}$ satisfies the Law of Iterated Logarithm with respect to the dynamical system (f, μ_t) and $A_g = \sqrt{J_{\mathscr{S}}}$.

In order to be able to apply this theorem, we need a technical result establishing (22) for a large class of functions from J_f to $\mathbb{R}$. This is the content of the following lemma.

Lemma 3. *Let $\psi = \psi_s : J_f \to \mathbb{R}$ be a loosely tame function. Then for every $\gamma > 0$,*

$$\int_{J_{\mathscr{S}}} |\hat{\psi}|^{\gamma} d\hat{m}_t < +\infty.$$

Proof. Since the measure $\hat{m}_t$ is proportional to m_t on $J_{\mathscr{S}}$, our equivalent task is to show that

$$\int_{J_{\mathscr{S}}} |\hat{\psi}|^{\gamma} dm_t < +\infty.$$

Fix $\varepsilon > 0$. Because of expanding properties of the function $f : \mathbb{C} \to \hat{\mathbb{C}}$ there exists a constant $C > 0$ such that

$$|\psi(z)| \leq C|f'(z)|_\tau^\varepsilon$$

for all $z \in J(f)$. Therefore, for every $e \in E$ and all $z \in J_{\mathscr{S}}$ we get,

$$\begin{aligned}|\hat{\psi}(\phi_e(z))| &= \left|\sum_{j=0}^{N_e-1} \psi(f^j(\phi_e(z)))\right| \le \sum_{j=0}^{N_e-1} \left|\psi(f^j(\phi_e(z)))\right| \\ &\le C\sum_{j=0}^{N_e-1} |f'(f^j(\phi_e(z))|_\tau^\varepsilon \le C\prod_{j=0}^{N_e-1} |f'(f^j(\phi_e(z))|_\tau^\varepsilon \\ &= C|(f^{N_e})'(\phi_e(z))|_\tau^\varepsilon.\end{aligned}$$

Thus

$$\begin{aligned}&|\hat{\psi}\phi_e(z))|^\gamma m_t(\phi_e(J_{\mathscr{S}})) \\ &\asymp |\hat{\psi}(\phi_e(z))|^\gamma \exp\bigl(S_{N_e}\psi_t(\phi_e(z)) - \mathrm{P}(\psi_t)N_e\bigr) \\ &\le C^\gamma \exp\bigl(\gamma\varepsilon \log|(f^{N_e})'(\phi_e(z))|_\tau - t\log|(f^{N_e})'(\phi_e(z))|_\tau + S_{N_e}k(z)) - \mathrm{P}(\psi_t)N_e\bigr) \\ &= C^\gamma \exp\bigl(S_{N_e}\psi_{t-\gamma\varepsilon}(\phi_e(z)) - \mathrm{P}(\psi_t)N_e\bigr) \\ &= C^\gamma \exp\bigl(S_{N_e}\psi_{t-\gamma\varepsilon}(\phi_e(z)) - \mathrm{P}(\psi_{t-\gamma\varepsilon})N_e\bigr)\exp\bigl((\mathrm{P}(\psi_{t-\gamma\varepsilon} - \mathrm{P}(\psi_t))N_e\bigr) \\ &\asymp C^\gamma \exp\bigl((\mathrm{P}(\psi_{t-\gamma\varepsilon}) - \mathrm{P}(\psi_t))N_e\bigr) m_{t-\gamma\varepsilon}(\phi_e(J_{\mathscr{S}})). \qquad (23)\end{aligned}$$

Since, by Lemma 7.5 in [9], the function $(t-\delta_1, t+\delta_1) \ni u \mapsto \mathrm{P}(\psi_u)$ ($\delta_1 > 0$ sufficiently small) is real-analytic, we get a constant $M > 0$ such that for all $\varepsilon > 0$ sufficiently small, we have that

$$|\mathrm{P}(\psi_{t-\gamma\varepsilon}) - \mathrm{P}(\psi_t)| \le M\varepsilon.$$

Formula (23) then yields

$$|\hat{\psi}(\phi_e(z))|^\gamma m_t(\phi_e(J_{\mathscr{S}}) \le C^\gamma \mathrm{e}^{M\varepsilon N_e} m_{t-\gamma\varepsilon}(\phi_e(J_{\mathscr{S}})). \qquad (24)$$

Now, for every $k \ge 1$, let

$$U_k^c = \bigcap_{j=0}^{k} f^{-j}(\mathbb{C}\setminus U).$$

Fixing $u > \rho/\alpha_2$, we have for every $n \ge 1$ that

$$\begin{aligned}m_u\Bigl(\bigcup_{e\in E: N_e = n} \phi_e(J_{\mathscr{S}})\Bigr) &= m_u\bigl(U \cap f^{-1}(U_{n-1}^c) \cap f^{-n}(U)\bigr) \le m_u(f^{-1}(U_{n-1}^c)) \\ &= m_u(f^{-1}(\mathbf{1}_{U_{n-1}^c})) = m_u(\mathbf{1}_{U_{n-1}^c} \circ f) \\ &= m_u\bigl(\mathrm{e}^{-\mathrm{P}(\psi_u)n}\mathscr{L}_u\bigl(\mathbf{1}_{U_{n-1}^c} \circ f\bigr)\bigr) \\ &= m_u\bigl(\mathrm{e}^{-\mathrm{P}(\psi_u)(n-1)}\mathscr{L}_u^{n-1}\bigl(\mathrm{e}^{-\mathrm{P}(\psi_u)}\mathscr{L}_u\bigl(\mathbf{1}_{U_{n-1}^c} \circ f\bigr)\bigr)\bigr)\end{aligned}$$

$$
\begin{aligned}
&= m_u\big(\mathrm{e}^{-\mathrm{P}(\psi_u)(n-1)}\mathscr{L}_u^{n-1}\big(\mathbf{1}_{U_{n-1}^c}(\mathrm{e}^{-\mathrm{P}(\psi_u)}\mathscr{L}_u\mathbf{1})\big)\\
&\le C_1 m_u\big(\mathrm{e}^{-\mathrm{P}(\psi_u)(n-1)}\mathscr{L}_u^{n-1}\big(\mathbf{1}_{U_{n-1}^c}\big)\big) \qquad (25)
\end{aligned}
$$

with some constant $C_1 > 0$. Looking at this moment at the proof of Proposition 6.3 in [14] and taking into account continuity properties of the Perron–Frobenius operator $\mathscr{L}_u$, we conclude that there exist $\kappa > 0$ and $c_2 > 0$ such that

$$
m_u\big(\mathscr{L}_u^{n-1}\big(\mathbf{1}_{U_{n-1}^c}\big)\big) \le C_2\mathrm{e}^{-\kappa n}\mathrm{e}^{\mathrm{P}(\psi_u)(n-1)}
$$

for all $u \in (t-\delta, t+\delta)$ with some $0 < \delta \le \delta_1$ small enough and all integers $n \ge 1$. Substituting this to (25) we get that

$$
m_u\left(\bigcup_{e\in E:N_e=n}\phi_e(J_{\mathscr{S}})\right) \le C_1C_2\mathrm{e}^{-\kappa n} \qquad (26)
$$

for all $\in (t-\delta, t+\delta)$. Fix $0 < \varepsilon < \min\{\delta/\gamma, \kappa/(2M)\}4$. Inserting then (26) into (24), we obtain

$$
\begin{aligned}
\int |\hat{\psi}|^\gamma \mathrm{d}m_t &= \sum_{n=1}^\infty \int_{\bigcup_{e\in E:N_e=n}\phi_e(J_{\mathscr{S}})} |\hat{\psi}|^\gamma \mathrm{d}m_t = \sum_{n=1}^\infty \sum_{N_e=n} \int_{\phi_e(J_{\mathscr{S}})} |\hat{\psi}|^\gamma \mathrm{d}m_t\\
&\le \sum_{n=1}^\infty \sum_{N_e=n} ||\hat{\psi}|_{\phi_e(J_{\mathscr{S}})}||_\infty^\gamma m_t(\phi_e(J_{\mathscr{S}}))\\
&\le C^\gamma \sum_{n=1}^\infty \mathrm{e}^{M\varepsilon n} \sum_{N_e=n} m_{t-\gamma\varepsilon}(\phi_e(J_{\mathscr{S}}))\\
&= C^\gamma \sum_{n=1}^\infty \mathrm{e}^{M\varepsilon n} m_{t-\gamma\varepsilon}\left(\bigcup_{N_e=n}\phi_e(J_{\mathscr{S}})\right)\\
&\le C^\gamma C_1C_2 \sum_{n=1}^\infty \mathrm{e}^{M\varepsilon n}\mathrm{e}^{-\kappa n}\\
&\le C^\gamma C_1C_2 \sum_{n=1}^\infty \mathrm{e}^{-\frac{1}{2}\kappa n}. \qquad (27)
\end{aligned}
$$

The proof is complete. □

Now we are in position to provide a short proof of the following theorem and its corollary, both forming the main results of this section.

Theorem 7. *Let $f : \mathbb{C} \to \hat{\mathbb{C}}$ be a dynamically semi-regular meromorphic function and fix $t > \rho/\alpha$. Let $\psi : J_f \to \mathbb{R}$ be a loosely tame function ($\psi(z) = \psi_s(z) = -s\log|f'(z)|_\tau + h(z)$ with $s \neq 0$). Then the asymptotic variance $\sigma_f^2(\psi)$ exists and, if $\sigma_f^2(\psi) > 0$, or equivalently if $\psi : J_f \to \mathbb{R}$ is not cohomologous to a constant in the* *class of Hölder continuous functions on J_f, then the function $\psi : J_f \to \mathbb{R}$ satisfies the Law of Iterated Logarithm with respect to the dynamical system (f, μ_t).*

Proof. Adding a constant to ψ we may assume without loss of generality that $\int \psi d\mu_t = 0$. The existence of the asymptotic variance $\sigma_f^2(\psi)$ was established in Theorem 6.17 of [9]. The fact that $\sigma_f^2(\psi) > 0$ if and only if $\psi : J_f \to \mathbb{R}$ is not cohomologous to a constant in the class of Hölder continuous functions on J_f is the content of Proposition 6.21 in [9]. It follows from Lemma 5.2 in [9] that the function $\hat{\psi}$ is Hölder continuous; more precisely, its composition with the canonical projection from $E^{\mathbb{N}}$ onto $J_{\mathscr{S}}$ is a Hölder continuous map. Along with Lemma 3 this implies (see Lemma 2.5.6 in [7] and the beginning of the p. 41 in [7]) that the asymptotic variance $\sigma_{\hat{f}}^2(\hat{\psi})$ exists and, in addition with Theorem 2.5.5 and Lemma 2.5.6, both in [7], that the function $\hat{\psi}$ satisfies the Law of Iterated Logarithm with respect to the dynamical system $(\hat{f}, \hat{\mu}_t)$ (with $A_{\hat{\psi}} = \sigma_{\hat{f}}^2(\hat{\psi})$), provided that $\sigma_{\hat{f}}^2(\hat{\psi}) > 0$. But since, by Lemma 7.11 in [9], the function ψ has all moments with respect to the measure μ_t, we in particular have that $\psi \in L_4(\mu_t)$. Then, using (21), Lemma 1 implies that $\hat{\psi}$ is not a coboundary in the class of bounded measurable functions on $J_{\mathscr{S}}$. It then directly follows from Lemma 4.8.8 in [7] that $\sigma_{\hat{f}}^2(\hat{\psi}) > 0$. Now, with the help of Lemma 3, a direct application of Theorem 6 completes the proof. □

As an immediate consequence of this theorem, with the help of Theorem 6.20 in [9], we get the following.

Corollary 1. *Let $f : \mathbb{C} \to \hat{\mathbb{C}}$ be a dynamically semi-regular meromorphic function and fix $t > \rho/\alpha$. If $\psi : J_f \to \mathbb{R}$ is a loosely tame function, then the function $\psi : J_f \to \mathbb{R}$ satisfies the Law of Iterated Logarithm with respect to the dynamical system (f, μ_t) with $A_\psi = \sqrt{2}\sigma_{\hat{f}}(\hat{\psi}) > 0$.*

5 Equilibrium States Versus Hausdorff Measures

Keep $f : \mathbb{C} \to \hat{\mathbb{C}}$ a dynamically semi-regular meromorphic function and $t > \rho/\alpha$. Let

$$D_t := \mathrm{HD}(\mu_t),$$

let the function $\zeta : J_{\mathscr{S}} \to (0, +\infty)$ be defined by the formula

$$\zeta(\phi_e(z)) = -\log|\phi_e'(z)|,$$

and let

$$\chi_t := \int \zeta d\mu_t.$$

The number χ_t is called *the Lyapunov exponent* of the measure μ_t. Let $h:(a,+\infty)\to(0,+\infty)$ ($a>0$ small enough) be a nondecreasing function. This function h is said to *belong to the lower class*, if

$$\int_a^{\infty}\frac{h(r)}{r}\exp\left(-\frac{1}{2}(h(r)^2\right)\,\mathrm{d}r<+\infty$$

and *to the upper class*, if

$$\int_a^{\infty}\frac{h(r)}{r}\exp\left(-\frac{1}{2}(h(r)^2\right)\,\mathrm{d}r=+\infty.$$

Let finally $\mathrm{H}_{\tilde{h}}$ be the Hausdorff measure on $\mathbb{C}$ induced by the gauge function $\tilde{h}$.

Now, let

$$\psi_t^*=\psi_t+D_t\zeta-\mathrm{P}(\psi_t)\ =-(t-D_t)\log|f'|_\tau+(h-\mathrm{P}(\psi_t)).$$

Associated to the function h and the function ψ_t is the function $\tilde{h}$ defined for all sufficiently small $t>0$ by the following formula:

$$\tilde{h}(r)=r^{D_t}\exp\left(\frac{\sigma(\psi_t^*)}{\sqrt{\chi_t}}h(-\log r)\sqrt{-\log r}\right).$$

Since the projection $\pi:E^{\mathbb{N}}\to J_{\mathscr{S}}$ and the function $\hat{\psi}_t^*:J_{\mathscr{S}}\to\mathbb{R}$ are both Hölder continuous, since all the integrals $\int|\hat{\psi}_t^*|^\gamma\mathrm{d}\mu_t$ ($\gamma>2$) are finite (see Lemma 3 where this is proved for all $\gamma>0$), and since the measures $\hat{m}_t$ and $\hat{\mu}_t$ are respectively F_t-conformal and invariant, Theorem 4.8.3 in [7] applies to give the following.

Theorem 8. *Suppose that $\sigma^2(\hat{\psi}_t^*)>0$ and that $h:(a,+\infty)\to(0,+\infty)$ is a slowly growing function. Then:*

(a) If h belongs to the upper class, then the measures $\hat{\mu}_t$ and $\mathrm{H}_{\tilde{h}}|_{J_{\mathscr{S}}}$ are mutually singular.
(b) If h belongs to the lower class, then $\hat{\mu}_t$ is absolutely continuous with respect to $\mathrm{H}_{\tilde{h}}$.

We shall now prove a sufficient condition for $\sigma^2(\hat{\psi}_t^*)$ to be positive. It is trivially verifiable. Let $J_{r,f}$ be the set of points in J_f that do not escape to infinity under the action of the map $f:\mathbb{C}\to\hat{\mathbb{C}}$. It is called in the literature the radial (or conical) Julia set of f.

Lemma 4. *If $t\neq\mathrm{HD}(J_{r,f})$, then the function $\psi_t^*=-(t-D_t)\log|f'|_\tau+(h-\mathrm{P}(\psi_t))$ is not cohomologous to a constant in the class of Hölder continuous functions on J_f and $\sigma^2(\hat{\psi}_t^*)>0$. In particular this is true for all $t\geq 2$.*

Proof. First observe that because of Theorem 8.1 (Volume Lemma) and Theorem 6.25 (Variational Principle), both in [9], we have

$$\begin{aligned}
\int \psi_t^* d\mu_t &= -t\chi_{\mu_t} + \frac{h_{\mu_t}(f)}{\chi_{\mu_t}}\chi_{\mu_t} + \int h d\mu_t - P(\psi_t) \\
&= h_{\mu_t}(f) - t\chi_{\mu_t} + \int h d\mu_t - P(\psi_t) \\
&= 0. \qquad (28)
\end{aligned}$$

We already know that ψ_t^* is cohomologous to a constant in the class of Hölder continuous functions on J_f if and only if $\sigma^2(\hat{\psi}_t^*) = 0$. So, assume that ψ_t^* is cohomologous to a constant. By (28) ψ_t^* is then a coboundary. By Theorem 6.20 in [9], we get that

$$t = D_t. \qquad (29)$$

The fact that ψ_t^* is a coboundary equivalently means that the function $-D_t \log|f'|_\tau$ is cohomologous to $\psi_t - P(\psi_t)$. But the topological pressure of the latter function vanishes; whence $P(-D_t \log|f'|_\tau) = 0$. Theorem 8.3 in [9] (Bowen's Formula) then implies that

$$D_t = \mathrm{HD}(J_{r,f}). \qquad (30)$$

This theorem was in fact in [9] formulated for dynamically regular functions only; however, apart from dynamical semi-regularity all what was needed there was the existence of a zero of the pressure function of potentials $-t\log|f'|_\tau$. Combining (29) and (30) yields $t = \mathrm{HD}(J_{r,f})$, proving the first part of our lemma. Knowing now however that Bowen's Formula holds and having the conformal measure $m_{\mathrm{HD}(J_{r,f}})$, the fact that $\mathrm{HD}(J_{r,f})) < 2$ follows in the same way as Corollary 1.4 in [8]. The proof is complete. □

As an immediate consequence of Theorem 8 and Lemma 4 we get the following main result of this chapter.

Theorem 9. *Let $f : \mathbb{C} \to \hat{\mathbb{C}}$ be a dynamically semi-regular meromorphic function and for every $t > \rho/\alpha$ let $\psi_t = -t\log|f'|_\tau + h$. Suppose that $\sigma^2(\hat{\psi}_t^*) > 0$ (this is in particular true if $t \neq \mathrm{HD}(J_{r,f})$, more particularly if $t \geq 2$) and that $h : (a, +\infty) \to (0, +\infty)$ is a slowly growing function. Then:*

(a) If h belongs to the upper class, then the measures μ_t and $\mathrm{H}_{\tilde{h}}|_{J_f}$ are mutually singular.
(b) If h belongs to the lower class, then μ_t is absolutely continuous with respect to $\mathrm{H}_{\tilde{h}}$.

Towards the end of the chapter note that the function $h_c(t) = c\sqrt{\log\log t}$, $c \geq 0$, belongs to the upper class if and only if $c \leq \sqrt{2}$. With the consistent notation

$$\begin{aligned}\tilde{h}_c(r) &= r^{D_t}\exp\left(\frac{\sigma(\hat{\psi})}{\sqrt{\chi_t}}h_c(-\log r)\sqrt{-\log r}\right)\\ &= r^{D_t}\exp\left(c\frac{\sigma(\hat{\psi})}{\sqrt{\chi_t}}\sqrt{\log(1/r)\log_3(1/r)}\right),\end{aligned}$$

we therefore immediately obtain the following consequence of Theorem 9.

Theorem 10. *Let $f:\mathbb{C}\to\hat{\mathbb{C}}$ be a dynamically semi-regular meromorphic function and for every $t>\rho/\alpha$, let $\psi_t=-t\log|f'|_\tau+h_c$. Suppose that $\sigma^2(\hat{\psi}_t^*)>0$; this is in particular true if $t\neq \mathrm{HD}(J_{r,f})$, more particularly if $t\geq 2$. Then:*

(a) The measures μ_t and $\mathrm{H}_{\tilde{h}_c}|_{J_f}$ are mutually singular for all $0\leq c\leq\sqrt{2}$.
(b) The measure μ_t is absolutely continuous with respect to $\mathrm{H}_{\tilde{h}_c}$ for all $c>\sqrt{2}$.

Given $\kappa>0$ let H_κ be the standard Hausdorff measure corresponding to the parameter κ, i.e., $\mathrm{H}_\kappa=\mathrm{H}_{r\mapsto r^\kappa}$ with the notation introduced above. Taking in Theorem 10 $c=0$, we obtain the following.

Corollary 2. *Let $f:\mathbb{C}\to\hat{\mathbb{C}}$ be a dynamically semi-regular meromorphic function and for every $t>\rho/\alpha$ let $\psi_t=-t\log|f'|_\tau+h$. Suppose that $\sigma^2(\hat{\psi}_t^*)>0$; this is in particular true if $t\neq\mathrm{HD}(J_{r,f})$, more particularly if $t\geq 2$. Then the measures μ_t and $\mathrm{H}_{\mathrm{HD}(\mu_t)}|_{J_f}$ are mutually singular.*

Acknowledgements The research of the Mariusz Urbański was supported in part by the NSF Grant DMS 0700831.

References

1. Badeńska, A.: Almost sure invariance principle for some hyperbolic meromorphic maps. Preprint (2008)
2. Billingsley, P.: Probability and Measure. Wiley, New York (1995)
3. Denker, M., Urbański, M.: Relating Hausdorff measures and harmonic measures on parabolic Jordan curves. Journal für die Reine und Angewandte Mathematik **450**, 181–201 (1994)
4. Dobbs, N.: Nice sets and invariant densities in complex dynamics. Math. Proc. Cambridge Philos. Soc. **150**, 157–165 (2011)
5. Makarov, N.: On the distortion of boundary sets under conformal mappings. Proc. London Math. Soc. **51**, 369–384 (1985)
6. Mauldin, D., Urbański, M.: Dimensions and measures in infinite iterated function systems. Proc. London Math. Soc. **73**(3), 105–154 (1996)
7. Mauldin, D., Urbański, M.: Graph Directed Markov Systems: Geometry and Dynamics of Limit Sets. Cambridge University Press, Cambridge (2003)
8. Mayer, V., Urbański, M.: Geometric thermodynamical formalism and real analyticity for meromorphic functions of finite order. Ergod. Theor. Dyn. Syst. **28**, 915–946 (2008)
9. Mayer, V., Urbański, M.: Thermodynamical formalism and multifractal analysis for meromorphic functions of finite order. Memoir. Am. Math. Soc. **203**, No. 954 (2010)
10. Philipp, W., Stout, W.: Almost sure invariance principles for partial sums of weakly dependent random variables. Memoir. Am. Math. Soc. **161**, No. (2) (1975)

11. Przytycki, F., Urbański, M.: Conformal fractals: ergodic theory methods. London Mathematical Society Lecture Note Series, vol. 371. Cambridge University Press, Cambridge (2010)
12. Przytycki, F., Zdunik, A.: Harmonic, Gibbs and Hausdorff measures on repellers for holomorphic maps I. Ann. Math. **130**, 1–40 (1989)
13. Rivera-Letelier, J.: A connecting lemma for rational maps satisfying a no-growth condition. Ergod. Theor. Dyn. Syst. **27**, 595–636 (2007)
14. Skorulski, B., Urbański, M.: Dynamical rigidity of transcendental meromorphic functions. Nonlinearity **25**(8), 2337–2348 (2012)
15. Skorulski, B., Urbański, M.: On Hausdorff dimension of radial julia sets of meromorphic function. Preprint (2011)
16. Urbański, M.: Hausdorff measures versus equilibrium states of conformal infinite iterated function systems. Periodica Math. Hung. **37**, 153–205 (1998)
17. Urbański, M., Szostakiewicz, M., Zdunik, A.: Fine inducing and equilibrium measures for rational functions of the Riemann sphere. Preprint (2011)

Cookie-Cutter-Like Sets with Graph-Directed Construction

Shen Fan, Qing-Hui Liu, and Zhi-Ying Wen

Abstract In this chapter, we extend the cookie-cutter-like construction introduced by Ma, Rao, and Wen to the case having the graph-directed construction which is introduced by Mauldin and Williams and obtain a new class of fractals, which can be used to study the dimensions of the spectrum of discrete Schrödinger operators. Under suitable assumptions we prove that this class of fractals possesses the properties of bounded variation, bounded distortion, bounded covariation, and the existence of Gibbs-like measures. With these properties we give expressions for the Hausdorff dimensions, box dimensions, and packing dimensions of the fractals. We also discuss the continuous dependence of the dimensions on the defining data.

1 Introduction

Cookie-cutter (CC) set is a typical fractal generated by a nonlinear *iterated function system (IFS)* which plays an important role in the fractals and dynamical systems (for a survey, see [1]), then this construction is extended to the case with a Moran structure, called cookie-cutter-like construction (CCL) [7] and has been used to study the dimensions of the spectrum of discrete Schrödinger operators [6]. Based on this construction, in this chapter, we will consider a further extension to the case having graph-directed construction, which is introduced in [9]. This new extension appears also in the structure of the spectrum of discrete Schrödinger operators [4]. After studying the properties of the associated limit fractal sets

S. Fan (✉) • Z.-Y. Wen
Department of Mathematics, Tsinghua University, Beijing 100084, P.R. China
e-mail: ai-hua.fan@u-picardie.fr; wenzy@mail.tsinghua.edu.cn

Q.-H. Liu
Department of Computer Science and Engineering, Beijing Institute of Technology, Beijing 100083, P.R. China
e-mail: qhliu@bit.edu.cn

J. Barral and S. Seuret (eds.), *Further Developments in Fractals and Related Fields*, Trends in Mathematics, DOI 10.1007/978-0-8176-8400-6_12,

in detail, we determine the Hausdorff dimensions, packing dimensions, and the box dimensions of these new fractals by explicit expressions; we discuss also the continuous dependence of the dimensions on the defining data.

We first recall some known facts and preliminaries starting with the cookie-cutter structure.

1.1 Cookie-Cutter and Cookie-Cutter-Like Constructions

Let $I = [0,1]$, I_1, I_2 be two disjoint subintervals of I, and f is an expanding continuous mapping from $I_1 \cup I_2$ to I, the restriction of f to each I_1, I_2 is 1–1 and onto. Let

$$E = \{x \in I \big| f^n(x) \text{ is defined}, \quad \forall n \in \mathbb{N}\}, \tag{1}$$

then E is called a *cookie-cutter set* generated by f. This classical construction comes from dynamical systems; see [1, 3, 10], for a good introduction.

Another way of defining cookie-cutter set is by IFS. Denote the corresponding branch inverse of f on I_1, I_2 by ϕ_1, ϕ_2, and let $\Sigma_2^n = \{i_1 i_2 \cdots i_n \big| i_k = 1 \text{ or } 2, 1 \leq k \leq n\}$, $\Sigma_2 = \{1,2\}^{\mathbb{N}}$, then

$$E = \bigcap_{n\geq 1} \quad \bigcup_{i_1\cdots i_n \in \Sigma_2^n} \quad \phi_{i_1} \circ \cdots \circ \phi_{i_n}(I). \tag{2}$$

Let σ be the shift mapping on Σ_2, i.e., $\sigma(i_1 i_2 i_3 \cdots) = i_2 i_3 \cdots$. Define $\pi : \Sigma_2 \to E$ as

$$\pi(i_1 i_2 \cdots) = \lim_{n\to\infty} \phi_{i_1} \circ \cdots \circ \phi_{i_n}(I),$$

then it is well known that (Σ_2, σ) is isomorphic to (E, f), i.e., the following diagram is commutable:

$$\mathrm{CD}\Sigma_2 @ > \sigma > > \Sigma_2 @V\pi VV @VV\pi V E @ > f > > E.$$

π is called the *coding mapping* of E.

In [7], Ma, Rao, and Wen generalized the cookie-cutter construction in the following way: let $I = [0,1]$, $\{f_k\}_{k\geq 1}$ be a sequence of expanding continuous mappings

$$f_k : \bigcup_{j=1}^{q_k} I_j^k \to I,$$

where $I_1^k, \ldots, I_{q_k}^k$ are disjoint subintervals of I. For each $k \geq 1$, the restriction of f_k to each initial interval I_j^k $(1 \leq j \leq q_k)$ is 1-1 and onto. The *CCL* set generated by $\{f_k\}_{k\geq 1}$ is defined as

$$E = \{x \in I \big| f_n \circ \cdots \circ f_1(x) \text{ is defined}, \quad \forall n \in \mathbb{N}\}. \tag{3}$$

By an analogous argument with the case of cookie-cutter construction, we can define it in another way: for any $k \geq 1$, denote the corresponding branch inverse of f_k by $\phi_1^{(k)}, \ldots, \phi_{q_k}^{(k)}$. Let $\Omega_n = \prod_{k=1}^{n} \{1, 2, \ldots, q_k\}$, then

$$E = \bigcap_{n \geq 1} \bigcup_{i_1 \cdots i_n \in \Omega_n} \phi_{i_1}^{(1)} \circ \cdots \circ \phi_{i_n}^{(n)}(I). \tag{4}$$

Let $\Omega = \prod_{k=1}^{\infty} \{1, 2, \ldots, q_k\}$, which is called the *coding space* of E; we may also define a coding map π from Ω to E.

1.2 Graph-Directed Construction

Recall now some preliminaries on the graph-directed construction introduced by Mauldin and Williams [9].

Let $\mathscr{V} = \{1, \ldots, n\}$ be the vertex set, $\mathscr{E} \subset \{(i, j) : i, j \in V\}$ be the set of edges, and denote $\mathscr{G} = (\mathscr{V}, \mathscr{E})$ the directed graph. Let $I_1, \ldots, I_n$ be disjoint intervals of $\mathbb{R}$.

Suppose for each $e = (i, j) \in \mathscr{E}$, there exists a contractive similitude $\phi_e : I_j \to I_i$ with contraction ratio $0 < t_e < 1$, and suppose that for each $i \in \mathscr{V}$, there is some $j \in \mathscr{V}$ such that $(i, j) \in \mathscr{E}$.

For $e = (i, j) \in \mathscr{E}$, we denote the initial point of e as $i(e) = i$ and the terminal point of e as $t(e) = j$. We say that $e_1 e_2 \cdots e_m$ is an admissible word of length m, if for each $1 \leq k \leq m - 1$, $t(e_k) = i(e_{k+1})$.

Let Ω_m be the set of admissible words of length m; then the graph-directed invariant set can be defined as

$$K = \bigcap_{m \geq 1} \bigcup_{e_1 \cdots e_m \in \Omega_m} \phi_{e_1} \circ \cdots \circ \phi_{e_m}(I_{t(e_m)}). \tag{5}$$

We can define K in another way: $K = \cup_{i=1}^{n} K_i$, where $(K_1, \ldots, K_n) \in \prod_{i=1}^{n} \mathscr{K}(I_i)$ $\left(\mathscr{K}(I_i) \text{ denotes the set of compact subsets of } I_i\right)$ is the unique vector of compact sets such that

$$K_i = \bigcup_{e \in \mathscr{E}, i(e) = i} \phi_e(K_{t(e)}), \quad \forall 1 \leq i \leq n. \tag{6}$$

Let $\Sigma = \{e_1 e_2 \cdots | t(e_k) = i(e_{k+1}),\ \forall k \geq 1\}$; there exists a surjection $\pi : \Sigma \to K$, π is also called coding mapping.

Example 1. Let $\mathscr{V} = \{0, 1\}, \mathscr{E} = \{e_1, e_2, e_3\}$ where $e_1 = (0, 0), e_2 = (0, 1), e_3 = (1, 0), \mathscr{G} = (V, \mathscr{E})$.

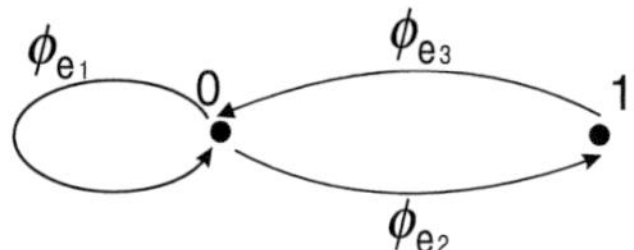

Remark 1. The above construction can be extended as follows: (1) for each pair of vertices (i, j), we allow several edges starting at i and ending at j; (2) Each similitude ϕ_e can be replaced by contractive mappings. In both cases, by an analogous argument with the case of graph-directed set, the corresponding graph-directed invariant set still exists and the equality (6) holds still. The construction corresponding to case (2) is called the graph-directed cookie-cutter set (GCC).

1.3 Cookie-Cutter-Like Sets with Graph-Directed Construction (GCCL)

We are going to extend further the GCC construction in Remark 1 in the following way, i.e., at each iteration, by keeping the vertex set invariant, we allow the set of edges change.

Let $I_1, \ldots, I_n$ be pairwise disjoint intervals in $[0,1]$, $\mathscr{V} = \{1, \ldots, n\}$ be the set of vertices, and $Q_k := [q_{i,j}^{(k)}]_{i,j\leq n} (k \geq 1)$ be nonnegative integer matrix of order n.

For each matrix Q_k there is a corresponding digraph $\mathscr{G}^{(k)} = (\mathscr{V}, \mathscr{E}_k)$ such that for any pair of vertices $i, j \in \mathscr{V}$, there exist $q_{i,j}^{(k)}$ edges starting at the vertex i and ending at the vertex j. We denote by $\big((i,j),l\big)$ the l-th edge in $\mathscr{E}_k$ starting at i and ending at j; if no confusion will happen, we will denote this edge as (e,l) where $e = (i, j)$ pointing out the initial and the terminal vertex of this edge, and we write $i(e) = i$ and $t(e) = j$, respectively, as the initial and the terminal vertex of the edge (e,l). Similar with the graph-directed construction, we suppose that:

(1) For each $k \geq 1$ and for any $i \in \mathscr{V}$, there exist $(e,l) \in \mathscr{E}_k$ such that $i(e) = i$ and $q_e^{(k)} \geq 1$, i.e., there exist edges start from the vertex i in each digraph $\mathscr{G}^{(k)}$.
(2) If $q_e^{(k)} \geqslant 1$, we suppose that there exist $q_e^{(k)}$ pairwise disjoint subintervals $\{I_{e,l}^{(k)}\}_{1\leqslant l\leqslant q_e^{(k)}}$ of $I_{i(e)}$ and an expanding continuous map $f_e^{(k)} : \bigcup_{l=1}^{q_e^{(k)}} I_{e,l}^{(k)} \to I_{t(e)}$, such that the restriction of $f_e^{(k)}$ to each initial interval $I_{e,l}^{(k)}$ $(1 \leq l \leq q_e^{(k)})$ is 1-1 and onto, and denote the corresponding inverse branches of $f_e^{(k)}$ as $\{\phi_{e,l}^{(k)}, l = 1, \ldots, q_e^{(k)}\}$.

For $(e_1,l_1) \in \mathscr{E}_1, \ldots, (e_m,l_m) \in \mathscr{E}_m$, we say that $(e_1,l_1)\cdots(e_m,l_m)$ is an *admissible path* (or *admissible word*) of length m, if $t(e_l) = i(e_{l+1})$.

Let

$$\Omega_m = \{(e_1,k_1)\cdots(e_m,k_m) | (e_1,k_1)\cdots(e_m,k_m) \text{ is an admissible path}\} \tag{7}$$

be the set of all the admissible path of length m.

Let $m \geq 1$ and $\sigma = (e_1,k_1)\cdots(e_m,k_m) \in \Omega_m$ be an admissible path of length m, and set

$$\phi_\sigma = \phi^{(1)}_{e_1,k_1} \circ \cdots \circ \phi^{(m)}_{e_m,k_m}, \quad \text{and} \quad I_\sigma = \phi_\sigma(I_{t(e_m)}).$$

Finally, define

$$E = \bigcap_{m \geqslant 1} \bigcup_{\sigma \in \Omega_m} I_\sigma; \tag{8}$$

then E is called the *cookie-cutter-like set with graph-directed construction (GCCL)*.

Example 2. From the graph-directed construction point of view, the CCL set can be obtained through a sequence of graphs $\{\mathscr{G}^{(k)} := (\mathscr{V}, \mathscr{E}_k)\}$ which have the same vertex set $\mathscr{V} = \{0\}$, and for each $k \geq 1$, $\mathscr{G}^{(k)}$ have q_k edges from vertex 0 to vertex 0,

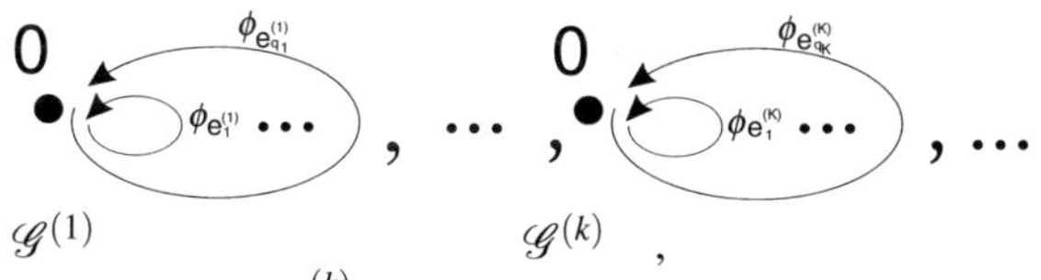

0 $\phi_{e^{(1)}_{q_1}}$ $\phi_{e^{(1)}_1}$ $\cdots$, $\cdots$, 0 $\phi_{e^{(K)}_{q_K}}$ $\phi_{e^{(K)}_1}$ $\cdots$, $\cdots$

$\mathscr{G}^{(1)}$ $\qquad$ $\mathscr{G}^{(k)}$,

where $\phi_{e^{(k)}_j} = \phi^{(k)}_j, \forall k \geq 1, j = 1, \ldots, q_k$. Then the CCL set

$$E = \bigcap_{n \geq 1} \bigcup_{i_1 \cdots i_n \in \Omega_n} \phi_{e^{(1)}_{i_1}} \circ \cdots \circ \phi_{e^{(n)}_{i_n}}(I). \tag{9}$$

Remark 2. (1) This new construction generalizes the graph-directed one in several aspects: (i) multi-edges are allowed for a pair of vertices; (ii) the involved maps can be nonlinear; (iii) the indexed graphs may change in each step.

(2) We can compare the generalization from CCL to GCCL with the generalization from full shift to subshift of finite type.

We are going to study the following problems under some regularity condition on the maps:

(1) The expressions of the Hausdorff dimension and the packing dimension of the set E defined by Eq. (8)
(2) Conditions such that the Hausdorff measure $H^{s_*}(E)$ (resp. packing measure $P^{s^*}(E)$) is finite and positive, where s_* and s^* are the Hausdorff and packing dimension of E, respectively
(3) The continuous dependence of the dimensions of E on the defining data of E

The statements of the answers of the above problems will be given in Sect. 2.

The remainder of this chapter is organized as follows. In Sect. 2, we present the regularity assumptions on the maps and state the main results. Section 3 is devoted to establishing four principal properties of GCCL, namely bounded variation, bounded distortion, bounded covariation, and the existence of Gibbs-like measures, which play the essential role in the proofs of the main theorems of this chapter. Section 4 will be devoted to the proofs of the main theorems.

2 Main Results

2.1 Basic Assumption

Definition 1. Let $I, I_1, \ldots, I_q \subset \mathbb{R}$ be bounded closed intervals and $I_1, \ldots, I_q$ be pairwise disjoints. $f : \cup_{j=1}^{q} I_j \to I$ is called a mapping of *cookie-cutter type* provided:

(1) $f : I_j \to I$ is 1-1 and onto for each $j = 1, \ldots, q$. (The corresponding branch inverse is denoted by $\phi_j = (f|_{I_j})^{-1} : I \to I_j$).
(2) f is differentiable with Hölder continuous derivative Df, i.e., there exist constant $c_f > 0$ and $\gamma_f \in (0, 1]$ such that for any $x, y \in I_j$, $j = 1, \ldots, q$,

$$|Df(x) - Df(y)| \leq c_f |x - y|^{\gamma_f}.$$

(3) f is boundly expanding in the sense that

$$1 < b_f := \inf_x \{|Df(x)|\} \leq \sup_x \{|Df(x)|\} := B_f < +\infty.$$

The array $[\cup_{j=1}^{q} I_j; c_f, \gamma_f, b_f, B_f]$ is called the *defining data* of f.

In the following, we suppose always the following two conditions are fulfilled:

(C1) If $q_e^{(k)} \geq 1$, then the mapping $f_e^{(k)}$ is of cookie-cutter type with the defining data $[\bigcup_{l=1}^{q_e^{(k)}} I_{e,l}^{(k)}; c_e^{(k)}, \gamma_e^{(k)}, b_e^{(k)}, B_e^{(k)}]$.
(C2) There exist constants $B \geq b > 1$, $1 \geq \gamma > 0$, and $c > 0$ such that

$$b \leq \inf\{b_e^{(k)} | q_e^{(k)} \geq 1\} \leqslant \sup\{B_e^{(k)} | q_e^{(k)} \geq 1\} \leq B,$$
$$\gamma \leq \inf\{\gamma_e^{(k)} | q_e^{(k)} \geq 1\} \leqslant \sup\{\gamma_e^{(k)} | q_e^{(k)} \geq 1\} \leqslant 1,$$
$$\sup\{c_e^{(k)} | q_e^{(k)} \geq 1\} \leq c.$$

Define $\mathscr{F}^{(k)} := \{f_e^{(k)} : q_e^{(k)} \geq 1\}$. We say that the GCCL set E is generated by $[(\mathscr{F}^{(k)})_{k \geq 1}, c, \gamma, b, B]$.

Remark 3. Let $a = \min\{|I_i| : i \in \mathscr{V}\}, A = \max\{|I_i| : i \in \mathscr{V}\}$; from the condition (C2), we see that

$$\sup\{q_e^{(k)} : q_e^{(k)} \geq 1\} \leqslant \frac{BA}{a} < \infty,$$

and we denote $M := \max\{q_e^{(k)} : q_e^{(k)} \geq 1\}$.

2.2 Main Theorems

For each $m \geqslant 1$, let s_m be the unique solution of the following equation:

$$\sum_{\omega\in\Omega_m} |I_\omega|^{s_m} = 1,$$

and define $s_* := \liminf s_m$, $s^* := \limsup s_m$.

Definition 2 ([6]). Let $\{M_k\}_{k\geqslant 1}$ be a sequence of nonnegative matrix of order n. We say that the sequence of matrices is primitive (with index p) if there exists $p \geqslant 0$ such that for any $h \geqslant 1$, the entries of the matrix $M_h \cdots M_{h+p}$ are all strictly positive.

Now we state the main results of this chapter.

Theorem 1. *Let E be a GCCL, then*

$$\overline{\dim}_B E = s^*.$$

Moreover, if the sequence of matrices $\{Q_k\}_{k\geqslant 1}$ is primitive, then

$$\dim_H E = s_*, \quad \dim_P E = \overline{\dim}_B E = s^*.$$

Theorem 2. *Let E be a GCCL and the sequence of matrices $\{Q_k\}_{k\geqslant 1}$ be primitive. Then $H^{s_*}(E)$ (resp. $P^{s^*}(E)$) and $\liminf_{l\to\infty} \sum_{\omega\in\Omega_l} |I_\omega|^{s_*}$ (resp. $\limsup_{l\to\infty} \sum_{\omega\in\Omega_l} |I_\omega|^{s^*}$) are simultaneously null, positive and finite, or infinite.*

Now let $\mathscr{V} := \{1,2,\ldots,n\}$, $I_1,\ldots,I_n$ and $\{Q_k\}_{k\geqslant 1}$ be fixed. Let E and $\tilde{E}$ be generated, respectively, by $[f_e^{(k)}, I_{e,l}^{(k)}, c, \gamma, b, B]$ and $[\tilde{f}_e^{(k)}, \tilde{I}_{e,l}^{(k)}, \tilde{c}, \tilde{\gamma}, \tilde{b}, \tilde{B}]$, and let s_* and $\tilde{s}_*$ (resp. s^* and $\tilde{s}^*$) be the Hausdorff dimension (resp. packing dimension) of GCCL's E and $\tilde{E}$.

Let $d_k := \max_{q_e^{(k)}\geq 1} \{\rho_H(\bigcup_{l=1}^{q_e^{(k)}} I_{e,l}^{(k)}, \bigcup_{l=1}^{q_e^{(k)}} \tilde{I}_{e,l}^{(k)}),\ \rho_U(f_e^{(k)}, \tilde{f}_e^{(k)})\}$, where ρ_H is the Hausdorff metric and ρ_U is the uniform metric on the function space $C^1[0,1]$, and define $\widetilde{d} := \widetilde{d}(E,\tilde{E}) := \sup_k d_k$.

Theorem 3 (Continuous Dependence of the Dimensions). *Let E be a GCCL. Then the Hausdorff dimension and the packing dimensions of E depend continuously on its defining data under the metric $\tilde{d}$.*

3 Four Properties of GCCL

For the convenience at first we introduce more notations.

3.1 More on Coding Space and Other Notations

We have defined Ω_m as the set of all admissible paths of length m in Eq. (7). Define $\Omega^* = \bigcup_{m\geqslant 1}\Omega_m$.

For $\sigma = (e_1,k_1)(e_2,k_2)\cdots(e_m,k_m)$ and $1 \leqslant l \leqslant l' \leqslant m$, we denote

$$|\sigma| = m,\ \sigma|_l = (e_1,k_1)(e_2,k_2)\cdots(e_l,k_l),\ \sigma|_l^{l'} = (e_l,k_l)\cdots(e_{l'},k_{l'})$$

and denote $\Omega_{l,m} = \{\sigma|_l^m :\ \sigma \in \Omega^*, |\sigma| \geqslant m\}$. Define

$$\Omega_\infty = \{(e_1,k_1)(e_2,k_2)\cdots \big| \forall m \geq 1, (e_1,k_1)\cdots(e_m,k_m) \text{ is admissible}\}.$$

For $\omega \in \Omega_l$ and $\sigma \in \Omega_{l+1,m}$ we can define the operation $\omega * \sigma$ by simply conjuncting two words. If $\omega * \sigma \in \Omega_m$, we say that ω and σ are compatible and denote by $\omega \hookleftarrow \sigma$.

For $\sigma = (e_1,k_1)\cdots(e_m,k_m) \in \Omega_m$ with $|\sigma| = m \geqslant 1$, set

$$F_\sigma = f_{e_m}^{(m)} \circ \cdots \circ f_{e_1}^{(1)}.$$

Recall that by the definitions ϕ_σ and I_σ, we have $|I_\sigma| \leqslant b^{-(m-1)}|I_{i_m}|$ and $|I_\sigma| \to 0$ as $|\sigma| \to \infty$.

It is easy to verify that these basic intervals $\{I_\sigma\}_{\sigma\in\Omega^*}$ have the following net properties:

- For $\sigma = (e_1,k_1)\cdots(e_m,k_m) \in \Omega_m$, $F_\sigma : I_\sigma \to I_{t(e_m)}$ is $1-1$ and onto.
- $I_{\sigma'} \subset I_\sigma$ if $|\sigma'| \geqslant |\sigma|$ and $\sigma = \sigma'|_{|\sigma|}$.
- $I_{\omega_1} \cap I_{\omega_2} = \emptyset$, if $\omega_1, \omega_2 \in \Omega_m$ and $\omega_1 \neq \omega_2$.

3.2 Proofs of Four Properties

Lemma 1 (Bounded Variation). *There exists a constant* $0 < \xi < \infty$, *such that for each* $m \geqslant 1, \omega \in \Omega_m$, *and* $x,y \in I_\omega$, *we have*

$$\xi^{-1} \leqslant \frac{|DF_\omega(x)|}{|DF_\omega(y)|} \leqslant \xi.$$

Proof. Fix $m \geqslant 1, \omega = (e_1,k_1)(e_2,k_2)\cdots(e_m,k_m) \in \Omega_m$, and $x,y \in I_\omega$. Notice that for each $1 \leqslant l \leqslant m$, $F_{\omega|_l}$ maps I_ω to the set $\phi_{e_{l+1},k_{l+1}}^{(l+1)} \circ \cdots \circ \phi_{e_m,k_m}^{(m)}(I_{t(e_m)})$; hence by the chain rule and the mean value theorem, we get

$$|F_{\omega|_l}(x) - F_{\omega|_l}(y)| \leqslant |\phi_{e_{l+1},k_{l+1}}^{(l+1)} \circ \cdots \circ \phi_{e_m,k_m}^{(m)}(I_{t(em)})| \leqslant b^{-(m-l)}.$$

Then we can get

$$\begin{aligned}&\big|\log|Df_{e_{l+1}}^{(l+1)}(F_{\omega|_l}(x))|-\log|Df_{e_{l+1}}^{(l+1)}(F_{\omega|_l}(y))|\big|\\&\leqslant\big||Df_{e_{l+1}}^{(l+1)}(F_{\omega|_l}(x))|-|Df_{e_{l+1}}^{(l+1)}(F_{\omega|_l}(y))|\big|\\&\leqslant cb^{-(m-l)\gamma};\end{aligned}$$

therefore, by the above inequality and chain rule,

$$\begin{aligned}&\big|\log|DF_\omega(x)|-\log|DF_\omega(y)|\big|\\&\leqslant\sum_{l=1}^{m-1}\big|\log|Df_{e_{l+1}}^{(l+1)}(F_{\omega|_l}(x))|-\log|Df_{e_{l+1}}^{(l+1)}(F_{\omega|_l}(y))|\big|\\&\leqslant\sum_{l=1}^{m-1}cb^{-(m-l)\gamma}\leqslant\frac{cb^{-\gamma}}{1-b^{-\gamma}}=\frac{c}{b^\gamma-1}.\end{aligned}$$

Taking $\xi=\exp\{\frac{c}{b^\gamma-1}\}$, we get this lemma. □

Lemma 2 (Bounded Distortion). *There exists a constant $0<\zeta<\infty$, such that for any $m\geqslant 1$, $\omega\in\Omega_m$, and $x\in I_\omega$, we have*

$$\zeta^{-1}\leqslant|I_\omega||DF_\omega(x)|\leqslant\zeta.$$

Moreover, for each $\omega\in\Omega_m$,

$$|I_\omega|\geqslant a\zeta^{-1}B^{-1}|I_{\omega|_{|\omega|-1}}|.$$

Proof. Let $\omega=(e_1,k_1)(e_2,k_2)\cdots(e_m,k_m)$, then $F_\omega(I_\omega)=I_{t(e_m)}$ and there exists some point $y\in I_\omega$ such that $|DF_\omega(y)||I_\omega|=|I_{t(e_m)}|$. By Lemma 1, we get

$$\xi^{-1}|I_{t(e_m)}|\leqslant|I_\omega||DF_\omega(x)|=|I_\omega||DF_\omega(y)|\frac{|DF_\omega(x)|}{|DF_\omega(y)|}\leqslant\xi|I_{t(e_m)}|\leqslant\xi.$$

Take $\zeta=\frac{\xi}{\min\{|I_i|:i\in V\}}$, then we get

$$\zeta^{-1}\leqslant|I_\omega||DF_\omega(x)|\leqslant\zeta,\quad\forall x\in I_\omega.\tag{10}$$

Moreover, since

$$|DF_\omega(y)|=|Df_{e_m}^{(m)}(F_{\omega|_{|\omega|-1}}(y))|\;|DF_{\omega|_{|\omega|-1}}(y)|$$

and

$$|DF_\omega(y)||I_\omega|=|I_{t(e_m)}|,$$

by Eq. (10), we get

$$
\begin{aligned}
\frac{|I_\omega|}{|I_{\omega|_{|\omega|-1}}|} &= \frac{|I_{t(e_m)}|}{|DF_\omega(y)|\,|I_{\omega|_{|\omega|-1}}|}\\
&= \frac{|I_{t(e_m)}|}{|Df_{e_m}^{(m)}(F_{\omega|_{|\omega|-1}}(y))|\,|DF_{\omega|_{|\omega|-1}}(y)|\,|I_{\omega|_{|\omega|-1}}|}\\
&\geqslant aB^{-1}\zeta^{-1}.
\end{aligned}
$$

Then

$$
|I_\omega| \geqslant a\zeta^{-1}B^{-1}|I_{\omega|_{|\omega|-1}}|. \tag{11}
$$

□

Lemma 3 (Bounded Covariation). *There exists a positive constant $\rho > 0$, such that for all $m > l \geqslant 1$, $\omega_1, \omega_2 \in \Omega_l, \sigma \in \Omega_{l+1,m}$, if $\omega_1 \hookleftarrow \sigma$ and $\omega_2 \hookleftarrow \sigma$, we have*

$$
\rho^{-1}\frac{|I_{\omega_2*\sigma}|}{|I_{\omega_2}|} \leqslant \frac{|I_{\omega_1*\sigma}|}{|I_{\omega_1}|} \leqslant \rho\frac{|I_{\omega_2*\sigma}|}{|I_{\omega_2}|}.
$$

Proof. Suppose that $\sigma = (e_{l+1}, k_{l+1})\cdots(e_m, k_m)$; we then denote

$$
J_\sigma = \phi_{e_{l+1},k_{l+1}}^{(l+1)} \circ \cdots \circ \phi_{e_m,k_m}^{(m)}(I_{t(e_m)}).
$$

Since $F_{\omega_1}(I_{\omega_1*\sigma}) = J_\sigma$ and $F_{\omega_2}(I_{\omega_2*\sigma}) = J_\sigma$, by the mean value theorem, we get $|J_\sigma| = |DF_{\omega_1}(z_1)|\,|(I_{\omega_1*\sigma})|$, for some $z_1 \in I_{\omega_1*\sigma}$.

By Eq. (10) there exists a constant $\zeta > 0$ such that $\zeta^{-1} \leqslant |I_{\omega_1}|\,|DF_{\omega_1}(z_1)| \leqslant \zeta$. Thus

$$
\zeta^{-1}|J_\sigma| \leqslant \frac{|I_{\omega_1*\sigma}|}{|I_{\omega_1}|} \leqslant \zeta|J_\sigma|.
$$

According to the same discussion, we also have

$$
\zeta^{-1}|J_\sigma| \leqslant \frac{|I_{\omega_2*\sigma}|}{|I_{\omega_2}|} \leqslant \zeta|J_\sigma|;
$$

hence we

$$
\zeta^{-2}\frac{|I_{\omega_2*\sigma}|}{|I_{\omega_2}|} \leqslant \frac{|I_{\omega_1*\sigma}|}{|I_{\omega_1}|} \leqslant \zeta^2\frac{|I_{\omega_2*\sigma}|}{|I_{\omega_2}|}.
$$

Let $\rho = \zeta^2$, we get the lemma. □

Now, with Remark 3 and Lemma 3, one can easily prove the following lemma.

Lemma 4. *If the matrix sequence $\{Q_k\}_{k\geqslant 1}$ is primitive with index p, then there exists a positive constant $\eta > 0$ determined by p, M, ρ, such that $\forall l \leqslant m$, and $\omega_0, \omega_1 \subset \Omega_l$*

$$\eta^{-1} \sum_{\substack{\sigma_1 \in \Omega_{l+1,m} \\ \omega_1 \hookleftarrow \sigma_1}} \frac{|I_{\omega_1 * \sigma_1}|^\beta}{|I_{\omega_1}|^\beta} \leqslant \sum_{\substack{\sigma_0 \in \Omega_{l+1,m} \\ \omega_0 \hookleftarrow \sigma_0}} \frac{|I_{\omega_0 * \sigma_0}|^\beta}{|I_{\omega_0}|^\beta} \leqslant \eta \sum_{\substack{\sigma_1 \in \Omega_{l+1,m} \\ \omega_1 \hookleftarrow \sigma_1}} \frac{|I_{\omega_1 * \sigma_1}|^\beta}{|I_{\omega_1}|^\beta}. \tag{12}$$

Lemma 5 (Existence of Gibbs-Like Measures). *If the sequence of matrices $\{Q_k\}_{k\geqslant 1}$ is primitive, then $\forall \beta > 0$, there exist $\eta > 0$ and a probability μ_β supported by E such that for all $l \geqslant 1$ and $\omega_0 \in \Omega_l$, we have*

$$\eta^{-1} \frac{|I_{\omega_0}|^\beta}{\sum_{\sigma \in \Omega_l} |I_\sigma|^\beta} \leqslant \mu_\beta(I_{\omega_0}) \leqslant \eta \frac{|I_{\omega_0}|^\beta}{\sum_{\sigma \in \Omega_l} |I_\sigma|^\beta}.$$

Proof. First, for each $m \geqslant 1$, we define a probability μ_m supported by E such that for any $\omega \in \Omega_m$

$$\mu_m(I_\omega) = \frac{|I_\omega|^\beta}{\sum_{\sigma \in \Omega_m} |I_\sigma|^\beta}.$$

Now we fix $l \geqslant 1, \omega_0 \in \Omega_l$.
For $m > l$, from Eq. (12), we get

$$\begin{aligned}
\eta |I_{\omega_0}|^\beta \sum_{\omega \in \Omega_m} |I_\omega|^\beta &= \eta |I_{\omega_0}|^\beta \sum_{\omega_1 \in \Omega_l} \sum_{\substack{\sigma_1 \in \Omega_{l+1,m} \\ \omega_1 \hookleftarrow \sigma_1}} |I_{\omega_1 * \sigma_1}|^\beta \\
&= \sum_{\omega_1 \in \Omega_l} \left(\eta |I_{\omega_0}|^\beta \sum_{\substack{\sigma_1 \in \Omega_{l+1,m} \\ \omega_1 \hookleftarrow \sigma_1}} |I_{\omega_1 * \sigma_1}|^\beta \right) \\
&\geqslant \sum_{\omega_1 \in \Omega_l} \left(|I_{\omega_1}|^\beta \sum_{\substack{\sigma_0 \in \Omega_{l+1,m} \\ \omega_0 \hookleftarrow \sigma_0}} |I_{\omega_0 * \sigma_0}|^\beta \right) \\
&= \left(\sum_{\omega_1 \in \Omega_l} |I_{\omega_1}|^\beta \right) \left(\sum_{\substack{\sigma_0 \in \Omega_{l+1,m} \\ \omega_0 \hookleftarrow \sigma_0}} |I_{\omega_0 * \sigma_0}|^\beta \right).
\end{aligned}$$

Then we have

$$\mu_m(I_{\omega_0}) = \sum_{\substack{\sigma_0 \in \Omega_{l+1,m} \\ \omega_0 \hookleftarrow \sigma_0}} \frac{|I_{\omega_0 * \sigma_0}|^\beta}{\sum_{\omega \in \Omega_m} |I_\omega|^\beta} \leqslant \eta \sum_{\omega_1 \in \Omega_l} \frac{|I_{\omega_0}|^\beta}{|I_{\omega_1}|^\beta}.$$

According the same discussion, we can get

$$\eta^{-1} \sum_{\omega_1 \in \Omega_l} \frac{|I_{\omega_0}|^\beta}{|I_{\omega_1}|^\beta} \leqslant \mu_m(I_{\omega_0}).$$

Let μ_β be a weak limit of the measure $\{\mu_m\}_{m \geqslant 1}$, and by Theorem 1.24 of [8], we get this lemma. □

4 Proofs of the Theorems

4.1 *Proof of Theorem 1*

Proof. One can get the first statement directly using Theorem 3 of [6]. We just need to discuss the case that $\{Q_k\}_{k \geq 1}$ is primitive.

(1) $\dim_H E = s_*$

Let $\beta > s_*$. Then by the definition of s_*, there exists a subsequence of integers m_l such that $s_{m_l} < \beta$; thus

$$H^\beta(E) \leqslant \liminf_{l \to \infty} \sum_{\omega \in \Omega_{m_l}} |I_\omega|^\beta \leqslant \liminf_{l \to \infty} \sum_{\omega \in \Omega_{m_l}} |I_\omega|^{s_{m_l}} = 1,$$

which yields $\dim_H E \leqslant \beta$ and so $\dim_H E \leqslant s_*$.

Now take any $\beta < s_*$, thus $s_l > \beta$ for l large enough; thus

$$\sum_{\omega \in \Omega_l} |I_\omega|^\beta > \sum_{\omega \in \Omega_l} |I_\omega|^{s_*} = 1.$$

Let μ_β be a Gibbs-like measure defined in Lemma 5. Then by Lemma 5, for large m and $\omega \in \Omega_m$, we have

$$\mu_\beta(I_\omega) \leqslant \eta |I_\omega|^\beta.$$

Take $r > 0$ small and let $\mathscr{M}_r = \{I_\omega : \ \omega \in \Omega^*, \ |I_\omega| < r \leqslant |I_{\omega|_{|\omega|-1}}|\}$. Then by Eq. (11), we get $\forall x \in E$

$$\mu_\beta(B(x,r)) \leqslant \sum_{\substack{I_\omega \in \mathscr{M}_r \\ I_\omega \cap B(x,r) \neq \emptyset}} \mu_\beta(I_\omega) \leqslant \eta \sum_{\substack{I_\omega \in \mathscr{M}_r \\ I_\omega \cap B(x,r) \neq \emptyset}} |I_\omega|^\beta \leqslant a^{-1} \eta \zeta B r^\beta.$$

By Frostman's Lemma [12], we know that $\dim_H E \geqslant \beta$; hence, $\dim_H E \geqslant s_*$.

(2) $\dim_P E = \overline{\dim}_B E = s^*$

Since the sequence of matrices $\{Q_k\}_{k\geqslant 1}$ is primitive, notice that for any ω, from the construction of E it is seen that $\overline{\dim}_B E = \overline{\dim}_B(E \cap I_\omega)$. Suppose an open set U with $U \cap E \neq \emptyset$, then U will contain a basic interval of some order I_ω, so $\overline{\dim}_B(U \cap E) = \overline{\dim}_B(E)$, by Corollary 3.9 in [2], $\dim_P E = \overline{\dim}_B E$.

Let $\beta > s^*$, then $s_l < \beta$ for l large enough; thus

$$\sum_{\omega\in\Omega_l} |I_\omega|^\beta \leqslant \sum_{\omega\in\Omega_l} |I_\omega|^{s_l} = 1.$$

Let μ_β be a Gibbs-like measure associated with E. Then by the above inequality and Lemma 5, for large l and each $\omega_0 \in \Omega_l$, we have

$$\mu_\beta(I_{\omega_0}) \geqslant \eta^{-1}|I_{\omega_0}|^\beta.$$

Let $r > 0$ and let $\mathscr{M}_r = \{I_\omega : \omega \in \Omega^*, |I_\omega| \leqslant r < |I_{\omega|_{|\omega|-1}}|\}$. Suppose $x \in E$ and $x \in I_{\omega_j} \in \mathscr{M}_r$, then $I_{\omega_j} \subset B(x,r)$; we have by Eq. (11)

$$\mu_\beta(B(x,r)) \geqslant \mu_\beta(I_{\omega_j}) \geqslant \eta^{-1}|I_{\omega_j}|^\beta \geqslant \eta^{-1}a^\beta\zeta^{-\beta}B^{-\beta}r^{-\beta}.$$

By Frostman's lemma again, we get $\dim_P E \leqslant \beta$ and hence $\dim_P E \leqslant s^*$.

For any $\beta < s^*$, there exists a subsequence of integers m_l such $s_{m_l} > \beta$; hence

$$Q^\beta(E) \geqslant \limsup_{l\to\infty} \sum_{\omega\in\Omega_l} |I_\omega|^\beta \geqslant \limsup_{l\to\infty} \sum_{\omega\in\Omega_{m_l}} |I_\omega|^{s_{m_l}} = 1;$$

by the definition of upper box-counting dimension introduced by Tricot [11], we know $\overline{\dim}_B E \geqslant \beta$; hence $\overline{\dim}_B E \geqslant s^*$.

By the above discussion, we get $\dim_P E = \overline{\dim}_B E = s^*$. □

4.2 Proof of Theorem 2

Proof. (1) Denote by $a_* = \liminf_{l\to\infty} \sum_{\omega\in\Omega_l} |I_\omega|^{s_*}$; then by the definition of Hausdorff dimension and s_*, it suffices to prove $a_* > 0 \Rightarrow H^{s_*}(E) > 0$ and $a_* = \infty \Rightarrow H^{s_*} = \infty$. Now we consider the two cases, respectively:

(i) $a_* > 0 \Rightarrow H^{s_*}(E) > 0$. Take l large such that

$$\sum_{\omega\in\Omega_l} |I_\omega|^{s_*} > \frac{a_*}{2}.$$

Then by Lemma 5, the Gibbs-like measure μ_{s_*} satisfies, for l and each $\omega_0 \in \Omega_l$,

$$\mu_{s_*}(I_{\omega_0}) \leqslant \frac{2\eta}{a_*}|I_{\omega_0}|^{s_*}.$$

Now for small $r>0$ and $\forall x\in E$, let $\mathscr{M}_r=\{I_\omega:\ \omega\in\Omega^*,\ |I_\omega|\leqslant r<|I_{\omega|_{|\omega|-1}}|\}$; by Lemma 11, we get

$$\mu_{s_*}(B(x,r))\leqslant\sum_{\substack{I_\omega\in\mathscr{M}_r\\ I_\omega\cap B(x,r)\neq\emptyset}}\mu_{s_*}(I_\omega)\leqslant\frac{2\eta}{a_*}\sum_{\substack{I_\omega\in\mathscr{M}_r\\ I_\omega\cap B(x,r)\neq\emptyset}}|I_\omega|^{s_*}\leqslant\frac{4\eta}{aB^{-1}\zeta^{-1}a_*}r^{s_*};$$

therefore

$$\limsup_{r\to 0}\frac{\mu_{s_*}(B(x,r))}{r^{s_*}}\leqslant\frac{4\eta}{aB^{-1}\zeta^{-1}a_*},$$

then we can get $H^{s_*}(E)\geqslant\frac{aB^{-1}\zeta^{-1}a_*}{4\eta}>0$.

(ii) $a_*=\infty\Rightarrow H^{s_*}(E)=\infty$.

Suppose that $a_*=\infty$; then for any $\varepsilon>0$, we have for large l

$$\sum_{\omega\in\Omega_l}|I_\omega|^{s_*}>\frac{1}{\varepsilon}.$$

By an analogous discussion as in part (i), we get

$$\limsup_{r\to 0}\frac{\mu_{s_*}(B(x,r))}{r^{s_*}}\leqslant\frac{4\eta\varepsilon}{aB^{-1}\zeta^{-1}},$$

so $H^{s_*}(E)\geqslant\dfrac{aB^{-1}\zeta^{-1}}{4\eta\varepsilon}$, it follows $H^{s_*}(E)=\infty$.

(2) For the case of packing dimension, let $a^*:=\limsup_{l\to\infty}\sum_{\omega\in\Omega_l}|I_\omega|^{s^*}$. We prove the proposition by the following steps:

Claim 1. There exists $\lambda>0$ such that $P_0^{s^*}(E)\geqslant\lambda a^*$. Since the sequence of matrices $\{Q_k\}_{k\geqslant 1}$ is primitive, with the condition of $f_{i,j}^{(k)}$, we know that there exists $q>0$, such that $\forall\omega\in\Omega^*$, there exist a subinterval of I_ω of order $|\omega|+q$, say $I_{\omega'}$ and a point $x_\omega\in E\cap I_{\omega'}$ such that $B(x,|I_{\omega'}|)\subset I_\omega$. First suppose $a^*<\infty$, then there exist infinitely many l's such that $\sum_{\omega\in\Omega_l}|I_\omega|^{s^*}>\frac{a^*}{2}$. For such l's and each $\omega\in\Omega_l$, we choose the balls $B(x,|I_{\omega'}|)$; such balls are disjoint and centered in E, so they form a packing of E, then we get

$$\sum_{\omega\in\Omega_l}|I_{\omega'}|^{s^*}\geqslant\sum_{\omega\in\Omega_l}(\zeta^{-q}a^qB^{-q}|I_\omega|)^{s^*}\geqslant\frac{a^*}{2}\left(\frac{a^q}{\zeta^qB^q}\right)^{s^*},$$

which yield $P_0^{s^*}(E)\geqslant\frac{a^*}{2}\left(\frac{a^q}{\zeta^qB^q}\right)^{s^*}$; take $\lambda=\frac{1}{2}\left(\frac{a^q}{\zeta^qB^q}\right)^{s^*}$, and we get the desired conclusion.

By the same way, we prove that if $a^*=\infty$, then $P_0^{s^*}(E)=\infty$.

Claim 2. $a^*=0\Rightarrow P_0^{s^*}=0$. Let $a^*=0$. Then for any $\varepsilon>0$, we have for large l enough

$$\sum_{\omega\in\Omega_l}|I_\omega|^{s^*}\leqslant\varepsilon.$$

Let $\mathscr{M}_r = \{I_\omega : \ \omega \in \Omega^*, \ |I_\omega| \leqslant r < |I_{\omega|_{|\omega|-1}}|\}$ and let μ_{s^*} be the Gibbs-like measure. Then by Lemma 5, we have for large l and each $\omega_0 \in \Omega_l$,

$$\mu_{s^*}(I_{\omega_0}) \geqslant (\eta\varepsilon)^{-1}|I_{\omega_0}|^{s^*}.$$

Suppose $x \in I_\omega \in \mathscr{M}_r$, then $I_\omega \subset B(x,r)$, we have

$$\mu_{s^*}(B(x,r)) \geqslant \mu_{s^*}(I_\omega) \geqslant (\eta\varepsilon)^{-1}|I_\omega|^{s^*} \geqslant (\varepsilon\eta)^{-1}a^{s^*}\zeta^{-s^*}B^{-s^*}r^{s^*},$$

then we know

$$P^{s^*}(E) \leqslant P_0^{s^*}(E) \leqslant 2^{s^*}\eta a^{-s^*}\zeta^{s^*}B^{s^*}\varepsilon,$$

from which follows $P^{s^*}(E) = P_0^{s^*}(E) = 0$.

Claim 3. $0 < P_0^{s^*}(E) < \infty \Rightarrow 0 < a^* < \infty$. For l large enough,

$$\sum_{\omega\in\Omega_l} |I_\omega|^{s^*} < 2a^*.$$

By Lemma 5, the Gibbs-like measure μ_{s^*} fulfills, for large k and each $\omega_0 \in \Omega_k$,

$$\mu_{s^*}(I_{\omega_0}) \geqslant (2a^*\eta)^{-1}|I_{\omega_0}|^{s^*}.$$

Let $\mathscr{M}_r = \{I_\omega : \ \omega \in \Omega^*, \ |I_\omega| \leqslant r < |I_{\omega|_{|\omega|-1}}|\}$; suppose $x \in I_\omega \in \mathscr{M}_r$, then $I_\omega \subset B(x,r)$ and $|I_\omega| \leqslant r \leqslant a^{-1}\zeta B|I_\omega|$, then we have

$$\mu_{s^*}(B(x,r)) \geqslant \mu_{s^*}(I_\omega) \geqslant (2a^*\eta)^{-1}|I_\omega|^{s^*} \geqslant (2a^*\eta a^{-s^*}\zeta^{s^*}B^{s^*})^{-1}r^{s^*},$$

so

$$\liminf_{r\to 0}\frac{\mu_{s^*}(B(x,r))}{(2r)^{s^*}} \geqslant 2^{-s^*}(2a^*\eta a^{-s^*}\zeta^{s*}B^{s^*})^{-1},$$

then $P^{s^*}(E) \leqslant P_0^{s^*}(E) \leqslant 2^{s^*+1}a^*\eta a^{-s^*}\zeta^{s^*}B^{s^*} < \infty$.

Claim 4. $P_0^{s^*}(E) = \infty \Rightarrow P^{s^*}(E) = \infty$. Suppose $P_0^{s^*}(E) = \infty$, then by Claim 3, $a^* = \infty$; thus for any $R > 0$, there exist infinitely many l's such that

$$\sum_{\omega\in\Omega_l} |I_\omega|^{s^*} > R. \tag{13}$$

Now take any open set U that meets E; it will contain a basic interval, say $I_{\omega_0}(\omega_0 \in \Omega_m \text{with some} m)$.

For any $\omega \in \Omega_m$ and $l > m$ satisfying Eq. (13), by Eq. (12), we have

$$|I_\omega|^{s^*}\sum_{\substack{\sigma\in\Omega_{m+1}^l\\ \omega_0\hookleftarrow\sigma}} |I_{\omega_0*\sigma}|^{s^*} \geqslant \eta^{-1}|I_{\omega_0}|^{s^*}\sum_{\substack{\sigma_1\in\Omega_{m+1}^l\\ \omega\hookleftarrow\sigma_1}} |I_{\omega*\sigma_1}|^{s^*},$$

then, we get

$$\begin{aligned}
&\sum_{\omega\in\Omega_m}|I_\omega|^{s^*}\sum_{\substack{\sigma\in\Omega^l_{m+1}\\ \omega_0\hookleftarrow\sigma}}|I_{\omega_0*\sigma}|^{s^*}\\
&\geqslant \eta^{-1}|I_{\omega_0}|^{s^*}\sum_{\omega\in\Omega_m}\sum_{\substack{\sigma_1\in\Omega^l_{m+1}\\ \omega\hookleftarrow\sigma_1}}|I_{\omega*\sigma'}|^{s^*},\\
&\geqslant \eta^{-1}|I_{\omega_0}|^{s^*}R,
\end{aligned}$$

which yields

$$\sum_{\substack{\sigma\in\Omega^l_{m+1}\\ \omega_0\hookleftarrow\sigma}}|I_{\omega_0*\sigma}|^{s^*}\geqslant\frac{R|I_{\omega_0}|^{s^*}}{\eta\sum_{\omega\in\Omega_m}|I_\omega|^{s^*}}.$$

An analogous argument as in Claim 1 asserts that

$$P_0^{s^*}(E\cap U)\geqslant P_0^{s^*}(E\cap I_{\omega_0})=\infty,$$

then $P^{s^*}(E)=\infty$.

Claim 5. $P_0^{s^*}(E)<\infty\Rightarrow P^{s^*}(E)<\infty$. This is Theorem 2.3 of [5].

With the above five claims, we get the proposition for packing dimension. □

4.3 *Proof of Theorem 3*

Proof. Let $\pi:\Omega_\infty\to E$, $\tilde\pi:\Omega_\infty\to\tilde E$ be the coding mappings. Then it is easy to see that they are homeomorphisms and $h:=\tilde\pi\circ\pi^{-1}$ is a homeomorphism from E to $\tilde E$. For $x\in E$, denote $\tilde x=h(x)$, and we have the following lemma.

Lemma 6. *With the notations as above, for $x=\pi(\omega)\in E$, $\tilde x=\tilde\pi(w)\in\tilde E$ and $m\geqslant 1$, we have*

$$|F_{\omega|_m}(x)-\tilde F_{\omega|_m}(\tilde x)|\leqslant\frac{\tilde d(E,\tilde E)}{b-1}.$$

Proof. Suppose that $\omega=(e_1,k_1)(e_2,k_2)(e_3,k_3)\cdots$; then for any $l\geqslant 1$, $a\in I_{t(e_l)}$, since $f_{e_l}^{(l)}(\phi_{e_l,k_l}^{(l)}(a))=\tilde f_{e_l}^{(l)}(\tilde\phi_{e_l,k_l}^{(l)}(a))$, by the definition of $\tilde d$

$$\begin{aligned}
&|f_{e_l}^{(l)}(\phi_{e_l,k_l}^{(l)}(a))-f_{e_l}^{(l)}(\tilde\phi_{e_l,k_l}^{(l)}(a))|\\
&=|\tilde f_{e_l}^{(l)}(\tilde\phi_{e_l,k_l}^{(l)}(a))-f_{e_l}^{(l)}(\tilde\phi_{e_l,k_l}^{(l)}(a))|\\
&\leqslant\tilde d_l\leqslant\tilde d(E,\tilde E),
\end{aligned}$$

we get thus by the mean value theorem

$$|\phi^{(l)}_{e_l,k_l}(a) \quad \tilde{\phi}^{(l)}_{e_l,k_l}(a)| \leqslant b^{-1}\tilde{d}(E,\tilde{E}). \tag{14}$$

For any $m \geqslant 1, l \geqslant 1$, set

$$\Phi_{\omega|_{m+1,m+l}} = \phi^{(m)}_{e_m,k_m} \circ \cdots \circ \phi^{(m+l)}_{e_{m+l},k_{m+l}},$$

$$\tilde{\Phi}_{\omega|_{m+1,m+l}} = \tilde{\phi}^{(m)}_{e_m,k_m} \circ \cdots \circ \tilde{\phi}^{(m+l)}_{e_{m+l},k_{m+l}},$$

then we have

$$F_{\omega|_m}(x) = \bigcap_{l=1}^{\infty} \Phi_{\omega|_{m+1,m+l}}(I_{t(e_{m+l})}), \tag{15}$$

$$\tilde{F}_{\omega|_m}(\tilde{x}) = \bigcap_{l=1}^{\infty} \tilde{\Phi}_{\omega|_{m+1,m+l}}(I_{t(e_{m+l})}). \tag{16}$$

For any $t \in I_{i_{m+l}}$, $m \geqslant 1, l \geqslant 1$, by Eq. (14), we get

$$\begin{aligned}
&\left|\Phi_{\omega|_{m+1,m+l}}(t) - \tilde{\Phi}_{\omega|_{m+1,m+l}}(t)\right| \\
&\leqslant \sum_{q=0}^{l-1} \left|\Phi_{\omega|_{m+1,m+l-q}} \circ \tilde{\Phi}_{\omega|_{m+l-q+1,m+l}}(t)\right. \\
&\quad \left. -\Phi_{\omega|_{m+1,m+l-q-1}} \circ \tilde{\Phi}_{\omega|_{m+l-q,m+l}}(t)\right| \\
&= \sum_{q=0}^{l-1} \left|D\Phi_{\omega|_{m+1,m+l-q-1}}(z_q)\right| \left|\phi^{(m+l-q)}_{e_{m+l-q-1},k_{m+l-q}} \circ \tilde{\Phi}_{\omega|_{m+l-q,m+l}}(t)\right. \\
&\quad \left. -\tilde{\phi}^{(m+l-q)}_{e_{m+l-q},k_{m+l-q}} \circ \tilde{\Phi}_{\omega|_{m+l-q,m+l}}(t)\right| \\
&\leqslant \sum_{q=0}^{l-1} b^{-q}(b^{-1}\tilde{d}(E,\tilde{E})) \\
&\leqslant \frac{\tilde{d}(E,\tilde{E})}{b-1},
\end{aligned}$$

where $z_q \in I$ and we take $\Phi_{\omega|_{l_1,l_2}} = id$ for $l_1 > l_2$. Then by the definition of the Hausdorff metric, we get

$$\rho_H\left(\Phi_{\omega|_{m+1,m+l}}(I_{t(e_{m+l})}), \tilde{\Phi}_{\omega|_{m+1,m+l}}(I_{t(e_{m+l})})\right) \leqslant \frac{\tilde{d}(E,\tilde{E})}{b-1}.$$

By Eqs. (15) and (16), we have

$$\begin{aligned}
&|F_{\omega|_m}(x) - \tilde{F}_{\omega|_m}(\tilde{x})| \\
&\quad \leqslant \rho_H\big(\Phi_{\omega|_{m+1,m+l}}(I_{t(e_{m+l})}), \tilde{\Phi}_{\omega|_{m+1,m+l}}(I_{t(e_{m+l})})\big) \\
&\qquad + |\Phi_{\omega|_{m+1,m+l}}(I_{t(e_{m+l})})| \\
&\quad \leqslant \frac{\tilde{d}(E,\tilde{E})}{b-1} + b^{-l};
\end{aligned}$$

letting $l \to \infty$, we get the conclusion of the lemma. □

Now, let us continue the proof of Theorem 3.

Let $s_m, \tilde{s}_m$ be the m-th pre-dimensions of E and $\tilde{E}$, respectively. First assume that $[c,\gamma,b,B] = [\tilde{c},\tilde{\gamma},\tilde{b},\tilde{B}]$.

For each $m \geqslant 1, \omega = (e_1,k_1)(e_2,k_2)\cdots(e_m,k_m) \in \Omega_m$, take $x \in E \cap I_\omega$ and $\tilde{x} \in \tilde{E} \cap \tilde{I}_\omega$; then by the chain rule and Lemma 6, we have

$$\begin{aligned}
&\left|\log \frac{|DF_\omega(x)|}{|D\tilde{F}_\omega(\tilde{x})|}\right| \\
&\quad \leqslant \sum_{k=1}^{m} \left|\log |Df_{e_k}^{(k)}(F_{\omega|_{k-1}}(x))| - \log |D\tilde{f}_{e_k}^{(k)}(\tilde{F}_{\omega|_{k-1}}(\tilde{x}))|\right| \\
&\quad \leqslant \sum_{k=1}^{m} |Df_{e_k}^{(k)}(F_{\omega|_{k-1}}(x)) - D\tilde{f}_{e_k}^{(k)}(\tilde{F}_{\omega|_{k-1}}(\tilde{x}))| \\
&\quad \leqslant \sum_{k=1}^{m} |Df_{e_k}^{(k)}(F_{\omega|_{k-1}}(x)) - Df_{e_k}^{(k)}(\tilde{F}_{\omega|_{k-1}}(\tilde{x}))| \\
&\qquad + \sum_{k=1}^{m} |Df_{e_k}^{(k)}(\tilde{F}_{\omega|_{k-1}}(x)) - D\tilde{f}_{e_k}^{(k)}(\tilde{F}_{\omega|_{k-1}}(\tilde{x}))| \\
&\quad \leqslant m\frac{c\tilde{d}^\gamma}{(b-1)^\gamma} + m\tilde{d}.
\end{aligned}$$

From the above inequality and Lemma 2, we get

$$\zeta^{-2}\exp\left\{m\left[\frac{c\tilde{d}^\gamma}{(b-1)^\gamma} + \tilde{d}\right]\right\} \leqslant \frac{|I_\omega|}{|\tilde{I}_\omega|} \leqslant \zeta^2 \exp\left\{m\left[\frac{c\tilde{d}^\gamma}{(b-1)^\gamma} + \tilde{d}\right]\right\}, \tag{17}$$

where $\zeta = \frac{1}{\min\{|I_i|:i\in\mathscr{V}\}}\exp\{\frac{c}{b^\gamma-1}\}$.

Put $\limsup_{m\to\infty} |s_m - \tilde{s}_m| = d$; then there exist infinite m such that $s_m \geqslant \tilde{s}_m + d$ or $s_m \leqslant \tilde{s}_m - d$. We only discuss the first case, and the second case can be treated in the same way. From Eq. (17), we have

$$
\begin{aligned}
1 &= \sum_{\omega\in\Omega_m} |I_\omega|^{s_m} \leqslant \sum_{\omega\in\Omega_m} |I_\omega|^{\tilde{s}_m+d} \\
&\leqslant b^{-md} \sum_{\omega\in\Omega_m} \left[\zeta^2 \exp\left\{m\left[\frac{c\tilde{d}^\gamma}{(b-1)^\gamma}+\tilde{d}\right]\right\}|\tilde{I}_\omega|\right]^{\tilde{s}_m} \\
&= b^{-md}\left[\zeta^2 \exp\left\{m\left[\frac{c\tilde{d}^\gamma}{(b-1)^\gamma}+\tilde{d}\right]\right\}\right]^{\tilde{s}_m}
\end{aligned}
$$

which gives

$$
d \leqslant \frac{\tilde{s}_m}{m\log b}\left\{\log\zeta^2 + m\left[\frac{c\tilde{d}^\gamma}{(b-1)^\gamma}+\tilde{d}\right]\right\},
$$

so

$$
d \leqslant \frac{c\tilde{s}^*}{(b-1)^\gamma \log b}\tilde{d}^\gamma + \frac{\tilde{s}^*}{\log b}\tilde{d}, \tag{18}
$$

where $\tilde{s}^* = \limsup_{m\to\infty} \tilde{s}_m \leq 1$.

From Eq. (18), we get the statement for the case of $[c,\gamma,b,B] = [\tilde{c},\tilde{\gamma},\tilde{b},\tilde{B}]$.

For the case that $[c,\gamma,b,B] \neq [\tilde{c},\tilde{\gamma},\tilde{b},\tilde{B}]$, by an analogous argument as the above, we get finally the conclusion of the theorem. □

Acknowledgments This work was supported by National Natural Science Foundation of China (Grant Nos. 10971013,61071066).

References

1. Bedford, T.: Application of dynamical systems theory to fractals—a study of cookie-cutter sets. In: Bélair, J., Dubue, S. (eds.) Fractal Geometry and Analysis, pp. 1–44. Kluwer, Amsterdam (1991)
2. Falconer, K.J.: Fractal Geometry-Mathematical Foundation and Applications. Wily, London (1990)
3. Falconer, K.J.: Techniques in Fractal Geometry. Wiley, New York (1997)
4. Fan, S., Liu, Q., Wen. Z.-Y.: Gibbs-Like Measure for Spectrum of a Class of One-Dimensional Schrödinger Operator with Sturm Potentials. http://arxiv.org/PS_cache/arxiv/pdf/0909/0909.2301v1.pdf
5. Feng, D., Hua, S., Wen, Z.-Y.: Some relations between pre-packing measure and packing measure. Bull. London Math. Soc. **31**, 665 (1999)
6. Liu, Q., Wen Z.-Y.: On the shape of Cantor sets. Math. Proc. Camb. Phil. Soc. **139**, 541–553 (2005)
7. Ma, J., Rao, H., Wen, Z.-Y.: Dimensions of cookie-cutter-like sets. Sci. China Ser. A. **44**(11), 1400–1412 (2001)
8. Mattila, P.: Geometry of Sets and Measures in Euclidean Spaces. Cambridge University Press, Cambridge (1995)

9. Mauldin, R.D., Williams, S.C.: Hausdorff dimension in graph directed constructions. Trans. Am. Math. Soc. **309**(2), 811–829 (1988)
10. Palis, J., Takens, F.: Hyperbolicity and sensitive chaotic dynamics at homoclinic bifurcations: Fractal dimensions and infinitely many attractors. In: Cambridge Studies in Advanced Mathematics, vol. 35. Cambridge University Press, Cambridge (1993)
11. Tricot, C.: Two definitions of fractal dimensions. Math. Proc. Camb. Phil. Soc. **91**(1), 57–74 (1982)
12. Wen, Z.-Y.: Mathematical Foundations of Fractal Geometry. Shang Scientific and Technological Education Publishing House, Shanghai (2000)

Recent Developments on Fractal Properties of Gaussian Random Fields

Yimin Xiao

Abstract We review some recent developments in studying fractal and analytic properties of Gaussian random fields. It is shown that various forms of strong local nondeterminism are useful for studying many fine properties of Gaussian random fields. A list of open questions is included.

1 Introduction

Fractal properties of Brownian motion and Lévy processes have been studied since the pioneering works of [66, 101] and have become significant part of the theory on stochastic processes. We refer to [102, 121], and the references therein for further information.

In recent years, there has been an increased interest in investigating various properties of random fields. On one hand, random fields arise naturally in probability theory, stochastic partial differential equations, and studies of Markov processes. On the other hand, they are extensively applied as stochastic models in various scientific areas such as image processing, physics, biology, hydrology, geostatistics, and spatial statistics. However, compared with the rich theory on fine properties of Brownian motion and Lévy processes, the progress in studying random fields has been relatively slow. One of the main difficulties is the lack of powerful technical tools such as Markov property and stopping times.

In this chapter we survey some recent studies on sample path properties of Gaussian random fields. We will mainly focus on results which are either based on various properties of strong local nondeterminism or on new concepts in fractal geometry (such as packing dimension profiles).

Y. Xiao (✉)
Department of Statistics and Probability, Michigan State University, A-413 Wells Hall, East Lansing, MI 48824, USA
e-mail: xiao@stt.msu.edu

J. Barral and S. Seuret (eds.), *Further Developments in Fractals and Related Fields*, Trends in Mathematics, DOI 10.1007/978-0-8176-8400-6_13,

The rest of this chapter is organized as follows. Section 2 is an introduction on Gaussian random fields in which we recall the notions of strong local nondeterminism and provide some typical examples. In Sect. 3, we discuss analytic properties of Gaussian random fields, such as exact modulus of continuity, law of the iterated logarithm (LIL), Chung's LIL, existence, and regularity of local times. Section 4 is on fractal properties of Gaussian random fields. We illustrate how anisotropy in the time variable and/or space variable may affect the fractal structures of the random fields. More specifically we provide Hausdorff and packing dimension results on the images, graphs, and inverse images and set of intersections, set of exceptional oscillations, and of Gaussian random fields. There are many open problems on analytic and geometric properties of Gaussian random fields. We list some of them in Sects. 3 and 4.

We end the introduction with some general notation. Throughout this chapter, we consider random fields $\{X(t), t \in \mathbb{R}^N\}$ which take values in $\mathbb{R}^d$, and we call them (N,d)-random fields. We use $|\cdot|$ and $\langle \cdot, \cdot \rangle$ to denote, respectively, the Euclidean norm and the inner product in $\mathbb{R}^N$ (or $\mathbb{R}^d$). The Lebesgue measure in $\mathbb{R}^N$ is denoted by λ_N. A point $t \in \mathbb{R}^N$ is written as $t = (t_1, \ldots, t_N)$, or $\langle t_j \rangle$ or as $\langle c \rangle$ if $t_1 = \cdots = t_N = c$. For any $s, t \in \mathbb{R}^N$ such that $s_j < t_j$ $(j = 1, \ldots, N)$, $[s,t] = \prod_{j=1}^N [s_j, t_j]$ is called a closed interval (or a rectangle). We will let $\mathscr{A}$ denote the class of all closed intervals in $\mathbb{R}^N$. For two functions f and g defined on $T \subseteq \mathbb{R}^N$, the notation $f(t) \asymp g(t)$ for $t \in T$ means that the function $f(t)/g(t)$ is bounded from below and above by positive constants that do not depend on $t \in T$.

We will use c to denote an unspecified positive and finite constant which may not be the same in each occurrence. More specific constants are numbered as $c_1, c_2, \ldots$.

2 Gaussian Random Fields

Two of the most important Gaussian random fields are, respectively, the Brownian sheet $W = \{W(t), t \in \mathbb{R}^N_+\}$ and fractional Brownian motion $B^H = \{B^H(t),\ t \in \mathbb{R}^N\}$ of index $H \in (0,1)$, and they have been under extensive investigations for several decades. Both of them are centered (N,d)-Gaussian random fields, the former has covariance function

$$\mathbb{E}[W_i(s) W_j(t)] = \delta_{ij} \prod_{k=1}^N s_k \wedge t_k, \quad \forall s, t \in \mathbb{R}^N_+, \tag{1}$$

and the latter has covariance function

$$\mathbb{E}[B_i^H(s) B_j^H(t)] = \frac{1}{2} \delta_{ij} \Big(|s|^{2H} + |t|^{2H} - |s-t|^{2H} \Big), \quad \forall s, t \in \mathbb{R}^N. \tag{2}$$

In the above $\delta_{ij} = 1$ if $i = j$ and 0 if $i \neq j$. The Brownian sheet W and fractional Brownian motion B^H are natural multiparameter extensions of Brownian motion in $\mathbb{R}^d$ and have played important roles in probability theory and in various applications. It is known that there are some fundamental differences between W and B^H. For example, it follows from Eq. (1) that the increments of W over disjoint intervals of

$\mathbb{R}_+^N$ are independent and, along each direction of the axis, W is a rescaled Brownian motion in $\mathbb{R}^d$. On the other hand, Eq. (2) implies that B^H is H-self-similar and has stationary increments. Moreover, B^H is isotropic in the sense that $B^H(s) - B^H(t)$ depends only on the Euclidean distance $|s-t|$. For further information on the Brownian sheet W and fractional Brownian motion, we refer to Adler [1], Kahane [43], Khoshnevisan [51], and Samorodnitsky and Taqqu [92].

Several classes of anisotropic Gaussian random fields have been introduced and studied for theoretical and application purposes. For instance, [46] introduced fractional Brownian sheets and studied some of their regularity properties. Bonami and Estrade [14], Biermé et al. [12], Li and Xiao [68], and Xue and Xiao [127] constructed several classes of anisotropic Gaussian random fields with stationary increments and certain operator-scaling properties. Anisotropic Gaussian random fields also arise naturally in stochastic partial differential equations (see, e.g., [20, 41, 82, 84]) and as spatial or spatiotemporal models in statistics (e.g., [17, 35, 95]).

It is known that, compared with isotropic Gaussian fields such as fractional Brownian motion, the probabilistic and geometric properties of anisotropic Gaussian random fields are much richer (see [4, 109, 110, 112, 123, 127]) and their estimation and prediction problems are more difficult to study.

There are three kinds of anisotropy, namely, time-variable anisotropy, space-variable anisotropy, and the anisotropy in both variables. Typical examples of time-anisotropic Gaussian random fields are fractional Brownian sheets introduced by [46], the solution of the stochastic heat equation driven by space–time white noise (see [82]), and operator-scaling Gaussian random fields with stationary increments constructed by [12]. Examples of space-anisotropic Gaussian random fields include those in [1, 113], Gaussian fields with fractional Brownian motion components in [115], and the operator-fractional Brownian motion studied by [26,74]. A large class of (N,d)-random fields which are anisotropic in both space and time variables have been constructed and studied by [68].

In the following two sections, we provide a brief discussion on the properties of anisotropy in the space and time variables of (N,d)-random fields.

2.1 Space-Anisotropic Gaussian Random Fields

Let $X = \{X(t), t \in \mathbb{R}^N\}$ be an (N,d)-Gaussian random field defined by

$$X(t) = (X_1(t), \ldots, X_d(t)), \tag{3}$$

where the coordinate processes $X_1, \ldots, X_d$ are assumed to be stochastically continuous. For every $i = 1, \ldots, d$, let

$$\sigma_i^2(t,h) = \mathbb{E}\Big[\big(X_i(t+h) - X_i(t)\big)^2\Big], \qquad t,h \in \mathbb{R}^N.$$

Then for every $t \in \mathbb{R}^N$, $\sigma_i(t,h) \to 0$ as $|h| \to 0$. Many probabilistic and geometric properties of X are determined by the asymptotic properties of $\sigma_i^2(t,h)$. If, as $h \to 0$, $\sigma_i^2(t,h) \to 0$ with different rates for $i = 1,\ldots,d$, then the coordinate processes $X_1,\ldots,X_d$ have different asymptotic properties which can affect the properties of X in various ways. In this case, we say that X is anisotropic in the space variable (or, simply, space anisotropic). An important class of space-anisotropic random fields are those satisfying the operator-self-similarity. An (N,d)-random field X is called *operator-self-similar in the space variable* if there exists a $d \times d$ matrix $D = (d_{ij})$ such that for all constants $c > 0$,

$$\{X(ct), t \in \mathbb{R}^N\} \stackrel{d}{=} \{c^D X(t), t \in \mathbb{R}^N\}. \tag{4}$$

In the above and in the sequel, $\stackrel{d}{=}$ denotes equality of all finite-dimensional distributions and c^D is the linear operator on $\mathbb{R}^d$ defined by

$$c^D = \sum_{n=0}^{\infty} \frac{(\ln c)^n D^n}{n!}.$$

The linear operator D is called a space-variable self-similarity exponent (which may not be unique).

Mason and Xiao [74] constructed a class of operator-self-similar Gaussian random fields with stationary increments, which are called operator-fractional Brownian motions, and they showed that the Hausdorff dimension of the image $X(E)$ is determined by the positive parts of the eigenvalues of D, the self-similarity exponent of X. To study the effect of space anisotropy on fractal properties of X, we can first work with the special example of $X = \{X(t), t \in \mathbb{R}^N\}$, which is defined by Eq. (3), where $X_1,\ldots,X_d$ are independent N-parameter fractional Brownian motions in $\mathbb{R}$ with indices $\alpha_1,\ldots,\alpha_d$, respectively. Then the (N,d)-Gaussian field X is operator-self-similar with exponent $D = (d_{ij})$ which is the diagonal matrix with $d_{ii} = \alpha_i$ for $i = 1,\ldots,d$. When $\alpha_1,\ldots,\alpha_d$ are not the same, X is anisotropic in the space variable. We call X a Gaussian random field with independent fractional Brownian motion components. Didier and Pipiras [26] consider more general framework and provide a characterization of all operator-fractional Brownian motions by means of their integral representations in the spectral and time domains.

2.2 Time-Anisotropic Gaussian Random Fields

Let $X = \{X(t), t \in \mathbb{R}^N\}$ be an (N,d)-random field defined by Eq. (3). We say that X is anisotropic in the time variable (or time anisotropic) if the coordinate processes are (approximately) identically distributed and for some $1 \le i \le d$, the random field $\{X_i(t), t \in \mathbb{R}^N\}$ has different distributional properties along different directions of $\mathbb{R}^N$.

Analogous to the space-anisotropy case, an (N,d)-random field $\{X(t), t \in \mathbb{R}^N\}$ is called *operator-self-similar in the time variable* if there exists an $N \times N$ matrix E such that for all constants $c > 0$,

$$\{X(c^E t), t \in \mathbb{R}^N\} \stackrel{d}{=} \{cX(t), t \in \mathbb{R}^N\}. \tag{5}$$

The linear operator E is called a time-variable self-similarity exponent (which may not be unique).

A typical example of Gaussian random fields which are operator-self-similar in the time variable is fractional Brownian sheets (cf. [46]). For a given vector $\mathbf{H} = (H_1, \ldots, H_N) \in (0,1)^N$, an $(N,1)$-fractional Brownian sheet $W_0^{\mathbf{H}} = \{W_0^{\mathbf{H}}(t), t \in \mathbb{R}^N\}$ with index $\mathbf{H}$ is a real-valued, centered Gaussian random field with covariance function given by

$$\mathbb{E}\big[W_0^{\mathbf{H}}(s) W_0^{\mathbf{H}}(t)\big] = \prod_{j=1}^{N} \frac{1}{2}\Big(|s_j|^{2H_j} + |t_j|^{2H_j} - |s_j - t_j|^{2H_j}\Big), \qquad s, t \in \mathbb{R}^N. \tag{6}$$

It follows from Eq. (6) that $W_0^{\mathbf{H}}$ is operator-self-similar in the time variable with exponent $E = (a_{ij})$, which is the $N \times N$ diagonal matrix with $a_{ii} = (NH_i)^{-1}$ for all $1 \le i \le N$ and $a_{ij} = 0$ if $i \ne j$. Note that, in the direction of ith coordinate, $W_0^{\mathbf{H}}(s)$ is a re-scaled fractional Brownian motion of index H_i. Hence, if $H_1, \ldots, H_N$ are not the same, then $W_0^{\mathbf{H}}$ is a time-anisotropic Gaussian field. Other important examples of time-anisotropic random fields include the solution to stochastic heat equation driven by space–time white noise [41, 82] and those with stationary increments constructed by [12, 123].

Recently Li and Xiao [68] have extended the notions of operator-self-similarity and operator scaling to multivariate random fields by allowing scaling of the random field in both "time" domain and state space by linear operators. For any given $N \times N$ matrix E and $d \times d$ matrix D, they construct a large class of (N,d)-Gaussian or stable random fields $X = \{X(t), t \in \mathbb{R}^N\}$ such that for all constants $c > 0$

$$\{X(c^E t), t \in \mathbb{R}^N\} \stackrel{d}{=} \{c^D X(t), t \in \mathbb{R}^d\}. \tag{7}$$

If $E = I$, the identity matrix, then Eq. (7) reduces to Eq. (4). If $D = I$, then Eq. (7) reduces to Eq. (5).

Similarly to [74], Xiao [123] proved that the Hausdorff and packing dimensions of the range, graphs and level sets of a Gaussian random field X which satisfies Eq. (5) are determined by the real parts of the eigenvalues of E. However, fractal properties of (N,d)-Gaussian random fields which satisfy Eq. (7) have not been studied in general.

2.3 Assumptions

Now let us specify the class of Gaussian random fields to be considered in this chapter.

Let $X = \{X(t), t \in \mathbb{R}^N\}$ be a Gaussian random field with values in $\mathbb{R}^d$ defined by Eq. (3). We assume that $X_1, \ldots, X_d$ are independent copies of a real-valued, centered Gaussian random field $X_0 = \{X_0(t), t \in \mathbb{R}^N\}$. Many sample path properties of X can be determined by the following function:

$$\sigma^2(s,t) = \mathbb{E}\big(X_0(s) - X_0(t)\big)^2, \qquad \forall s,t \in \mathbb{R}^N. \tag{8}$$

Let $I \in \mathscr{A}$ be a fixed closed interval and we will consider various sample path properties of $X(t)$ when $t \in I$. We say that X_0 is approximately isotropic on I if there is a function $g : \mathbb{R}_+ \to \mathbb{R}_+$ such that

$$\sigma^2(s,t) \asymp g(|s-t|), \qquad \forall\, s,t \in I.$$

For simplicity we will mostly assume $I = [\varepsilon, 1]^N$, where $\varepsilon \in (0,1)$ is fixed, or $I = [0,1]^N$.

Let $\mathbf{H} = (H_1, \ldots, H_N) \in (0,1)^N$ be a fixed vector. Let ρ be the metric on $\mathbb{R}^N$ defined by

$$\rho(s,t) = \sum_{j=1}^{N} |s_j - t_j|^{H_j}, \qquad \forall\, s,t \in \mathbb{R}^N. \tag{9}$$

For any $r > 0$ and $t \in \mathbb{R}^N$, we denote by $B_\rho(t,r) = \{s \in \mathbb{R}^N : \rho(s,t) \le r\}$ the closed ball in the metric ρ.

As in [123], we will make use of the following general conditions on X_0:

(C1) There exists a positive constant $c \ge 1$ such that

$$c^{-1} \rho(s,t)^2 \le \sigma^2(s,t) \le c\,\rho(s,t)^2, \qquad \forall\, s,t \in I. \tag{10}$$

(C2) There exists a constant $c > 0$ such that for all $s,t \in I$,

$$\mathrm{Var}\big(X_0(t) \big| X_0(s)\big) \ge c\,\rho(s,t)^2.$$

Here and in the sequel, $\mathrm{Var}\big(Y\big|Z\big)$ denotes the conditional variance of Y given Z.

(C3) There exists a constant $c > 0$ such that for all integers $n \ge 1$ and all u, $t^1, \ldots, t^n \in I$ (and $u \ne 0$ if $0 \in I$),

$$\mathrm{Var}\Big(\, X_0(u) \,|\, X_0(t^1), \ldots, X_0(t^n)\Big) \ge c \sum_{j=1}^{N} \min_{0 \le k \le n} \big|u_j - t_j^k\big|^{2H_j}, \tag{11}$$

where $t_j^0 = 0$ for every $j = 1, \ldots, N$.

(C4) There exists a constant $c > 0$ such that for all integers $n \ge 1$ and all u, $t^1, \ldots, t^n \in I$ (and $u \ne 0$ if $0 \in I$),

$$\mathrm{Var}\big(\, X_0(u) \,|\, X_0(t^1), \ldots, X_0(t^n)\big) \ge c \min_{0 \le k \le n} \rho(u, t^k)^2, \tag{12}$$

where $t^0 = 0$.

Remark 1. The following are some remarks about the above conditions:

- Under Condition (C1), X_0 has a version which has continuous sample functions on I almost surely. Henceforth we will assume that the Gaussian random field X has continuous sample paths.
- Condition (C2) is referred to as "two-point" local nondeterminism in the metric ρ [or with indices $\mathbf{H} = (H_1, \dots, H_N) \in (0,1)^N$]. Together with (C1), it is useful for determining the fractal dimensions of many random sets generated by X.
- Condition (C3) is weaker than (C4). Following the terminology in [57], (C3) is called the property of sectorial local nondeterminism. Condition (C4) is called the strong local nondeterminism in the metric ρ. These conditions are important for establishing sharp results on modulus of continuity, Chung's LIL, sharp Hölder conditions for the local times, and exact Hausdorff measure functions, among others.
- Pitt [89] proved that multiparameter fractional Brownian motion B^H satisfies (C4) with $\mathbf{H} = lH\rho$. Wu and Xiao [109] proved that a fractional Brownian sheet $W^{\mathbf{H}}$ satisfies condition (C3). We refer to [123, 127] for more examples of anisotropic Gaussian random fields which satisfy condition (C4).

  Recently, Luan and Xiao [71] have provided a general condition for a Gaussian random field X_0 with stationary increments to satisfy (C4) in terms of the spectral measure Δ of X_0. This condition can be applied even when Δ is singular, supported on a fractal set or on a discrete set such as $\mathbb{Z}^N$. Their theorem can be applied to prove that the solution of a fractional stochastic heat equation on the circle $\mathbb{S}_1$ (see [83, 105]) has the property of strong local nondeterminism in the space variable (at fixed time t). Hence fine properties of the sample functions of the solution can be obtained by using the results discussed below. Similarly, we can show that the spherical fractional Brownian motion on $\mathbb{S}_1$ introduced by Istas [42] is also strongly locally nondeterministic. These processes share local properties with ordinary fractional Brownian motion with appropriate indices.

3 Analytic Results

Sample functions of a Gaussian random field may present various fine properties such as continuity and differentiability. See [2, 72, 99].

For an anisotropic Gaussian random field, its sample function may be differentiable in certain directions, but not differentiable in other directions (see [127] for explicit criterion for Gaussian random fields with stationary increments) and may have rich (sometimes complicated) geometric structures. Our main objective is to characterize the analytic and geometric properties of a Gaussian random field C in terms of its parameter $\mathbf{H} = (H_1, \dots, H_N)$, if X satisfies some of the conditions (C1)–(C4).

Geometric properties of a Gaussian random field X are very closely related to the regularities (or irregularities) of the sample functions of X. In this section, we discuss analytic properties such as uniform and local moduli of continuity and local times of Gaussian random fields.

3.1 Exact Modulus of Continuity and LIL

Sample path continuity and Hölder regularity of Gaussian random fields have been studied by many authors. A powerful chaining argument leads to sharp upper bounds for uniform and local moduli of continuity of Gaussian processes in terms of metric entropy or majorizing measures. Here "sharp" means logarithmic correction factors can be obtained. See [2,72,99]. Sharp lower bounds for local and uniform moduli of continuity of Gaussian processes are discussed in [72, Chapter 7]. However, except for a few special cases such as certain one-parameter Gaussian processes [72], the Brownian sheet [85], and fractional Brownian motion [8], there have not been many explicit results on sharp lower bounds for uniform and local moduli of Gaussian random fields.

The following theorem on uniform modulus of continuity is proved in [76].

Theorem 1. *Let $X = \{X(t), t \in \mathbb{R}^N\}$ be real-valued and centered Gaussian random field which satisfies conditions (C1) and (C3). Then for every compact interval $I \subseteq \mathbb{R}^N$, there exists a positive and finite constant c_1, depending only on I and H_j, $(j = 1, \ldots, N)$ such that*

$$\lim_{\varepsilon \to 0+} \sup_{s,t \in I, \sigma(s,t) \le \varepsilon} \frac{|X(t) - X(s)|}{\sigma(s,t)\sqrt{\log(1 + \sigma(s,t)^{-1})}} = c_1 \quad a.s., \tag{13}$$

where $\sigma(s,t)$ is as in Eq. (8).

Notice that the limit in Eq. (13) exists almost surely due to monotonicity. So the real issue is to prove that the limit is a (nonrandom) constant which is positive and finite. Condition (C1) allows us to apply standard method (e.g., the entropy method) to derive an upper bound for $\sup_{s,t \in I, \sigma(s,t) \le \varepsilon} |X(t) - X(s)|$. Thus the 0-1 law in [72, Chapter 7] implies that the limit is nonrandom and finite. The hard part is to prove $c_1 > 0$; this is where the sectorial local nondeterminism (C3) plays an important role.

For the local modulus of continuity, [76] proved the following LIL.

Theorem 2. *Let $\{X(t), t \in \mathbb{R}^N\}$ be a real-valued, centered Gaussian random field with stationary increments and $X(0) = 0$. If X satisfies condition (C1) for $I = [0,1]^N$, then there is a positive constant c_2 such that for every $t_0 \in \mathbb{R}^N$ we have*

$$\limsup_{|\varepsilon| \to 0+} \sup_{s:|s_j| \le \varepsilon_j} \frac{|X(t_0 + s) - X(t_0)|}{\sigma(s)\sqrt{\log\log\left(1 + \prod_{j=1}^N |s_j|^{-H_j}\right)}} = c_2 \quad a.s., \tag{14}$$

where $\sigma^2(s) = \mathbb{E}\big[X(s)^2\big]$.

For the local oscillation of $X(t)$ with $t \in B_\rho(t_0, r)$, [76] proved the following result.

Theorem 3. *Let $X = \{X(t), t \in \mathbb{R}^N\}$ be a real-valued, centered Gaussian random field with stationary increments and $X(0) = 0$. If X satisfies condition (C1) for $I = [0,1]^N$, then there is a positive and finite constant c_3 such that for every $t_0 \in \mathbb{R}^N$ we have*

$$\lim_{r\to 0+} \sup_{s:\sigma(s)\le r} \frac{|X(t_0+s)-X(t_0)|}{\sigma(s)\sqrt{\log\log(1+\sigma(s)^{-1})}} = c_3 \quad a.s. \tag{15}$$

Remark 2. Noticed that the logarithmic factors in Theorems 2 and 3 are quite different, since (C1) implies $\sigma(s) \asymp \sum_{j=1}^N |s_j|^{H_j}$ as $s \to 0$ in Theorem 3, and the corresponding term $\prod_{j=1}^N |s_j|^{-H_j}$ in Theorem 2 is much bigger. This is due to the fact that the supremum in Eq. (14) is taken over a larger domain.

Theorems 2 and 3 cannot be applied directly to fractional Brownian sheet $W^{\mathbf{H}}$ because it does not have stationary increments in the ordinary sense. However, [76, Theorem 6.4] shows that Eqs. (14) and (15) still hold for all $t_0 \in [a,\infty)^N$ (where $a > 0$ is a constant). The oscillation behavior of $W^{\mathbf{H}}(t)$ at the origin $t_0 = 0$ is very different and is characterized by [108]. Together these results reveal the subtlety of the influence of anisotropy on fine properties of random fields.

Several interesting questions can be raised. Comparing Theorems 1–3, one can show as in [86] that there exists a random point t at which the local oscillation $\sup_{s:\sigma(s)\le r}|X(t+s)-X(t)|$ is unusually large, say, of the order $\sigma(s)\sqrt{\log(1+\sigma(s)^{-1})}$. Motivated by this, we define the following sets of "fast points":

$$F_1(\gamma) = \left\{ t \in I : \limsup_{|s|\to 0+} \frac{|X(t+s)-X(t)|}{\sigma(s)\sqrt{\log\left(1+\prod_{j=1}^N |s_j|^{-H_j}\right)}} \ge \gamma \right\} \tag{16}$$

and

$$F_2(\gamma) = \left\{ t \in I : \limsup_{|s|\to 0+} \frac{|X(t+s)-X(t)|}{\sigma(s)\sqrt{\log(1+\sigma(s)^{-1})}} \ge \gamma \right\}. \tag{17}$$

It follows from Theorems 2 and 3 and Fubini's theorem that both $F_1(\gamma)$ and $F_2(\gamma)$ have zero Lebesgue measure. It is interesting to study their Hausdorff and packing dimensions. In the case of $N = 1$ and X is Brownian motion, $F_1(\gamma) = F_2(\gamma)$ and its Hausdorff dimension was determined by [86] and further refined by [49]. Khoshnevisan et al. [53] developed a general method for studying limsup-type random fractals which is applicable not only to the set of fast points of Brownian motion but also to many other random sets defined by exceptional oscillation or growth, including the set of fast points of the Brownian sheet (see also [27] for another treatment of the set of fast points of two-parameter Brownian sheet W) and fractional Brownian motion, thick points of the sojourn measure of Brownian motion [25].

In the current random field setting, however, it is not known whether $F_1(\gamma)$ and $F_2(\gamma)$ have different fractal properties. Hence we formulate the following problem.

Problem 1. Determine the Hausdorff and packing dimensions of $F_1(\gamma)$ and $F_2(\gamma)$. For a given Borel set $E \subseteq \mathbb{R}^N$, when do we have $F_1(\gamma) \cap E \neq \varnothing$ and $F_2(\gamma) \cap E \neq \varnothing$?

Following [107], a point t_0 is called a singularity of X if the LIL fails at t_0. Hence the points in $F_1(\gamma)$ and $F_2(\gamma)$ are singularities of X. Walsh [107] studied the propagation of the singularities of the Brownian sheet and proved the following theorem. See also [129] for part (i).

Theorem 4. *Let $W = \{W(s,t), (s,t) \in \mathbb{R}_+^2\}$ be the real-valued Brownian sheet:*

(i) For each fixed $s \geq 0$,

$$\mathrm{P}\left\{ \limsup_{r\to 0} \frac{|W(s+r,t) - W(s,t)|}{\sqrt{2r\log\log(1/r)}} = \sqrt{t} \quad \text{for all } t \geq 0 \right\} = 1. \tag{18}$$

(ii) Let $t_0 > 0$ be fixed and let $S \geq 0$ be a random variable which is measurable with respect to $\sigma\{W(s,t) : s \geq 0, 0 \leq t \leq t_0\}$. Then for almost every $\omega \in \Omega$,

$$\limsup_{r\to 0} \frac{|W(S(\omega),t_0+r) - W(S(\omega),t_0)|}{\sqrt{2r\log\log(1/r)}} = \infty$$
$$\Longleftrightarrow \limsup_{r\to 0} \frac{|W(S(\omega),t+r) - W(S(\omega),t)|}{\sqrt{2r\log\log(1/r)}} = \infty \quad \text{for all } t \geq t_0. \tag{19}$$

Part (i) says that, at every fixed time $s \geq 0$, the law of iterated logarithm for the (rescaled) Brownian motion $W(\cdot,t)$ holds uniformly for all $t \geq 0$. This is a refinement of LIL for Brownian motion. Part (ii) is a converse of (i) and means that, if $S(\omega)$ is a random singularity of the Brownian motion $W(\cdot,t_0)$, then every point on the vertical ray $\{(S(\omega),t) : t \geq t_0\}$ is a singularity of W. This reveals specifically that the singularities of W propagate parallel to the coordinate axes.

Blath and Martin [13] have extended Theorem 4 to the two-parameter fractional Brownian sheet $W^{\mathbf{H}}$ with $H_1 = 1/2$ and $H_2 \in (0,1)$ (which they call semi-fractional Brownian sheet). The fact that, for any $t_2 > 0$, $\{W^{\mathbf{H}}(t_1,t_2), t_1 \geq 0\}$ is a Brownian motion plays a crucial role in their proofs.

Walsh [107] asked whether an analogous property holds for other Gaussian random fields such as Lévy's multiparameter Brownian motion $B^{1/2}$. As far as I know, this problem has not been solved. Hence we formulate the following problem.

Problem 2. We say that t_0 is a singularity of Gaussian random field X if the limsup in Eq. (15) is infinity. How do the singularities of X propagate?

3.2 *Chung's LIL and Modulus of Nondifferentiability*

The results in Sect. 3.1 are about large oscillations. For small local oscillation of X at $t_0 \in \mathbb{R}^N$, [70] proved the following Chung-type law of iterated logarithm. For earlier results on Chung's LIL for isotropic Gaussian random fields, as well as their connections to small ball probabilities, we refer to the excellent survey of [67].

Theorem 5. *Let $X = \{X(t), t \in \mathbb{R}^N\}$ be a Gaussian random field with stationary increments and $X(0) = 0$. If X satisfies conditions (C1) and (C4) on an interval $I \subseteq \mathbb{R}^N$, then there exists a positive and finite constant c_4 such that for every $t_0 \in \mathbb{R}^N$,*

$$\liminf_{r\to 0} \frac{\max_{\rho(t,t_0)\le r} |X(t) - X(t_0)|}{r(\log\log 1/r)^{-1/Q}} = c_4, \qquad a.s., \tag{20}$$

where $Q = \sum_{i=1}^N H_i^{-1}$.

When $N = 1$ this extends a result of [79]. Theorem 5 assumes that X has stationary increments and satisfies (C4) and thus is not applicable to other Gaussian random fields such as fractional Brownian sheets. Talagrand [96] and Zhang [128] proved the Chung's LIL for the Brownian sheet $W = \{W(s,t), (s,t) \in \mathbb{R}^2_+\}$ as "time" goes to ∞. Their arguments also prove the following Chun's LIL at $t_0 = 0$, which is quite different from Eq. (20).

Theorem 6. *There is a positive and finite constant c_5 such that*

$$\liminf_{r\to 0} \frac{(\log\log 1/r)^{1/2}}{r(\log\log\log 1/r)^{3/2}} \max_{0\le s,t\le r} |W(s,t)| = c_5 \qquad a.s. \tag{21}$$

The problem for proving a Chung's LIL for a general fractional Brownian sheet $W^{\mathbf{H}}$ with index $\mathbf{H} \in (0,1)^N$ has not been solved completely. For some interesting partial results, see [73]. By applying the result of [79] to the restriction of $W^{\mathbf{H}}$ to the direction of the jth coordinate, say $\{W^{\mathbf{H}}(1,\ldots,1,t_j,1,\ldots,1), t_j \in \mathbb{R}\}$, one can see that the Chung's LIL is analogous to that of a one-parameter fractional Brownian motion of Hurst index H_j.

It is also an open problem to establish a uniform version of Eq. (5) for Gaussian random fields. In the special case of fractional Brownian motion, we believe (cf. [116]) that there is a positive and finite constant c_6 such that

$$\liminf_{r\to 0} \sup_{t\in[0,1]^N} \frac{\max_{|s-t|\le r} |B^H(t) - B^H(s)|}{r^H(\log 1/r)^{-H/N}} = c_6, \qquad \text{a.s.} \tag{22}$$

Even though the lower bound can be easily proved (cf. [116]), the upper bound is more difficult. For the special case of $N = 1$, Eq. (22) has been proved recently by [39].

Similarly to Eqs. (16) and (17) one can define the set of exceptionally small oscillation

$$S(\gamma) = \left\{ t \in I : \liminf_{r\to 0+} \frac{\max_{\rho(t,t_0)\le r} |X(t) - X(t_0)|}{r(\log 1/r)^{-1/Q}} \le \gamma \right\} \tag{23}$$

and ask similar questions for $S(\gamma)$ as in Problem 1. The method of limsup random fractals developed by [53] should be useful in studying these problems.

3.3 Regularity of Local Times

The roughness or irregularity of sample functions of X can be reflected in the regularity (or smoothness) of the local times of X. This was first observed by Berman [9] who developed Fourier analytic method for studying the existence and continuity of local times of Gaussian processes. Berman [10] introduced the notion of "local nondeterminism" for Gaussian processes to overcome many difficulties caused by the lack of Markov property and to unify his methods for studying local times. Berman's work has been extended and strengthened in various ways. See [34, 122, 123] for more information.

We recall briefly the definition of local times. Let $Y(t)$ be a Borel vector field on $\mathbb{R}^p$ with values in $\mathbb{R}^q$. For any Borel set $T \subseteq \mathbb{R}^p$, the occupation measure of Y on T is defined as the following measure on $\mathbb{R}^q$:

$$\mu_T(\cdot) = \lambda_p\{t \in T : Y(t) \in \cdot\}.$$

If μ_T is absolutely continuous with respect to the Lebesgue measure λ_q, we say that $Y(t)$ has *local times* on T, and define its local time, $L(\cdot, T)$, as the Radon–Nikodým derivative of μ_T with respect to λ_q, i.e.,

$$L(x,T) = \frac{d\mu_T}{d\lambda_q}(x), \qquad \forall x \in \mathbb{R}^q.$$

In the above, x is called the *space variable*, and T is the *time variable*. Note that if Y has local times on T then for every Borel set $S \subseteq T$, $L(x,S)$ also exists.

Suppose we fix a rectangle $T = \prod_{i=1}^p [a_i, a_i + h_i] \subseteq \mathbb{R}^p$, where $a \in \mathbb{R}^p$ and $h \in \mathbb{R}^p_+$. If we can choose a version of the local time, still denoted by $L(x, \prod_{i=1}^p [a_i, a_i + t_i])$, such that it is a continuous function of $(x, t_1, \ldots, t_p) \in \mathbb{R}^q \times \prod_{i=1}^p [0, h_i]$, Y is said to have a *jointly continuous local time* on T. When a local time is jointly continuous, $L(x, \cdot)$ can be extended to a finite Borel measure supported on the level set

$$Y_T^{-1}(x) = \{t \in T : Y(t) = x\}; \tag{24}$$

see [1] for details. This makes local times a useful tool in studying fractal properties of Y.

When $X = \{X(t), t \in \mathbb{R}^N\}$ is an (N,d)-Gaussian random field with approximately isotropic increments (e.g., fractional Brownian motion), [116] proved sharp local and uniform modulus of continuity for the local time $L(x, \cdot)$ in the set variable. For simplicity, we focus on fractional Brownian motion $B^H = \{B^H(t), t \in \mathbb{R}^N\}$ with values in $\mathbb{R}^d$. The following theorem was proved by [5, 7]. See also [15] for the case of $N = 1$, where large deviation results for the local times and intersection local times are proved.

Theorem 7. *Let $B^H = \{B^H(t),\ t \in \mathbb{R}^N\}$ be a fractional Brownian motion in $\mathbb{R}^d$ with index $H \in (0,1)$. If $N > Hd$, then there exists a positive and finite constant c_7 such that for every $a \in \mathbb{R}^N$,*

$$\limsup_{r\to 0} \frac{L^*(B(a,r))}{r^{N-Hd}(\log\log(1/r))^{Hd/N}} = c_7, \quad \text{a.s.} \tag{25}$$

and for every $T > 0$, there is a positive and finite constant c_8 such that

$$\limsup_{r\to 0} \sup_{a\in[-T,T]^N} \frac{L^*(B(a,r))}{r^{N-Hd}(\log(1/r))^{Hd/N}} = c_8, \quad \text{a.s.} \tag{26}$$

In the above, $L^*(B(a,r)) = \max_{x\in\mathbb{R}^d} L(x, B(a,r))$.

Equation (25) gives the LIL for $L^*(B(a,r))$ and is used to derive an exact Hausdorff measure function for the level set of fractional Brownian motion in [6]. Their result significantly improves that in [116].

The existence and joint continuity of local times of a fractional Brownian sheet $W^{\mathbf{H}}$ with values in $\mathbb{R}^d$ and index $\mathbf{H} = (H_1,\ldots,H_N) \in (0,1)^N$ were studied by [126]. Ayache et al. [3] proved that the optimal condition for the joint continuity of the local times of $W^{\mathbf{H}}$ is $\sum_{j=1}^N H_j^{-1} > d$. Xiao [123] proved similar results for a class of Gaussian random fields with stationary increments which satisfy conditions (C1) and (C3). Wu and Xiao [112] provided a unified treatment by applying sectorial local nondeterminism to estimate high moments of local times and improved significantly the results in [3, 123].

When X is anisotropic, the problems for finding sharp local and uniform modulus of continuity for $L(x,\cdot)$ and $L^*(\cdot) = \max_{x\in\mathbb{R}^d} L(x,\cdot)$ in the set variable are more complicated and have not been solved. In the following we state the main result in [112].

First we give some notation. Henceforth we assume that $\mathbf{H} = (H_1,\ldots,H_N) \in (0,1)^N$ is fixed and

$$0 < H_1 \le \cdots \le H_N < 1. \tag{27}$$

When $\sum_{j=1}^N \frac{1}{H_j} > d$, there exists $\tau \in \{1,2,\ldots,N\}$ such that

$$\sum_{j=1}^{\tau-1} \frac{1}{H_j} \le d < \sum_{j=1}^{\tau} \frac{1}{H_j},$$

with the convention that $\sum_1^0(\cdot) \equiv 0$. We denote

$$\alpha := \sum_{j=1}^N \frac{1}{H_j} - d, \qquad \eta_\tau := \tau + H_\tau d - \sum_{j=1}^{\tau} \frac{H_\tau}{H_j} \tag{28}$$

and we will distinguish three cases:

Case 1. $\sum_{j=1}^{\tau-1} \frac{1}{H_j} < d < \sum_{j=1}^{\tau} \frac{1}{H_j}$

Case 2. $\sum_{j=1}^{\tau-1} \frac{1}{H_j} = d < \sum_{j=1}^{\tau} \frac{1}{H_j}$ and $H_{\tau-1} = H_\tau$

Case 3. $\sum_{j=1}^{\tau-1} \frac{1}{H_j} = d < \sum_{j=1}^{\tau} \frac{1}{H_j}$ and $H_{\tau-1} < H_\tau$

The first result is on local Hölder condition for $L^*(\cdot)$.

Theorem 8. *Let $X = \{X(t),\, t \in \mathbb{R}^N\}$ be an anisotropic Gaussian random field with values in $\mathbb{R}^d$ which satisfies conditions (C1) and (C3) on an interval $I \in \mathscr{A}$. Then there exist positive constants c_9 and c_{10} such that for every $a \in I$,*

$$\begin{aligned} &\limsup_{r\to 0} \frac{L^*(B_\rho(a,r))}{\varphi_1(r)} \le c_9, \quad \text{a.s.} \quad \textit{in Cases 1 and 2,} \\ &\limsup_{r\to 0} \frac{L^*(B_\rho(a,r))}{\varphi_2(r)} \le c_{10}, \quad \text{a.s.} \quad \textit{in Case 3,} \end{aligned} \tag{29}$$

where

$$\varphi_1(r) = r^\alpha \left(\log\log(1/r)\right)^{\eta_\tau},$$
$$\varphi_2(r) = r^\alpha \left(\log\log(1/r)\right)^{\eta_\tau} \log\log\log\left(1/r\right).$$

The second result is on uniform Hölder condition for $L^*(\cdot)$.

Theorem 9. *Under the same conditions as in Theorem 8, we have*

$$\begin{aligned} &\limsup_{r\to 0} \sup_{a\in I} \frac{L^*(B_\rho(a,r))}{\Phi_1(r)} \le c_{11}, \quad \text{a.s.} \quad \textit{in Cases 1 and 2,} \\ &\limsup_{r\to 0} \sup_{a\in I} \frac{L^*(B_\rho(a,r))}{\Phi_2(r)} \le c_{12}, \quad \text{a.s.} \quad \textit{in Case 3,} \end{aligned} \tag{30}$$

where c_{11} and c_{12} are positive and finite constants and

$$\Phi_1(r) = r^\alpha \left(\log(1/r)\right)^{\eta_\tau},$$
$$\Phi_2(r) = r^\alpha \left(\log(1/r)\right)^{\eta_\tau} \log\log\left(1/r\right).$$

In the special case of $H_1 = \cdots = H_N := H$, we have $\alpha = \frac{N}{H} - d$ and $\eta_\tau = Hd$. The above theorems give local and uniform Hölder conditions for $L^*(\cdot)$ in the Euclidean metric:

$$\limsup_{r\to 0} \frac{L^*(B(a,r))}{r^{N-Hd}(\log\log(1/r))^{Hd}} \le c_9, \quad \text{a.s.} \tag{31}$$

and

$$\limsup_{r\to 0}\sup_{a\in I}\frac{L^*(B_\rho(a,r))}{r^{N-Hd}(\log(1/r))^{Hd}}\leq c_{11},\qquad \text{a.s.} \tag{32}$$

Notice that the powers of $\log\log 1/r$ in Eqs. (25) and (31) are different. This is due to the different forms of strong local nondeterminism.

Unlike in Theorem 7, we do not know whether the results in Theorems 8 and 9 are sharp. Hence we raise the following question.

Problem 3. Under what conditions can one establish exact local and uniform moduli of continuity for the local time $L(x,\cdot)$ and $L^*(\cdot)$?

4 Fractal Properties

Now we turn to fractal properties of Gaussian random fields, which include Hausdorff and packing dimensions of random sets such as the images, graph, level sets, and the set of intersections. We also present uniform dimension results as well as exact Hausdorff and packing measure functions for the image sets. These latter results depend on properties of strong local nondeterminism. We refer to [30] or [75] for definitions and basic properties of Hausdorff and packing measures and corresponding dimensions.

Given an (N,d)-random field X, the following random sets generated by X are often random fractals:

1. Range or image set $X(E)=\{X(t):t\in E\}$, where $E\subseteq\mathbb{R}^N$
2. Graph $\mathrm{Gr}X(E)=\{(t,X(t)):t\in E\}$
3. Level set $X^{-1}(x)=\{t\in E:X(t)=x\}$, where $x\in\mathbb{R}^d$
4. Excursion set (or inverse image)

$$X^{-1}(F)=\big\{t\in\mathbb{R}^N: X(t)\in F\big\},$$

   where $F\subseteq\mathbb{R}^d$.
5. Set of k-multiple times

$$L_k=\big\{(t^1,\ldots,t^k)\in\mathbb{R}^{Nk}_{\neq}: X(t^1)=\cdots=X(t^k)\big\},$$

   where $\mathbb{R}^{Nk}_{\neq}=\big\{(t^1,\ldots,t^k)\in\mathbb{R}^{Nk}: t^1,\ldots,t^k \text{ are distinct}\big\}$.
6. Set of k-multiple points

$$M_k=\big\{x\in\mathbb{R}^d:\exists\,(t^1,\ldots,t^k)\in\mathbb{R}^{Nk}_{\neq}\ \text{such that}\ x=X(t^1)=\cdots=X(t^k)\big\}.$$

4.1 Hausdorff Dimension Results

Let $X = \{X(t), t \in \mathbb{R}^N\}$ be an (N,d)-Gaussian field defined by Eq. (3) such that the coordinate processes $X_1, \ldots, X_d$ are independent copies of a real-valued, centered Gaussian random field $X_0 = \{X_0(t), t \in \mathbb{R}^N\}$. Recall that we have assumed that $\mathbf{H} = (H_1, \ldots, H_N) \in (0,1)^N$ satisfies Eq. (27).

The Hausdorff and packing dimensions of the range, graph, level sets of fractional Brownian sheets, and, more generally, anisotropic Gaussian random fields which satisfy conditions (C1) and (C2) have been established in [4, 123].

Theorem 10. *Let $X = \{X(t), t \in \mathbb{R}^N\}$ be an (N,d)-Gaussian field defined in the above. Assume that X_0 satisfies condition (C1) with $I = [0,1]^N$. Then, with probability 1,*

$$\dim_{\mathrm{H}} X\big([0,1]^N\big) = \dim_{\mathrm{P}} X\big([0,1]^N\big) = \min\left\{d;\quad \sum_{j=1}^N \frac{1}{H_j}\right\} \tag{33}$$

and

$$\begin{aligned} &\dim_{\mathrm{H}} \mathrm{Gr}X\big([0,1]^N\big) = \dim_{\mathrm{P}} \mathrm{Gr}X\big([0,1]^N\big) \\ &\quad = \min\left\{ \sum_{j=1}^k \frac{H_k}{H_j} + N - k + (1-H_k)d,\ 1 \le k \le N;\ \sum_{j=1}^N \frac{1}{H_j} \right\} \\ &\quad = \begin{cases} \sum_{j=1}^N \frac{1}{H_j}, & \text{if } \sum_{j=1}^N \frac{1}{H_j} \le d, \\ \sum_{j=1}^k \frac{H_k}{H_j} + N - k + (1-H_k)d, & \text{if } \sum_{j=1}^{k-1} \frac{1}{H_j} \le d < \sum_{j=1}^k \frac{1}{H_j}, \end{cases} \end{aligned} \tag{34}$$

where $\sum_{j=1}^0 \frac{1}{H_j} := 0$.

Theorem 11. *Let $X = \{X(t), t \in \mathbb{R}^N\}$ be an (N,d)-Gaussian random field defined as above. Assume that X_0 satisfies conditions (C1) and (C2) on $I = [\varepsilon, 1]^N$. Then the following statements hold:*

(i) If $\sum_{j=1}^N \frac{1}{H_j} < d$, then for every $x \in \mathbb{R}^d$, $X^{-1}(x) \cap I = \emptyset$ a.s.
(ii) If $\sum_{j=1}^N \frac{1}{H_j} > d$, then for every $x \in \mathbb{R}^d$, with positive probability,

$$\begin{aligned} \dim_{\mathrm{H}} \big(X^{-1}(x) \cap I\big) &= \dim_{\mathrm{P}} \big(X^{-1}(x) \cap I\big) \\ &= \min\left\{ \sum_{j=1}^k \frac{H_k}{H_j} + N - k - H_k d,\ 1 \le k \le N \right\} \\ &= \sum_{j=1}^k \frac{H_k}{H_j} + N - k - H_k d, \quad \text{if } \sum_{j=1}^{k-1} \frac{1}{H_j} \le d < \sum_{j=1}^k \frac{1}{H_j}. \end{aligned} \tag{35}$$

For the inverse images, [11,123] provided conditions on F such that $\mathrm{P}\{X^{-1}(F)\cap I\neq\varnothing\}>0$ and proved the following result.

Theorem 12. *Let $X=\{X(t),t\in\mathbb{R}^N\}$ be an (N,d)-Gaussian random field as in Theorem 11 and let $F\subseteq\mathbb{R}^d$ be a Borel set such that* $\dim F\geq d-Q$. *Then*

$$\big\|\dim_{\mathrm{H}}\big(X^{-1}(F)\cap I\big)\big\|_{L^\infty(\mathrm{P})}=\min_{1\leq k\leq N}\left\{\sum_{j=1}^{k}\frac{H_k}{H_j}+N-k-H_k\big(d-\dim_{\mathrm{H}}F\big)\right\},\quad(36)$$

where, for any function $Y:\Omega\to\mathbb{R}_+$, $\|Y\|_{L^\infty(\mathrm{P})}$ is defined as

$$\|Y\|_{L^\infty(\mathrm{P})}=\sup\big\{\theta:Y\geq\theta\ \text{ on an event } E \text{ with } \mathrm{P}(E)>0\big\}.$$

Under an extra condition on F, Theorem 2.5 in [11] shows that with positive probability,

$$\begin{aligned}\dim_{\mathrm{H}}\big(X^{-1}(F)\cap I\big)&=\min_{1\leq k\leq N}\left\{\sum_{j=1}^{k}\frac{H_k}{H_j}+N-k-H_k(d-\dim_{\mathrm{H}}F)\right\}\\&=\sum_{j=1}^{k}\frac{H_k}{H_j}+N-k-H_k(d-\dim_{\mathrm{H}}F),\quad\text{if }\sum_{j=1}^{k-1}\frac{1}{H_j}\leq d-\dim_{\mathrm{H}}F<\sum_{j=1}^{k}\frac{1}{H_j}.\end{aligned}\quad(37)$$

There are several possible ways to strengthen and extend Theorems 10–12 and to ask further questions about these random sets. For example, it would be interesting to determine the exact Hausdorff and packing measure functions for the range $X([0,1]^N)$, graph $\mathrm{Gr}X([0,1]^N)$, and the level set $X^{-1}(x)$ and to characterize the hitting probabilities of these random sets. Further information on exact Hausdorff and packing measure functions will be provided in Sects. 4.5 and 4.6. Testard [104], Xiao [119,123], Bierme et al. [11], and Chen and Xiao [16] have provided necessary conditions and sufficient conditions on $E\subseteq\mathbb{R}^N$ or/and $F\subseteq R^d$ for $X([0,1]^N)\cap F\neq\varnothing$, or $\mathrm{Gr}X([0,1]^N)\cap(E\times F)\neq\varnothing$ with positive probability. See Sect. 4.7 below. However, except in a few special cases, the following questions are still open.

Problem 4. Find necessary and sufficient conditions on $F\subseteq\mathbb{R}^d$ or $E\subseteq\mathbb{R}^N$ for $X([0,1]^N)\cap F\neq\varnothing$, $\mathrm{Gr}X([0,1]^N)\cap(E\times F)\neq\varnothing$ with positive probability.

For results on the Brownian sheet, the hitting probabilities of the range and level sets have been completely characterized in [55,57]. The corresponding problem for the graph set is more complicated and has only been solved for Brownian motion; see [60] and the references therein for further information.

In the following, we consider the natural questions to find the Hausdorff, Fourier and packing dimensions of the image set $X(E)$, where $E\subseteq\mathbb{R}^N$ is an arbitrary Borel set (typically, a fractal set). It is not hard to see that, due to the anisotropy of X, the Hausdorff dimension of $X(E)$ cannot be determined by $\dim_{\mathrm{H}}E$ and the index $\mathbf{H}$ alone (see [109]). This is in contrast with the cases of fractional Brownian motion or the Brownian sheet.

To determine the Hausdorff dimension of $X(E)$ for an arbitrary Borel set E, [109, 123] make use of Hausdorff dimension $\dim_{\mathrm{H}}^{\rho}$ on the metric space $(\mathbb{R}^N, \rho)$, where ρ is defined in Eq. (9), and prove the following theorem.

Theorem 13. *Let $X = \{X(t), t \in \mathbb{R}^N\}$ be an (N,d)-Gaussian field as in Theorem 10. Then for every Borel set $E \subseteq \mathbb{R}^N$,*

$$\dim_{\mathrm{H}} X(E) = \min\left\{d, \dim_{\mathrm{H}}^{\rho} E\right\}, \qquad a.s. \tag{38}$$

The Fourier and packing dimensions of $X(E)$ will be discussed in Sects. 4.2 and 4.3 below. We end this section with the following remark.

Remark 3. Note that the results in this section and in most of the subsequent sections are concerned with Gaussian random fields which may be time anisotropic, but not space anisotropic. Hausdorff dimensions for the range, graph, level sets, and other properties of Gaussian random fields which are space anisotropic have been considered in [1, 18, 19, 114, 115]. The results are different, and there are still many open questions. For example, for space anisotropic Gaussian random fields, the Hausdorff dimension of $X^{-1}(F)$ and the Fourier dimensions (see Sect. 4.2 below) of the images have not been determined. Moreover, for (N,d)-random fields which are anisotropic in both time and space, little has been known about their fractal properties.

4.2 The Fourier Dimension and Salem Sets

Besides Hausdorff and packing dimensions, one can define the Fourier dimension of a set $F \subseteq \mathbb{R}^d$, which is related to the asymptotic behavior of the Fourier transforms of the probability measures carried by F.

Let us recall from [43] the definitions of Fourier dimension and Salem set. Given a constant $\beta \geq 0$, a Borel set $F \subseteq \mathbb{R}^d$ is said to be an M_β-set if there exists a probability measure ν on F such that

$$\left|\widehat{\nu}(\xi)\right| = o(|\xi|^{-\beta}) \quad \text{as } \xi \to \infty, \tag{39}$$

where $\widehat{\nu}$ denotes the Fourier transform of ν. The asymptotic behavior of $\widehat{\nu}(\xi)$ at infinity carries some information about the geometry of F. It can be verified that (i) if $\beta > d/2$ in Eq. (39), then $\widehat{\nu} \in L^2(\mathbb{R}^d)$ and, consequently, F has positive d-dimensional Lebesgue measure; (ii) if $\beta > d$, then $\widehat{\nu} \in L^1(\mathbb{R}^d)$. Hence ν has a continuous density function which implies that F has interior points.

For any Borel set $F \subseteq \mathbb{R}^d$, the Fourier dimension of F, denoted by $\dim_{\mathscr{F}} F$, is defined as

$$\dim_{\mathscr{F}} F = \sup\left\{\gamma \in [0,d] : \ F \text{ is an } \ M_{\gamma/2}\text{-set}\right\}. \tag{40}$$

It follows from Frostman's theorem that $\dim_{\mathscr{F}} F \le \dim_{\mathrm{H}} F$ for all Borel sets $F \subseteq \mathbb{R}^d$. The strict inequality may hold. For example, the Fourier dimension of triadic Cantor set is 0, but its Hausdorff dimension is $\log 2/\log 3$. It has been known that the Hausdorff dimension $\dim_{\mathrm{H}} F$ describes a metric property of F, whereas the Fourier dimension measures an arithmetic property of F. As a further example of this aspect, we mention that every set $F \subseteq \mathbb{R}^d$ with positive Fourier dimension generates $\mathbb{R}^d$ as a group (cf. [43]).

A Borel set $F \subseteq \mathbb{R}^d$ is called a *Salem set* if $\dim_{\mathscr{F}} F = \dim_{\mathrm{H}} F$. Such sets are of importance in studying the problem of uniqueness and multiplicity for trigonometric series (cf. [130, Chapter 9] and [45]) and the restriction problem for the Fourier transforms (cf. [77]).

The images of many random fields are Salem sets. For fractional Brownian motion $B^H = \{B^H(t), t \in \mathbb{R}^N\}$ with values in $\mathbb{R}^d$, [43, 44] proved that, for every Borel set $E \subseteq \mathbb{R}^N$ with $\dim_{\mathrm{H}} E \le Hd$, $B^H(E)$ is almost surely a Salem set with Fourier dimension $\frac{1}{H}\dim_{\mathrm{H}} E$.

The Fourier dimensions of the images of various Gaussian random fields have been studied by [56] for the Brownian sheet, [93] for a large class of approximately isotropic Gaussian random fields, and by [109] for fractional Brownian sheets. We mention that the argument in [109] is based on the property of sectorial local nondeterminism and can be applied more generally.

It would be interesting to know whether other random sets such as the graph, level sets, or inverse images of a Gaussian random field are Salem sets. Recently [33] showed that the zero set of one-dimensional Brownian motion is a Salem set of Fourier dimension 1/2 by studying the Fourier transform of the local times of Brownian motion. Their result is related to that of [44] for the images of stable Lévy processes because the zero set of Brownian motion equals, up to a countable set, the image of a stable subordinator of index $\frac{1}{2}$. However, for a Gaussian random field X, no direct connection between its level set $X^{-1}(x)$ and the image of a tractable random field has been established.

4.3 Packing Dimension Results

Packing measure and packing dimension were introduced by [103, 106] as dual concepts to Hausdorff measure and dimension. Since Hausdorff and packing dimensions of a set (or a measure) are determined by different geometric aspects of the set (or the measure), many random sets have different values for their Hausdorff and packing dimensions. To understand better the fractal nature of a set, it is important to determine both Hausdorff and packing dimensions of the set.

As we have seen in Sect. 4.1, for a Gaussian random field X which satisfies Conditions (C1) and (C2), the Hausdorff and packing dimensions of $X([0,1]^N)$, $\mathrm{Gr}X([0,1]^N)$, and $X^{-1}(x)$ coincide. However, [29, 122, 124] have shown that, for many Gaussian random fields, the Hausdorff and packing dimensions of these random sets may differ.

In this section, we mainly consider the packing dimension of $X(E)$, where X is a Gaussian field as in Sect. 4.1 and $E \subseteq \mathbb{R}^N$ is an arbitrary set.

In the special case of Brownian motion $W = \{W(t), t \in \mathbb{R}_+\}$ in $\mathbb{R}^d$, [88] proved that if $d \geq 2$, then with probability 1,

$$\dim_{\rm P} W(E) = 2\dim_{\rm P} E \quad \text{ for every Borel set } \; E \subseteq \mathbb{R}_+. \tag{41}$$

This not only determines the packing dimension of the image $W(E)$ but also says that the exceptional null probability event [on which Eq. (41) fails] does not depend on E. Hence Eq. (41) is called a *uniform dimension result*; see Sect. 4.4 for more information. However, when $d = 1$, [100] constructed a compact set $E \subseteq \mathbb{R}_+$ such that $\dim_{\rm P} W(E) < 2\dim_{\rm P} E$ a.s.; they also obtained the best possible lower bound for $\dim_{\rm P} W(E)$. Xiao [117] solved the problem of finding $\dim_{\rm P} W(E)$ by proving

$$\dim_{\rm P} W(E) = 2{\rm Dim}_{1/2} E \quad \text{ a.s.}, \tag{42}$$

where ${\rm Dim}_s E$ is the packing dimension profile of E defined by [31] [which is defined by replacing ρ in Eq. (43) below by the Euclidean metric]. We mention that [54] have recently introduced more general notion of packing dimension profiles and determined the packing dimension of the images of an arbitrary Lévy process.

Xiao [117] proved results analogous to Eq. (42) for fractional Brownian motion B^H and the Brownian sheet. Khoshnevisan and Xiao [59] provided a connection between $B^H(E)$ and the packing dimension profile of [38] and thus showed that the packing dimension profile of [31] coincides with that of [38].

In order to determine the packing dimension of $X(E)$ for time-anisotropic Gaussian field X as in Sect. 4.1, [29] extended the packing dimensional profile of [31] to the metric space $(\mathbb{R}^N, \rho)$ and define, for any finite Borel measure μ on $\mathbb{R}^N$, the s-dimensional packing dimension profile of μ in the metric ρ as

$${\rm Dim}_s^{\rho} \mu = \sup\left\{\beta \geq 0 : \; \liminf_{r \to 0} \frac{F_{s,\rho}^{\mu}(x,r)}{r^{\beta}} = 0 \; \text{ for } \mu\text{-a.a. } x \in \mathbb{R}^N \right\}, \tag{43}$$

where, for any $s > 0$, $F_{s,\rho}^{\mu}(x,r)$ is the s-dimensional potential of μ in metric ρ defined by

$$F_{s,\rho}^{\mu}(x,r) = \int_{\mathbb{R}^N} \min\left\{1, \; \frac{r^s}{\rho(x,y)^s}\right\} {\rm d}\mu(y). \tag{44}$$

For any Borel set $E \subseteq \mathbb{R}^N$, the s-dimensional packing dimension profile of E in the metric ρ is defined by

$${\rm Dim}_s^{\rho} E = \sup\left\{{\rm Dim}_s^{\rho} \mu : \; \mu \in \mathcal{M}_c^+(E)\right\}, \tag{45}$$

where $\mathcal{M}_c^+(E)$ denotes the family of finite Borel measures with compact support in E.

The following packing dimension analogue of Theorem 13 is a special case of Theorem 4.5 in [29].

Theorem 14. *Let $X = \{X(t), t \in \mathbb{R}^N\}$ be an (N,d)-Gaussian field as in Theorem 10. Then for every Borel set $E \subseteq \mathbb{R}^N$,*

$$\dim_{\rm P} X(E) = {\rm Dim}_d^{\rho} E, \qquad a.s. \tag{46}$$

In the aforementioned references, the measure-theoretic approach to packing dimension and packing dimension profiles has been essential. Even though in this chapter we focus on fractal properties of random sets, similar questions can be investigated for related random measures as well. We refer to [32, 94] for some recent results.

The packing dimensions of other random sets such as ${\rm Gr}X(E)$ and $X^{-1}(F)$ are not known in general. The former is related to the following problem.

Problem 5. What is the packing dimension of $X(E)$ if X is space anisotropic? In particular, what is $\dim_{\rm P} {\rm Gr} B^H(E)$?

Motivated by [54], we expect that an answer to the above question is to use the packing dimension profile associated with the kernels $\kappa = \{\kappa_r, r > 0\}$ defined by

$$\kappa_r(s,t) = {\rm P}\big\{\|X(s) - X(t)\| \le r\big\}, \quad \forall s,t \in \mathbb{R}^N. \tag{47}$$

When the coordinate processes of X are approximately independent and have approximately scaling properties, κ is comparable with the kernel

$$\widetilde{\kappa}_r(s,t) = \prod_{i=1}^{d} \min\left\{1, \frac{r}{|s_i - t_i|^{\alpha_i}}\right\}.$$

Details will be given elsewhere.

4.4 Uniform Dimension Results

We note that the exceptional null probability events in Eqs. (38), (46), and (37) depend on $E \subseteq \mathbb{R}^N$ and $F \subseteq \mathbb{R}^d$, respectively. In many applications, we have a random time set $E(\omega)$ or $F(\omega) \subseteq \mathbb{R}^d$ and wish to know the fractal dimensions of $X(E(\omega), \omega)$ and $X^{-1}(F(\omega), \omega)$. For example, for any Borel set $F \subseteq \mathbb{R}^d$, we can write the intersection $X(\mathbb{R}_+) \cap F$ as $X(X^{-1}(F))$, the set M_k of k-multiple points of X as $X(L_k')$, where L_k' is the projection of L_k into $\mathbb{R}^N$. For such problems, the results of the form Eqs. (38), (46), and (37) give no information.

Kaufman [47] was the first to show that if W is the planar Brownian motion, then

$${\rm P}\Big\{\dim_{\rm H} W(E) = 2\dim_{\rm H} E \ \text{ for all Borel sets } E \subseteq \mathbb{R}_+\Big\} = 1. \tag{48}$$

Since the exceptional null probability event in Eq. (48) does not depend on E, it is referred to as a *uniform dimension result*. For Brownian motion in $\mathbb{R}$, Eq. (48) does not hold. This can be seen by taking $E = W^{-1}(0)$.

Several authors, including J. Hawkes, W.E. Pruitt, E.A. Perkins, and S.J. Taylor, have studied the problems on uniform Hausdorff and packing dimension results for the ranges and level sets of stable Lévy processes. See [102] or [121] for more information.

For approximately isotropic Gaussian random fields, [78] established a uniform Hausdorff dimension result for the images under the condition of strong local nondeterminism. In the special case of fractional Brownian motion, their result gives: If $N \le Hd$, then a.s.

$$\dim_{\mathrm{H}} B^H(E) = \frac{1}{H}\dim_{\mathrm{H}} E \ \text{ for all Borel sets } E \subseteq \mathbb{R}^N.$$

A similar result for the Brownian sheet was established by [81] by using a very different method, which relies on special properties of the Brownian sheet. Khoshnevisan et al. [56] gave an alternative proof for Mountford's result by applying the sectorial local nondeterminism (C3), and their argument is similar in spirit to that in [78].

Recently, [110] have shown that, while the anisotropy in the space variable destroys the uniform dimension result for the images, the uniform Hausdorff dimension result still holds for the image sets of time-anisotropic Gaussian random fields.

Theorem 15. *Let $X = \{X(t), t \in \mathbb{R}^N\}$ be an (N,d)-Gaussian random field defined by Eq. (3) whose coordinate processes are independent copies of X_0. If X_0 satisfies conditions (C1) and (C3), and $\sum_{j=1}^N H_j^{-1} \le d$, then with probability 1,*

$$\dim_{\mathrm{H}} X(E) = \dim_{\mathrm{H}}^{\rho} E \quad \textit{for all Borel sets } \ E \subseteq (0,\infty)^N, \tag{49}$$

where $\dim_{\mathrm{H}}^{\rho}$ is Hausdorff dimension on the metric space $(\mathbb{R}^N, \rho)$.

In light of Theorem 15 and the results in Sect. 4.2, one can ask naturally whether the Hausdorff dimension $\dim_{\mathrm{H}} X(E)$ in Eq. (49) can be replaced by the Fourier dimension $\dim_{\mathscr{F}} X(E)$. Such a uniform result would be useful when E is a random set. This question is open even for Brownian motion.

Next we turn to the uniform dimension problem on the inverse images. It follows from [78] that for fractional Brownian motion B^H with $N > Hd$,

$$\dim_{\mathrm{H}} \big(B^H\big)^{-1}(F) = N - Hd + H\dim_{\mathrm{H}} F \quad \text{ for all Borel sets } F \subseteq \mathbb{R}^d. \tag{50}$$

More generally, if $X = \{X(t), t \in \mathbb{R}^N\}$ is an (N,d)-Gaussian random field which satisfies Conditions (C1) and (C3) with $H_1 = \cdots = H_N := H$ and $N > Hd$, then one can modify the proofs in [78] to prove that Eq. (50) still holds for X and for all $F \subseteq \mathscr{O}$, where

$$\mathscr{O} = \bigcup_{a<b:\, a,b\in\mathbb{Q}} \big\{x \in \mathbb{R}^d : L(x,[a,b]) > 0\big\}.$$

However, it is not known whether similar results still hold for Gaussian random fields which are anisotropic either in the space variable or in the time variable.

Problem 6. Do uniform Hausdorff and packing dimension results hold for the inverse images of time-anisotropic or space-anisotropic Gaussian random fields?

For the time-anisotropic Gaussian fields in Theorem 12 which also satisfy (C3), we can prove that if $\sum_{\ell=1}^{N} H_\ell^{-1} > d$, then almost surely

$$\dim_{\mathrm{H}} X^{-1}(F) \le \min_{1\le k\le N} \left\{ \sum_{j=1}^{k} \frac{H_k}{H_j} + N - k - H_k\big(d - \dim_{\mathrm{H}} F\big) \right\} \tag{51}$$

holds for all Borel sets $F \subseteq \mathbb{R}^d$.

We end this section with the following problem which is related to Problem 4. Note that, when $E \subseteq [0,1]^N$, $\mathrm{Gr}X\big([0,1]^N\big) \cap (E \times F) \ne \varnothing$ is equivalent to $X(E) \cap F \ne \varnothing$. When this happens, it is of interest to determine the Hausdorff and packing dimensions of the random sets $X(E) \cap F$ and $E \cap X^{-1}(F)$.

In the previous sections, Theorems 11, 13, and 14 consider the special cases of $E = I$ or $F = \mathbb{R}^d$, respectively. When $E \subseteq \mathbb{R}_+$ and $F \subseteq \mathbb{R}^d$ are both arbitrary Borel sets, there have only been a few partial results. Some upper and lower bounds for the Hausdorff dimension $\dim_{\mathrm{H}}\big(X(E) \cap F\big)$ have been obtained by [48] for Brownian motion, [37] for stable Lévy processes, and [104] for fractional Brownian motion. Recently, Khoshnevisan and Xiao [59] have determined the Hausdorff dimension $\dim\big(W(E) \cap F\big)$, where W is a Brownian motion in $\mathbb{R}^d$. Similar problems for Gaussian random fields and the packing dimension of $X(E) \cap F$ (even when X is Brownian motion) are open. Regarding the latter problem, we expect that a new form of packing dimension profile may be needed in order to determine the packing dimension of $X(E) \cap F$.

4.5 Exact Hausdorff Measure Functions

In Sect. 4.1, Hausdorff and packing dimensions of the range $X\big([0,1]^N\big)$, graph $\mathrm{Gr}X\big([0,1]^N\big)$, and level sets are obtained for time-anisotropic Gaussian random fields. It is a natural question to determine exact Hausdorff and packing measure functions for these random sets. Recall that a measure function $\varphi : (0,1) \to \mathbb{R}_+$ is called an exact Hausdorff measure function for a set $F \subseteq \mathbb{R}^d$ if $0 < \varphi\text{-}m(F) < \infty$. Here $\varphi\text{-}m$ denotes the φ-Hausdorff measure. In Sect. 4.6, we will use $\varphi\text{-}p$ to denote the φ-packing measure. A measure function φ is called an exact packing measure function for F if $0 < \varphi\text{-}p(F) < \infty$.

Investigating exact Hausdorff and packing measure functions for the random sets generated by a random field X not only provides more precise information about the fractal properties of the sample functions of X but also stimulates deep understanding of the probability properties such as small ball probabilities, large deviations, and dependence structures of X. These latter questions have proved to be significant and sometimes challenging.

The problems on finding exact Hausdorff measure functions for the range and graph of the Brownian sheet and fractional Brownian motion have been considered in [28, 97, 98, 115, 116]. Here is a brief summary of the known results on the ranges and graph sets:

1. Let $W = \{W(t), t \in \mathbb{R}^N_+\}$ be the Brownian sheet in $\mathbb{R}^d$. Ehm [28] proved the following results. If $2N < d$, then $\varphi_3(r) = r^{2N}\big(\log\log 1/r\big)^N$ is an exact Hausdorff measure function for the range and graph of W. If $2N > d$, then $W([0,1]^N)$ a.s. has interior points and $\varphi_4(r) = r^{N+\frac{d}{2}}(\log\log 1/r)^{\frac{d}{2}}$ is an exact Hausdorff measure function for the graph of W.

   When $2N = d$, the problems for finding exact Hausdorff measure functions for the range and graph of W are open.
2. Let $B^H = \{B^H(t), t \in \mathbb{R}^N\}$ be an (N,d)-fractional Brownian motion of index H. Talagrand [97] proved that if $N < Hd$ then $\varphi_5(r) = r^{N/H}\log\log 1/r$ is an exact Hausdorff measure function for the range and graph of B^H. Pitt [89] showed that if $N > Hd$, then $B^H([0,1]^N)$ a.s. has positive Lebesgue measure and interior points. Xiao [115] showed that $\varphi_6(r) = r^{N+(1-H)d}\big(\log\log 1/r\big)^{Hd/N}$ is an exact Hausdorff measure function for the graph of B^H.

   In the case of $N = Hd$, [98] showed that φ_7-$m\big(B^H([0,1]^N)\big)$ is σ-finite almost surely, where $\varphi_7(r) = r^d\log(1/r)\log\log\log 1/r$. The same is also true for the Hausdorff measure of the graph set of B^H. However, the corresponding lower bound problems for the Hausdorff measure of the range and graph have remained open.

It is interesting to notice the subtle differences in the exact Hausdorff functions for the range and graph sets of fractional Brownian motion and the Brownian sheet, respectively. The differences are a reflection of the two different types of strong local nondeterminism [i.e., (C3) and (C4)] that they satisfy.

The exact Hausdorff measure of the level sets of a class of approximately isotopic Gaussian random fields was obtained in [116]. In the case of fractional Brownian motion, [6] established the following result which improves Theorem 1.3 in [116] significantly.

Theorem 16. *Let $B^H = \{B^H(t), t \in \mathbb{R}^N\}$ be a d-dimensional fractional Brownian motion of index H. For every $I \in \mathscr{A}$, there exists a finite constant $c_{13} > 0$ (depending only on H, N, and d) such that, with probability 1,*

$$\varphi_8\text{-}m\Big(\big(B^H\big)^{-1}(\{0\}) \cap I\Big) = c_{13}\,L(0,I), \tag{52}$$

where $L(0,I)$ is the local time of B^H at 0 and $\varphi_8(r) = r^{N-Hd}(\log\log 1/r)^{Hd/N}$.

For the Brownian sheet $W = \{W(t), t \in \mathbb{R}^N_+\}$ with values in $\mathbb{R}^d$, [69] proved that $\varphi_9(r) = r^{N-d/2}(\log\log 1/r)^{d/2}$ is an exact Hausdorff measure function for $W^{-1}(\{0\})$. His method is based on moment estimates of the local times of W. Notice that there is a subtle difference between $\varphi_8(r)$ and $\varphi_9(r)$.

Xiao [115] provided the exact Hausdorff measure functions for the ranges of a class of space-anisotropic Gaussian random fields. The following result for the range of a time-anisotropic Gaussian random field is proved by [71].

Theorem 17. *Let $X = \{X(t), t \in \mathbb{R}^N\}$ be an (N,d)-Gaussian random field defined by Eq. (3), where $X_1, \ldots, X_d$ are independent copies of a centered, real-valued Gaussian field X_0 with stationary increments and $X_0(0) = 0$. We assume that X_0 satisfies conditions (C1) and (C4). If $d > \sum_{j=1}^N H_j^{-1}$, then we have*

$$0 < \varphi_{10}\text{-}m(X([0,1]^N)) < \infty \quad a.s., \tag{53}$$

where $\varphi_{10}(r) = r^{\sum_{j=1}^N H_j^{-1}} \log\log(1/r)$.

Many questions on exact Hausdorff measure functions for anisotropic Gaussian random fields remain unsolved. In particular, we ask:

Problem 7. What are the exact Hausdorff measure functions for the graph and level sets of Gaussian random fields in Theorem 17?

4.6 *Exact Packing Measure Functions*

The exact packing measure function for random sets was first considered by [103] who proved that $\psi_1(r) = r^2/(\log|\log r|)$ is an exact packing measure function for the range of Brownian motion in $\mathbb{R}^d$ $(d \geq 3)$. The situation for $d = 2$ is very different. Le Gall and Taylor [65] proved that the range of the planar Brownian motion $W^{(2)}$ does not have an exact packing measure function. More precisely they showed that, for any measure function of the form $\psi(r) = r^2 \log(1/r) h(r)$, where $h(r)$ is monotone increasing but $\log(1/r)h(r)$ is decreasing, almost surely,

$$\psi - p(W^{(2)}([0,1])) = \begin{cases} 0 \\ \infty \end{cases} \text{according as } \sum_{n=1}^{\infty} h(2^{-2^n}) \begin{cases} < \infty \\ = \infty. \end{cases} \tag{54}$$

Furthermore, [62] proved that, for every integer $k \geq 2$, the set of k-multiple points $M_k^{(2)}$ of $W^{(2)}$ does not have an exact packing measure function. Recently, [80] proved a similar result for the set of double points of Brownian motion in $\mathbb{R}^3$ and established an integral test in terms of the intersection exponent $\xi_3(2,2)$ of two packets of two independent Brownian motions in $\mathbb{R}^3$.

So far no exact packing measure results have been established for Gaussian random fields other than those mentioned above and fractional Brownian motion considered by [113, 120].

Theorem 18. *Let $B^H = \{B^H(t),\ t \in \mathbb{R}^N\}$ be a fractional Brownian motion in $\mathbb{R}^d$ of index H. If $N < Hd$, then there exist positive constants c_{14} and c_{15} such that, with probability 1,*

$$c_{14} \leq \psi_2 - p(X([0,1]^N)) \leq \psi_2 - p(\mathrm{Gr}X([0,1]^N)) \leq c_{15}, \tag{55}$$

where $\psi_2(s) = s^{N/H}/(\log\log 1/s)^{N/(2H)}$.

The proof of Theorem 18 in [120] relies on the liminf properties of the occupation measures of B^H and the delayed hitting probability estimates. Such results have not been established for other Gaussian fields including the Brownian sheet. The main difficulty lies in dealing with their complicated dependence structures. I think it would be interesting to investigate the following problems.

Problem 8. Determine the exact packing measure functions for the range, graph set, and level sets of the Brownian sheet and anisotropic Gaussian random fields.

4.7 *Hitting Probabilities and Intersections of Gaussian Random Fields*

Many authors have investigated intersections of the trajectories of stochastic processes. For Brownian motion, the questions have been studied by A. Dvoretzky, P. Erdös, S. Kakutani, S. J. Taylor, and J.-F. Le Gall. See [52] for historical accounts and a very nice proof for the existence theorem using an elementary argument based on the self-similarity and Markov property of Brownian motion. The results on intersections of Brownian motion have been extended to Lévy processes, Gaussian processes, and other processes. We refer to the survey papers of [102,121] for further information on intersections of Markov processes.

In this section, we give some recent results on intersections of two independent Gaussian random fields obtained in [16]. These results are established based on refining the hitting probability estimates for Gaussian random fields obtained in [11, 119, 123]; see also [21, 22, 24] for related results. This approach is different from those based on intersection local times in [40, 90, 91, 111], where fractional Brownian motions are considered.

The following theorem from [16] extends the results on hitting probabilities in the aforementioned references.

Theorem 19. *Let $X = \{X(t), t \in \mathbb{R}^N\}$ be an (N,d)-Gaussian random field defined by Eq. (3) and assume that X_0 satisfies conditions (C1) and (C2) on a closed interval I. If $E \subseteq I$ and $F \subseteq \mathbb{R}^d$ are Borel sets, then there is a constant $c_{16} \geq 1$, which depends on I, F, and $\mathbf{H}$ only, such that*

$$c_{16}^{-1}\,\mathscr{C}_{\widetilde{\rho},d}(E\times F) \leq \mathrm{P}\Big\{X(E)\cap F \neq \varnothing\Big\} \leq c_{16}\,\mathscr{H}^d_{\widetilde{\rho}}(E\times F), \tag{56}$$

where $\mathscr{C}_{\widetilde{\rho},d}$ and $\mathscr{H}^d_{\widetilde{\rho}}$ denote, respectively, the d-dimensional capacity and Hausdorff measure in the metric space $(\mathbb{R}^{N+d}, \widetilde{\rho})$ and where

$$\widetilde{\rho}\big((s,x),(t,y)\big) = \max\big\{\rho(s,t),\ |x-y|\big\}, \qquad \forall (s,x),(t,y) \in \mathbb{R}^N \times \mathbb{R}^d.$$

As we have mentioned in Problem 4, except in the case of Brownian motion, it is an open problem to provide a necessary and sufficient condition for $\mathrm{P}\{X(E)\cap F\neq\varnothing\}>0$.

Next we apply Theorem 19 to study intersections of two independent Gaussian random fields. Let $X^{\mathbf{H}}=\{X^{\mathbf{H}}(s), s\in\mathbb{R}^{N_1}\}$ and $X^{\mathbf{K}}=\{X^{\mathbf{K}}(t), t\in\mathbb{R}^{N_2}\}$ be two independent Gaussian random fields taking values in $\mathbb{R}^d$, defined as in Eq. (3). More specifically, we assume that $X^{\mathbf{H}}$ is defined as

$$X^{\mathbf{H}}(s)=(X_1^{\mathbf{H}}(s),\ldots,X_d^{\mathbf{H}}(s)),\qquad s\in\mathbb{R}^{N_1}, \tag{57}$$

where $X_1^{\mathbf{H}},\ldots,X_d^{\mathbf{H}}$ are independent copies of real-valued, centered Gaussian random field $X_0^{\mathbf{H}}$. The Gaussian random field $X^{\mathbf{K}}$ is defined in the same way. Here $\mathbf{H}\in(0,1)^{N_1}$ and $\mathbf{K}\in(0,1)^{N_2}$ are constant vectors.

We say the two Gaussian fields $X^{\mathbf{H}}$ and $X^{\mathbf{K}}$ intersect if there exist $s\in\mathbb{R}^{N_1}$ and $t\in\mathbb{R}^{N_2}$ such that $X^{\mathbf{H}}(s)=X^{\mathbf{K}}(t)$. The following problems are concerned with existence of intersections:

(i) When do $X^{\mathbf{H}}$ and $X^{\mathbf{K}}$ intersect (with positive probability)?
(ii) Let $E_1\subseteq\mathbb{R}^{N_1}$ and $E_2\subseteq\mathbb{R}^{N_2}$ be arbitrary Borel sets. When do $X^{\mathbf{H}}$ and $X^{\mathbf{K}}$ intersect if we restrict the "time" $s\in E_1$ and $t\in E_2$? That is, when is

$$\mathrm{P}\Big\{X^{\mathbf{H}}(E_1)\cap X^{\mathbf{K}}(E_2)\neq\varnothing\Big\}>0? \tag{58}$$

(iii) Given a Borel set $F\subseteq\mathbb{R}^d$, when does F contain intersection points of $X^{\mathbf{H}}(s)$ $(s\in E_1)$ and $X^{\mathbf{K}}(t)$, $(t\in E_2)$? That is, when is

$$\mathrm{P}\Big\{X^{\mathbf{H}}(E_1)\cap X^{\mathbf{K}}(E_2)\cap F\neq\varnothing\Big\}>0? \tag{59}$$

Clearly, Question (i) is a special case of Question (ii), which is a special case of Question (iii). For answering Questions (i) and (ii), consider the Gaussian random field $Z=\{Z(s,t),(s,t)\in\mathbb{R}^{N_1+N_2}\}$ with values in $\mathbb{R}^d$ defined by

$$Z(s,t)\equiv X^H(s)-X^K(t),\qquad s\in\mathbb{R}^{N_1}, t\in\mathbb{R}^{N_2}. \tag{60}$$

Then Eq. (58) is equivalent to $\mathrm{P}\big(Z(E_1\times E_2)\cap\{0\}\neq\varnothing\big)>0$. Hence sufficient conditions and necessary conditions for this to hold can be obtained from hitting probability estimates for Gaussian field Z, which is done in Theorem 2.1 of [16]. Instead of giving more details, we content with the following simpler result which provides an answer to Question (i). In the following, we let $Q:=\sum_{j=1}^{N_1}H_j^{-1}+\sum_{j=1}^{N_2}K_j^{-1}$.

Theorem 20. *Let $X^{\mathbf{H}}=\{X^{\mathbf{H}}(s), s\in\mathbb{R}^{N_1}\}$ and $X^{\mathbf{K}}=\{X^{\mathbf{K}}(t), t\in\mathbb{R}^{N_2}\}$ be two independent Gaussian random fields with values in $\mathbb{R}^d$ such that $X_0^{\mathbf{H}}$ and $X_0^{\mathbf{K}}$ satisfy*

(C1) and (C2), respectively, on interval $I_1 \subseteq \mathbb{R}^{N_1}$ *with indices* $H = (H_1, \ldots, H_{N_1})$ *and on interval* $I_2 \subseteq \mathbb{R}^{N_2}$ *with indices* $K = (K_1, \ldots, K_{N_2})$*:*

(i) If $d < Q$*, then* $\mathrm{P}\big\{X^H(I_1) \cap X^K(I_2) \neq \varnothing\big\} = 0$.
(ii) If $d > Q$*, then* $\mathrm{P}\big\{X^H(I_1) \cap X^K(I_2) \neq \varnothing\big\} > 0$.
(iii) If, in addition, we assume that $X_0^{\mathbf{H}}$ *has stationary increments and satisfies (C4) on interval* $I_1 \subseteq \mathbb{R}^{N_1}$*, then* $d = Q$ *implies* $\mathrm{P}\big\{X^H(I_1) \cap X^K(I_2) \neq \varnothing\big\} = 0$.

In order to answer Question (iii), we consider the Gaussian random field $Y = \{Y(s,t), (s,t) \in \mathbb{R}^{N_1+N_2}\}$ with values in $\mathbb{R}^{2d}$ defined by

$$Y(s,t) = \big(X^H(s), X^K(t)\big), \qquad \forall (s,t) \in \mathbb{R}^{N_1+N_2}.$$

Then Eq. (59) holds if and only if

$$\mathrm{P}\Big\{Y(E_1 \times E_2) \cap \widetilde{F} \neq \varnothing\Big\} > 0, \tag{61}$$

where $\widetilde{F} = \{(x,x) : x \in F\} \subseteq \mathbb{R}^{2d}$. This hitting probability is also estimated by [16] in terms of Bessel–Riesz-type capacity and Hausdorff measure of $E_1 \times E_2 \times F$, under appropriate metric on $\mathbb{R}^{N_1+N_2+d}$.

It follows from Theorem 20 that, when $\sum_{i=1}^{N_1} H_i^{-1} + \sum_{j=1}^{N_2} K_j^{-1} > d$, we have $\mathrm{P}\big(X^{\mathbf{H}}(I_1) \cap X^{\mathbf{K}}(I_2) \neq \varnothing\big) > 0$. It is of interest to determine the Hausdorff dimensions of the set of intersection times $L_2 := \{(s,t) \in I_1 \times I_2 : X^{\mathbf{H}}(s) = X^{\mathbf{K}}(t)\}$ and the set of intersections $M_2 = X^{\mathbf{H}}(I_1) \cap X^{\mathbf{K}}(I_2)$. Since L_2 is the level set of the Gaussian random field $Z(s,t) = X^{\mathbf{H}}(s) - X^{\mathbf{K}}(t)$, the Hausdorff and packing dimensions of L_2 can be obtained from Theorem 11. However, the Hausdorff and packing dimensions of M_2 have not been determined in their full generality.

In the following, we provide a partial answer for the intersection set of two independent fractional Brownian motions obtained by [111].

Let $B^{\alpha_1} = \{B^{\alpha_1}(s), s \in \mathbb{R}^{N_1}\}$ and $B^{\alpha_2} = \{B^{\alpha_2}(t), t \in \mathbb{R}^{N_2}\}$ be two independent fractional Brownian motions with values in $\mathbb{R}^d$ and indices α_1 and α_2, respectively. Let

$$\begin{aligned} M_2 &= \big\{x \in \mathbb{R}^d : x = B^{\alpha_1}(s) = B^{\alpha_2}(t) \ \text{ for some } (s,t) \in \mathbb{R}^{N_1+N_2}\big\} \\ &= B^{\alpha_1}(\mathbb{R}^{N_1}) \cap B^{\alpha_2}(\mathbb{R}^{N_2}). \end{aligned}$$

Theorem 21. *If* $\frac{N_1}{\alpha_1} + \frac{N_2}{\alpha_2} > d$*, then with probability 1,*

$$\dim_{\mathrm{H}} M_2 = \dim_{\mathrm{P}} M_2 = \begin{cases} d & \text{if } N_1 > \alpha_1 d \text{ and } N_2 > \alpha_2 d, \\ \frac{N_2}{\alpha_2} & \text{if } N_1 > \alpha_1 d \text{ and } N_2 \le \alpha_2 d, \\ \frac{N_1}{\alpha_1} & \text{if } N_1 \le \alpha_1 d \text{ and } N_2 > \alpha_2 d, \\ \frac{N_1}{\alpha_1} + \frac{N_2}{\alpha_2} - d & \text{if } N_1 \le \alpha_1 d \text{ and } N_2 \le \alpha_2 d. \end{cases} \tag{62}$$

Besides intersections of independent Gaussian random fields, one can study analogous questions for self-intersections of a Gaussian random field. The arguments described above are still applicable for studying the existence of self-intersections. For related work on fractional Brownian motion and the Brownian sheet, we refer to [23, 36, 61, 90, 98]. The first three papers provide sufficient conditions for the existence of k-multiple points and the last two papers show that the corresponding conditions are also necessary. We mention that the methods in [23, 98] are very different and [23] only prove necessity for $k = 2$.

Finally we remark that, while the Hausdorff and packing dimensions of the set of k-multiple points of fractional Brownian motion and the Brownian sheet have been obtained (cf. [56]), no results on exact Hausdorff or packing measure functions have been obtained for any Gaussian random fields.

Acknowledgments The author thanks the referee for his/her helpful comments which have led to improvement of the manuscript. Research partially supported by the NSF grant DMS-1006903.

References

1. Adler, R.J.: The Geometry of Random Fields. Wiley, New York (1981)
2. Adler, R.J., Taylor, J.E.: Random Fields and Geometry. Springer, New York (2007)
3. Ayache, A., Wu, D., Xiao, Y.: Joint continuity of the local times of fractional Brownian sheets. Ann. Inst. H. Poincaré Probab. Statist. **44**, 727–748 (2008)
4. Ayache, A., Xiao, Y.: Asymptotic properties and Hausdorff dimensions of fractional Brownian sheets. J. Fourier Anal. Appl. **11**, 407–439 (2005)
5. Baraka, D., Mountford, T.S.: A law of iterated logarithm for fractional Brownian motions. In: Séminaire de Probabilités XLI. Lecture Notes in Mathematics, vol. 1934, pp. 161–179. Springer, Berlin (2008)
6. Baraka, D., Mountford, T.S.: The exact Hausdorff measure of the zero set of fractional Brownian motion. J. Theor. Probab. **24**, 271–293 (2011)
7. Baraka, D., Mountford, T.S., Xiao, Y.: Hölder properties of local times for fractional Brownian motions. Metrika **69**, 125–152 (2009)
8. Benassi, A., Jaffard, S., Roux, D.: Elliptic Gaussian random processes. Rev. Mat. Iberoamericana **13**, 19–90 (1997)
9. Berman, S.M.: Gaussian sample function: uniform dimension and Hölder conditions nowhere. Nagoya Math. J. **46**, 63–86 (1972)
10. Berman, S.M.: Local nondeterminism and local times of Gaussian processes. Indiana Univ. Math. J. **23**, 69–94 (1973)
11. Biermé, H., Lacaux, C., Xiao, Y.: Hitting probabilities and the Hausdorff dimension of the inverse images of anisotropic Gaussian random fields. Bull. London Math. Soc. **41**, 253–273 (2009)
12. Biermé, H., Meerschaert, M.M., Scheffler, H.-P.: Operator scaling stable random fields. Stoch. Process. Appl. **117**, 313–332 (2007)
13. Blath, J., Martin, A.: Propagation of singularities in the semi-fractional Brownian sheet. Stoch. Process. Appl. **118**, 1264–1277 (2008)
14. Bonami, A., Estrade, A.: Anisotropic analysis of some Gaussian models. J. Fourier Anal. Appl. **9**, 215–236 (2003)
15. Chen, X., Li, W.V., Rosinski, J., Shao, Q.-M.: Large deviations for local times and intersection local times of fractional Brownian motions and Riemann-Liouville processes. Ann. Probab. **39**, 729–778 (2011)

16. Chen, Z., Xiao, Y.: On intersections of independent anisotropic Gaussian random fields. Sci. China Math. **55**, 2217–2232 (2012)
17. Christakos, M.J.: Modern Spatiotemporal Geostatistics. Oxford University Press, Oxford (2000)
18. Cuzick, J.: Some local properties of Gaussian vector fields. Ann. Probab. **6**, 984–994 (1978)
19. Cuzick, J.: Multiple points of a Gaussian vector field. Z. Wahrsch. Verw. Gebiete. **61**, 431–436 (1982)
20. Dalang, R.C.: Extending martingale measure stochastic integral with applications to spatially homogeneous s.p.d.e.'s. Electron. J. Probab. **4**(6), 1–29 (1999) (Erratum in Electron. J. Probab. **6**(6), 1–5 (2001))
21. Dalang, R.C., Khoshnevisan, D., Nualart, E.: Hitting probabilities for systems of non-linear stochastic heat equations with additive noise. Latin Am. J. Probab. Statist. (Alea) **3**, 231–271 (2007)
22. Dalang, R.C., Khoshnevisan, D., Nualart, E.: Hitting probabilities for the non-linear stochastic heat equation with multiplicative noise. Probab. Theor. Relat. Fields **117**, 371–427 (2009)
23. Dalang, R.C., Khoshnevisan, D., Nualart, E., Wu, D., Xiao, Y.: Critical Brownian sheet does not have double points. Ann. Probab. **40**, 1829–1859 (2012)
24. Dalang, R.C., Sanz-Solé, M.: Criteria for hitting probabilities with applications to systems of stochastic wave equations. Bernoulli **16**, 1343–1368 (2010)
25. Dembo, A., Peres, P., Rosen, J., Zeitouni, O.: Thick points for spatial Brownian motion: multifractal analysis of occupation measure. Ann. Probab. **28**, 1–35 (2000)
26. Didier, G., Pipiras, V.: Integral representations of operator fractional Brownian motions. Bernoulli **17**, 1–33 (2011)
27. Dindar, Z.: On the Hausdorff dimension of the set generated by exceptional oscillations of a two-parameter Wiener process. J. Multivar. Anal. **79**, 52–70 (2001)
28. Ehm, E.: Sample function properties of multi-parameter stable processes. Z. Wahrsch. Verw. Gebiete **56**, 195–228 (1981)
29. Estrade, A., Wu, D., Xiao, Y.: Packing dimension results for anisotropic Gaussian random fields. Commun. Stoch. Anal. **5**, 41–64 (2011)
30. Falconer, K.J.: Fractal Geometry—Mathematical Foundations and Applications, 2nd edn. Wiley, NJ (2003)
31. Falconer, K.J., Howroyd, J.D.: Packing dimensions of projections and dimension profiles. Math. Proc. Cambridge Philos. Soc. **121**, 269–286 (1997)
32. Falconer, K.J., Xiao, Y.: Generalized dimensions of images of measures under Gaussian processes (2011). Submitted
33. Fouché, W.L., Mukeru, S.: Fourier structure of the zero set of Brownian motion (2011). Preprint
34. Geman, D., Horowitz, J.: Occupation densities. Ann. Probab. **8**, 1–67 (1980)
35. Gneiting, T.: Nonseparable, stationary covariance functions for space-time data. J. Am. Statist. Assoc. **97**, 590–600 (2002)
36. Goldman, A.: Points multiples des trajectoires de processus gaussiens. Z. Wahrsch. Verw. Gebiete **57**, 481–494 (1981)
37. Hawkes, J.: Measures of Hausdorff type and stable processes. Mathematika **25**, 202–212 (1978)
38. Howroyd, J.D.: Box and packing dimensions of projections and dimension profiles. Math. Proc. Cambridge Phil. Soc. **130**, 135–160 (2001)
39. Hwang, K.-S., Wang, W., Xiao, Y.: The modulus of non-differentiability of a fractional Brownian motion (2012). Preprint
40. Hu, H., Nualart, D.: Renormalized self-intersection local time for fractional Brownian motion. Ann. Probab. **33**, 948–983 (2005)
41. Hu, Y., Nualart, D.: Stochastic heat equation driven by fractional noise and local time. Probab. Theor. Relat. Fields **143**, 285–328 (2009)
42. Istas, J.: Spherical and hyperbolic fractional Brownian motion. Electron. Comm. Probab. **10**, 254–262 (2005)

43. Kahane, J.-P.: Some Random Series of Functions, 2nd edn. Cambridge University Press, Cambridge (1985a)
44. Kahane, J.-P.: Ensembles aleatoires et dimensions. In: Recent Progress in Fourier Analysis (El Escorial, 1983), pp. 65–121. North-Holland, Amsterdam (1985b)
45. Kahane, J.-P., Salem, R.: Ensembles Parfaits et Series Trigonometriques, 2nd edn. Hermann, Paris (1994)
46. Kamont, A.: On the fractional anisotropic Wiener field. Probab. Math. Statist. **16**, 85–98 (1996)
47. Kaufman, R.: Une propriété métrique du mouvement brownien. C. R. Acad. Sci. Paris **268**, 727–728 (1968)
48. Kaufman, R.: Measures of Hausdorff-type, and Brownian motion. Mathematika **19**, 115–119 (1972)
49. Kaufman, R.: Large increments of Brownian motion. Nagoya Math. J. **56**, 139–145 (1975)
50. Kaufman, R.: Temps locaux et dimensions. C. R. Acad. Sci. Paris Sér. I Math. **300**, 281–282 (1985)
51. Khoshnevisan, D.: Multiparameter Processes: An Introduction to Random Fields. Springer, New York (2002)
52. Khoshnevisan, D.: Intersections of Brownian motions. Expos. Math. **21**, 97–114 (2003)
53. Khoshnevisan, D., Peres, Y., Xiao, Y.: Limsup random fractals. Electron. J. Probab. **5**(4), 1–24 (2000)
54. Khoshnevisan, D., Schilling, R., Xiao, Y.: Packing dimension profiles and Lévy processes. Bull. London Math. Soc. (2012). doi:10.1112/blms/bds022
55. Khoshnevisan, D., Shi, Z.: Brownian sheet and capacity. Ann. Probab. **27**, 1135–1159 (1999)
56. Khoshnevisan, D., Wu, D., Xiao, Y.: Sectorial local non-determinism and the geometry of the Brownian sheet. Electron. J. Probab. **11**, 817–843 (2006)
57. Khoshnevisan, D., Xiao, Y.: Images of the Brownian sheet. Trans. Am. Math. Soc. **359**, 3125–3151 (2007)
58. Khoshnevisan, D., Xiao, Y.: Packing dimension of the range of a Lévy process. Proc. Am. Math. Soc. **136**, 2597–2607 (2008a)
59. Khoshnevisan, D., Xiao, Y.: Packing dimension profiles and fractional Brownian motion. Math. Proc. Cambridge Philos. Soc. **145**, 205–213 (2008b)
60. Khoshnevisan, D., Xiao, Y.: Brownian motion and thermal capacity (2012). Submitted
61. Kõno, N.: Double points of a Gaussian sample path. Z. Wahrsch. Verw. Gebiete **45**, 175–180 (1978)
62. Le Gall, J.-F.: The exact Hausdorff measure of Brownian multiple points. In: Cinlar, E., Chung, K.L., Getoor, R.K. (eds.) Seminar on Stochastic Processes (Charlottesville, Va., 1986), pp. 107–137. Progress in Probability and Statistics, vol. 13. Birkhäuser, Boston (1987a)
63. Le Gall, J.-F.: Temps locaux d'intersection et points multiples des processus de Lévy. In: Séminaire de Probabilités, XXI. Lecture Notes in Mathematics, vol. 1247, pp. 341–374. Springer, Berlin (1987b)
64. Le Gall, J.-F.: The exact Hausdorff measure of Brownian multiple points II. In: Seminar on Stochastic Processes (Gainesville, FL, 1988), pp. 193–197. Progress in Probability, vol. 17. Birkhäuser, Boston (1989)
65. Le Gall, J.-F., Taylor, S.J.: The packing measure of planar Brownian motion. In: Seminar on Stochastic Processes (Charlottesville, Va., 1986), pp. 139–147. Progress in Probability and Statistics, vol. 13. Birkhäuser, Boston (1987)
66. Lévy, P.: La mesure de Hausdorff de la courbe du mouvement brownien. Giorn. Ist. Ital. Attuari. **16**, 1–37 (1953)
67. Li, W.V., Shao, Q.-M.: Gaussian processes: inequalities, small ball probabilities and applications. In: Rao, C.R., Shanbhag, D. (eds.) Stochastic Processes: Theory and Methods. Handbook of Statistics, vol. 19, pp. 533–597. North-Holland, Amsterdam (2001)
68. Li, Y., Xiao, Y.: Multivariate operator-self-similar random fields. Stoch. Process. Appl. **121**, 1178–1200 (2011)

69. Lin, H.: The local times and Hausdorff measure for level sets of a Wiener sheet. Sci. China Ser. A **44**, 696–708 (2001)
70. Luan, N., Xiao, Y.: Chung's law of the iterated logarithm for anisotropic Gaussian random fields. Statist. Probab. Lett. **80**, 1886–1895 (2010)
71. Luan, N., Xiao, Y.: Spectral conditions for strong local nondeterminism and exact Hausdorff measure of ranges of Gaussian random fields. J. Fourier Anal. Appl. **18**, 118–145 (2012)
72. Marcus, M.B., Rosen, J.: Markov Processes, Gaussian Processes, and Local Times. Cambridge University Press, Cambridge (2006)
73. Mason, D.M., Shi, Z.: Small deviations for some multi-parameter Gaussian processes. J. Theor. Probab. **14**, 213–239 (2001)
74. Mason, D.J., Xiao, Y.: Sample path properties of operator self-similar Gaussian random fields. Theor. Probab. Appl. **46**, 58–78 (2002)
75. Mattila, P.: Geometry of Sets and Measures in Euclidean Spaces. Cambridge University Press, Cambridge (1995)
76. Meerschaert, M.M., Wang, W., Xiao, Y.: Fernique-type inequalities and moduli of continuity of anisotropic Gaussian random fields. Trans. Am. Math. Soc. **365**, 1081–1107 (2013)
77. Mockenhaupt, G.: Salem sets and restriction properties of Fourier transforms. Geom. Funct. Anal. **10**, 1579–1587 (2000)
78. Monrad, D., Pitt, L.D.: Local nondeterminism and Hausdorff dimension. In: Cinlar, E., Chung, K.L., Getoor, R.K. (eds.) Progress in Probability and statistics. Seminar on Stochastic Processes 1986, pp. 163–189. Birkhäuser, Boston (1987)
79. Monrad, D., Rootzén, H.: Small values of Gaussian processes and functional laws of the iterated logarithm. Probab. Theor. Relat. Fields **101**, 173–192 (1995)
80. Mörters, P., Shieh, N.-R.: The exact packing measure of Brownian double points. Probab. Theor. Relat. Fields **143**, 113–136 (2009)
81. Mountford, T.S.: Uniform dimension results for the Brownian sheet. Ann. Probab. **17**, 1454–1462 (1989)
82. Mueller, C., Tribe, R.: Hitting probabilities of a random string. Electron. J. Probab. **7**(10), 1–29 (2002)
83. Nualart, E., Viens, F.: The fractional stochastic heat equation on the circle: time regularity and potential theory. Stoch. Process. Appl. **119**, 1505–1540 (2009)
84. Oksendal, B., Zhang, T.: Multiparameter fractional Brownian motion and quasi-linear stochastic partial differential equations. Stoch. Stoch. Rep. **71**, 141–163 (2000)
85. Orey, S., Pruitt, W.E.: Sample functions of the N-parameter Wiener process. Ann. Probab. **1**, 138–163 (1973)
86. Orey, S., Taylor, S.J.: How often on a Brownian path does the law of the iterated logarithm fail? Proc. London Math. Soc. **28**, 174–192 (1974)
87. Perkins, E.A.: The exact Hausdorff measure of the level sets of Brownian motion. Z. Wahrsch. Verw. Gebiete **58**, 373–388 (1981)
88. Perkins, E.A., Taylor, S.J.: Uniform measure results for the image of subsets under Brownian motion. Probab. Theor. Relat. Fields **76**, 257–289 (1987)
89. Pitt, L.D.: Local times for Gaussian vector fields. Indiana Univ. Math. J. **27**, 309–330 (1978)
90. Rosen, J.: Self-intersections of random fields. Ann. Probab. **12**, 108–119 (1984)
91. Rosen, J.: The intersection local time of fractional Brownian motion in the plane. J. Multivar. Anal. **23**, 37–46 (1987)
92. Samorodnitsky, G., Taqqu, M.S.: Stable non-Gaussian processes: stochastic models with infinite variance. Chapman and Hall, (1994)
93. Shieh, N.-R., Xiao, Y.: Images of Gaussian random fields: Salem sets and interior points. Studia Math. **176**, 37–60 (2006)
94. Shieh, N.-R., Xiao, Y.: Hausdorff and packing dimensions of the images of random fields. Bernoulli **16**, 926–952 (2010)
95. Stein, M.L.: Space-time covariance functions. J. Am. Statist. Assoc. **100**, 310–321 (2005)
96. Talagrand, M. The small ball problem for the Brownian sheet. Ann. Probab. **22**, 1331–1354 (1994)

97. Talagrand, M.: Hausdorff measure of trajectories of multiparameter fractional Brownian motion. Ann. Probab. **23**, 767–775 (1995)
98. Talagrand, M.: Multiple points of trajectories of multiparameter fractional Brownian motion. Probab. Theor. Relat. Fields **112**, 545–563 (1998)
99. Talagrand, M.: Generic Chaining. Springer, New York (2006)
100. Talagrand, M., Xiao, Y.: Fractional Brownian motion and packing dimension. J. Theor. Probab. **9**, 579–593 (1996)
101. Taylor, S.J.: The Hausdorff α-dimensional measure of Brownian paths in n-space. Proc. Cambridge Philo. Soc. **49**, 31–39 (1953)
102. Taylor, S.J.: The measure theory of random fractals. Math. Proc. Cambridge Philo. Soc. **100**, 383–406 (1986)
103. Taylor, S.J., Tricot, C.: Packing measure and its evaluation for a Brownian path. Trans. Am. Math. Soc. **288**, 679–699 (1985)
104. Testard, F.: Polarité, points multiples et géométrie de certains processus gaussiens. Publ. du Laboratoire de Statistique et Probabilités de l'U.P.S. Toulouse, 01–86 (1986)
105. Tindel, S., Tudor, C.A., Viens, F.: Sharp Gaussian regularity on the circle, and applications to the fractional stochastic heat equation. J. Funct. Anal. **217**, 280–313 (2004)
106. Tricot, C.: Two definitions of fractional dimension. Math. Proc. Cambridge Philo. Soc. **91**, 57–74 (1982)
107. Walsh, J.B.: Propagation of singularities in the Brownian sheet. Ann. Probab. **10**, 279–288 (1982)
108. Wang, W.: Almost-sure path properties of fractional Brownian sheet. Ann. Inst. H. Poincaré Probab. Statist. **43**, 619–631 (2007)
109. Wu, D., Xiao, Y.: Geometric properties of the images fractional Brownian sheets. J. Fourier Anal. Appl. **13**, 1–37 (2007)
110. Wu, D., Xiao, Y.: Uniform Hausdorff dimension results for Gaussian random fields. Sci. in China, Ser. A **52**, 1478–1496 (2009)
111. Wu, D., Xiao, Y.: Regularity of intersection local times of fractional Brownian motions. J. Theor. Probab. **23**, 972–1001 (2010)
112. Wu, D., Xiao, Y.: On local times of anisotropic Gaussian random fields. Comm. Stoch. Anal. **5**, 15–39 (2011)
113. Xiao, Y.: Dimension results for Gaussian vector fields and index-α stable fields. Ann. Probab. **23**, 273–291 (1995)
114. Xiao, Y.: Packing measure of the sample paths of fractional Brownian motion. Trans. Amer. Math. Soc. **348**, 3193–3213 (1996)
115. Xiao, Y.: Hausdorff dimension of the graph of fractional Brownian motion. Math. Proc. Cambridge Philo. Soc. **122**, 565–576 (1997)
116. Xiao, Y.: Hölder conditions for the local times and the Hausdorff measure of the level sets of Gaussian random fields. Probab. Theor. Relat. Fields **109**, 129–157 (1997)
117. Xiao, Y.: Packing dimension of the image of fractional Brownian motion. Statist. Probab. Lett. **333**, 379–387 (1997)
118. Xiao, Y.: Hausdorff-type measures of the sample paths of fractional Brownian motion. Stoch. Process. Appl. **74**, 251–272 (1998)
119. Xiao, Y.: Hitting probabilities and polar sets for fractional Brownian motion. Stoch. Stoch. Rep. **66**, 121–151 (1999)
120. Xiao, Y.: The packing measure of the trajectories of multiparameter fractional Brownian motion. Math. Proc. Cambridge Philo. Soc. **135**, 349–375 (2003)
121. Xiao, Y.: Random fractals and Markov processes. In: Lapidus, M.L., van Frankenhuijsen, M. (eds.) Fractal Geometry and Applications: A Jubilee of Benoit Mandelbrot, pp. 261–338. American Mathematical Society, New York (2004)
122. Xiao, Y.: Strong local nondeterminism and the sample path properties of Gaussian random fields. In: Lai, T.L, Shao, Q., Qian, L. (eds.) Asymptotic Theory in Probability and Statistics with Applications, pp. 136–176. Higher Education Press, Beijing (2007)

123. Xiao, Y.: Sample path properties of anisotropic Gaussian random fields. In: Khoshnevisan, D., Rassoul-Agha, F. (eds.) A Minicourse on Stochastic Partial Differential Equations. Lecture Notes in Mathematics, vol. 1962, pp. 145–212. Springer, New York (2009)
124. Xiao, Y.: A packing dimension theorem for Gaussian random fields. Statist. Probab. Lett. **79**, 88–97 (2009)
125. Xiao, Y.: Properties of strong local nondeterminism and local times of stable random fields. In: Dalang, R.C., Dozzi, M., Russo, F. (eds.) Seminar on Stochastic Analysis, Random Fields and Applications VI, Progress in Probability, vol. 63, pp. 279–310. Birkhäuser, Basel (2011)
126. Xiao, Y., Zhang, T.: Local times of fractional Brownian sheet. Probab. Theor. Relat. Fields **124**, 204–226 (2002)
127. Xue, Y., Xiao, Y.: Fractal and smoothness properties of anisotropic Gaussian models. Frontiers Math. China **6**, 1217–1246 (2011)
128. Zhang, L.X.: Two different kinds of liminfs on the LIL for two-parameter Wiener processes. Stoch. Process. Appl. **63**, 175–188 (1996)
129. Zimmerman, G.J.: Some sample function properties of the two-parameter Gaussian process. Ann. Math. Statist. **43**, 1235–1246 (1972)
130. Zygmund, A.: Trigonometric Series, vol. I. Cambridge University Press, Cambridge (1959)

MIX
Papier aus verantwortungsvollen Quellen
Paper from responsible sources
FSC® C105338

If you have any concerns about our products,
you can contact us on
ProductSafety@springernature.com

In case Publisher is established outside the EU,
the EU authorized representative is:
Springer Nature Customer Service Center GmbH
Europaplatz 3, 69115 Heidelberg, Germany

Printed by Libri Plureos GmbH
in Hamburg, Germany